MW00848765

RADAR
BRITAIN'S SHIELD
AND THE DEFEAT OF THE
LUFTWAFFE

RADAR
BRITAIN'S SHIELD
AND THE DEFEAT OF THE
LUFTWAFFE

DAVID ZIMMERMAN

AMBERLEY

This edition first published 2013

First published 2001

Amberley Publishing
The Hill, Stroud
Gloucestershire, GL5 4EP

www.amberley-books.com

British Library Cataloguing in Publication Data.
A catalogue record for this book is available from the British Library.

ISBN 978-1-4456-0859-4

Typesetting and Origination by Amberley Publishing.
Printed in Great Britain.

To Elaine Zimmerman.
Thanks for everything, mom.
Love David

Skip from pg 22 - chpt 3.

Contents

Foreword

In the six years of the Second World War there were many great battles on land, sea and in the air that led to the ultimate Allied victory. Of these the Battle of Britain in 1940 remains the most celebrated more than half a century after the end of the great conflict. This battle is the centrepiece of David Zimmerman's scholarly work, Britain's Shield.

Anyone reading this book must be left gasping at the narrowness of the victory. In that famous phrase of Churchill's speech when he reminded us how much we owed to the 'few', he was rightly paying tribute to the pilots of RAF Fighter Command, who, against the overwhelming odds of the Luftwaffe bombers, scored that crucial victory that led Hitler to cancel his invasion plans and thereby save Europe from Fascist domination.

It is the essence of Professor Zimmerman's work that he both destroys and supports the many mythical beliefs that have become prominent about the events leading up to that historic victory in the summer of 1940. The radar chain around the coast was one essential component, for without that early warning system, Fighter Command would, undoubtedly, have been overwhelmed. That fact is common knowledge, but Zimmerman shows that it was the detailed integration of this early warning with the command structure of the RAF fighter squadrons that led to the ultimate success of fighter pilots. The filtering of the radar information, the allocation of action to the appropriate fighter group and the radio telephonic communication between fighters and ground control were essential elements in the critical years and months leading to the first great clashes of the air battle in July 1940.

The close association of the radar scientists with RAF personnel is revealed as another important element in the complex organisation leading to the ultimate encounter of fighter and bomber. Indeed, nowhere is that better illustrated than in the failure of the German command to recognize the importance of the radar chain. Goering refused to accept the advice of his intelligence chief General Martini that the radar stations should be destroyed. The relatively few attacks on the coastal radar chain reveal that had Goering accepted Martini's advice the outcome of the battle might well have been victory to the Luftwaffe. Indeed, the close integration of the British and Allied scientists with the operational commands, in contrast with that in Germany, was at that time and throughout the war, a significant factor in conflict.

Another refreshing feature of Professor Zimmerman's book is the description of the evolution of the early warning system from the acoustic mirror and the abrupt change to the development of the radar system in the 1930s. Shortly after Baldwin's warning that 'the bomber would always get through', Watson-Watt staged his famous Daventry demonstration to show that radio waves could be detected after reflection from a bomber. Wimperis, the scientific advisor to the Air Ministry, had recently convened the committee with Sir Henry Tizard to advise about the possibility of defence against the bomber. Zimmerman is the first historian to describe in detail the years of political intrigue in which the Tizard Committee became involved in their support of the development of radar.

The personnel conflict between Tizard and Professor Lindemann (later Lord Cherwell) has often been described. Here we find the detailed scientific and political facts underlying that conflict. Lindemann was Churchill's wealthy friend and in his intrigue against the government Churchill ruthlessly used Lindemann's advice. It is truly an amazing story and Zimmerman shows how narrowly Tizard and scientific sanity prevailed over the machinations of the Churchill-Lindemann axis in those pre-war years.

Zimmerman ends his book on a rather sad note – the failure of the RAF to defeat the blitz on London and other cities, when the Luftwaffe turned to night bombing after their defeat in the daytime battle of 1940. The sequence had been foreseen by Tizard and RAF Command and development of the airborne AI (Air Interception) radar for night fighters had commenced. The early prototypes rushed into action were a total failure. Zimmerman places the blame on the failure to engage the British electronics industries at this early stage. In fact, there were other fundamental problems that required another two years to overcome. It was not until the spring of 1941 before the scientific-industrial collaboration produced operational equipment that defeated the night bombers.

Meanwhile, there had been tragic loss of life in London and in other cities and the reader of this book may well feel regret that the RAF Commander-in-Chief and the scientists, who had been the inspiration and core of the shield that saved Britain in 1940, were either dismissed or neglected.

Sir Bernard Lovell, 2001

Acknowledgments

I would like to thank Simon and Carolyn Chenery for accompanying me on several jaunts around the British coast visiting sites related to the history of the air defences. The staffs of the archives at Nuffield College, Churchill College, the Public Records Office and the Imperial War Museum provided invaluable assistance in providing me with access to the documentation and photographs required for this study. A special thanks to Sebastian Cox of the Air Historical branch for taking time out of his busy schedule to provide some useful sources and priceless advice. Thanks to David Hall for conducting research for me at the PRO. Also a special mention to Paul Mackenzie for his advice and his acquisition of a valuable photograph from the PRO. The Science Museum, after a long delay, provided me with photographs of the Daventry equipment. The *Daily Mail* gave permission to publish the drawing by Emmwood. Eric Gilson provided a wonderful tour of Bawdsey and several rare photographs of the radar research laboratory. Alexanders International School was kind enough to allow me to visit Bawdsey. Grant Lohoar, Head Warden of The National Trust's Orford Ness site, provided an informative guided tour of the remains of the radar research facilities. Ian Brown and the Centre for the History of Defence Electronics provided invaluable advice. Faith Guildenhuys, Rob Alexander and Wendy Muscat-Tyler gave sterling service as editors and critics for my rough drafts. Financial support for this research was provided by the Social Sciences and Humanities Research Council of Canada and the University of Victoria. Finally, my thanks goes to the team at Amberley Publishing and to Alan Sutton in particular for publishing the book.

Introduction

RDF might well be the title of a colourful and romantic novel. Its introduction to the RAF, the opening of the first AME station at Dover, and its subsequent growth to the first line of Britain's Defence would make a most interesting and delightful reading. The present situation however does not permit romance, it is plain hard fact we have to deal with.

W. H. Lawrence, Sqn Ldr commanding No. 7 Radio Maintenance Unit, RDF Organization – Technical and Administrative, 28 October 1940.

The Introduction of Radar into Fighter Command between 1937 and 1940 is a textbook example of the successful application of science to war. It ought to be studied in staff colleges: and it ought to be studied by anyone who thinks that scientific war is an affair of bright ideas.

C. P. Snow, Appendix to *Science and Government*, 1962.

Almost exactly seventy years ago the Royal Air Force defeated the Luftwaffe in the Battle of Britain. As a result Britain and perhaps all liberal democracies were saved from the evil and despotic Nazi regime of Germany. Churchill spoke in awe of the valiant few, the fighter pilots who in their Spitfires and Hurricanes fought against terrible odds and often made the ultimate sacrifice in defence of their nation. The story of the battle has been told and retold, yet one vital element of the battle has almost been completely overlooked – the development of the radar-based air defence system. The individual heroism of the pilots would have certainly proven futile if they were simply romantic warriors, fighting like medieval knights as individuals in a deadly aerial joust. Instead, the pilots of Fighter Command were in fact warriors of the machine age, an integrated human component in what was then the most sophisticated and complex weapon system ever

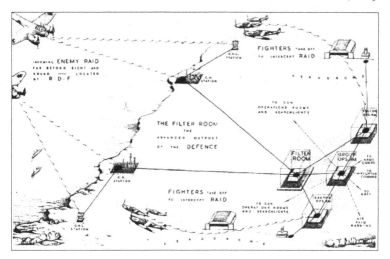

The Air Defence System.

developed. This is the first full narrative account of the development of this air defence system using all available archival and published sources.

At the heart of the air defence system was a technology that was invented just five years before the battle – radar or RDF, as it was known in British Commonwealth service until 1943. The importance of radar and Fighter Command's command and control system has long been recognized. What has not been told is the story of how this scientific and technological miracle was accomplished in such a remarkably short period of time. Some writers have dealt with the discovery of radar, mainly using often inadequate memoirs as their main sources. Almost all of the histories of radar begin in 1935 and primarily concentrate on the technical details of the device. Few have placed the story into the broader picture of air defence technological development and the political situation in Britain in the 1930s.

Radar was only one, albeit a major, part of a much larger story which has its roots in the First World War, when Britain began developing air defences to counter the first strategic bombing campaign in history. It continues in the 1920s with the re-establishment of strategic air defence and failed efforts to develop aircraft early warning prior to the discovery of radar. The discovery of radar itself can only be understood in terms of the political, strategic, and scientific context of the mid-1930s.

The attitude of the Royal Air Force towards air defence technology will be one of the major themes of this study. Far from being obsessed with the concept of strategic bombing, senior air force officers and Air Ministry officials embraced radar and did everything in their power to promote its development as an integral feature of the nation's defences. The crucial role of Air Marshal Sir Hugh Dowding in the creation of the new air defence system will emerge. Dowding and others worked closely with the scientists that supervised, managed, and conducted air defence research.

These scientists, whose toil is all too often forgotten in histories of the period, are the true heroes of this story. They include Henry Wimperis, the Director of Scientific Research at the Air Ministry from 1924 to 1937; Sir Henry Tizard, chairman of the Committee for the Scientific Survey of Air Defence (the Tizard Committee); Robert Watson-Watt, the discoverer of British radar; E. G. Bowen, who developed airborne radar; and, perhaps most unfairly overlooked, Albert Percival Rowe, the civil servant who was at the centre of air defence research policy development until he succeeded Watson-Watt as head of the Air Ministry's radar research laboratory at Bawdsey Manor in 1938, a position he would hold throughout the Second World War.

This is first detailed study of the often tumultuous but magical research programme undertaken at Orfordness in 1935 and than at Bawdsey from 1936 to 1939. The work on the daytime defences was eventually successful,

Sir Hugh Dowding.

but there were more than a few failures along the way. One of the most surprising aspects of the history of radar is that the decision to proceed with the Chain Home early warning system was made at a time when the technology was far from proven. This decision can only be understood in the context of the entire programme of research supervised by the Tizard Committee. From aerial minefields to unguided rockets no stone was left unturned in the quest for means to defeat the bomber threat. All of these other research projects failed and, with no other alternatives to defend Britain from the expected aerial onslaught of the German air force, the Air Ministry and the government put their faith and trust in the abilities of the Bawdsey staff.

Robert Watson-Watt.

The effectiveness of the radar system was ensured by two major research programmes conducted outside the confines of the laboratory. The development of interception techniques by a group at Fighter Command's air station at Biggin Hill was instrumental to success in the summer of 1940. The Biggin Hill team linked radar with two other technologies – radio-telephony and radio direction finding – which together made effective air-to-air interception techniques possible. The bringing together of all of the components of the air defences into a smooth running system was in large measure the product of the first extensive and systematic programme of operational research (OR). Modern OR originated at Bawdsey in 1938.

The development of the radar defence system is an integral part of the politics of rearmament. Central to this part of the story were the attempts of Winston Churchill and his scientific confidant, Frederick Lindemann, to seize control of the research agenda. Often viewed as simply a personality clash between Tizard and Lindemann, what emerges is a more unsavoury tale in which Churchill threatened to reveal the radar secret in public debate in order to force the government to accede to his demands. Churchill was completely misguided by Lindemann, whose single-minded obsession with aerial minefields prevented either of them from seeing the revolution in aerial defences that was being overseen by the Tizard Committee.

Much of the rapid early progress in the earlier years was a direct result of the drive, energy and leadership of Watson-Watt. Paradoxically, it would be Watson-Watt's erratic, almost manic, behaviour and lack of administrative skills which would be a significant factor in the failure to have effective night defences ready in time for the Blitz. Watson-Watt was primarily responsible for the disastrous move from Bawdsey to Dundee in September 1939, which led to the virtual collapse of radar research. Another cause of the failure to develop effective Airborne Interception radar in time was the decision in the late 1930s to not draw on the talents of the British electronics industry.

Despite the inadequacy of the night defences, this colourful and romantic story is one of triumph, since what resulted is one of the most remarkable military technological accomplishments of the twentieth century. Not only did the radar air defence system save Britain in its time of greatest need, but the long-term ramifications of the programme were felt throughout the Second World War and beyond.

Note on Spelling and Terminology

The British term for radar until 1943 was RDF. The term was always confusing, particularly since radio direction finding was a separate technology of great importance to the air defence system. I have explained the origins of the term RDF in chapter six. For purposes of clarity the modern term radar is used throughout the text except in direct quotations, in which case the original terminology is preserved.

I

The Origins of Strategic Air Defence and Aircraft Early Warning

On the day before Christmas 1914 a small German seaplane flew over the coastline of southern England and dropped a bomb, which landed in a garden near Dover. No one was hurt and minor material damage was done, but it marked the beginning of the first German strategic bombing offensive against Great Britain. In the next four years, in a series of fifty-one raids by Zeppelin dirigibles, and sixty-three by Gotha and Giant bombers, 1,415 people were killed, and 3,367 were wounded. Material damage was measured in hundreds of thousands of pounds, but more economic loss was caused by disruption of work, which occurred when people sought the safety of air-raid shelters.

In Britain the immediate response to this new form of warfare was the creation of the first strategic air defence system. The origins of the British air defence system which proved so vital in the defeat of the Luftwaffe in 1940-41 must be traced back to this earlier campaign. The system did not emerge overnight, and did not mature as an effective response to aerial assault until early 1918. Given that air warfare was itself brand new, however, the progress was remarkably rapid. By the end of the First World War some of the basic principles and components of the air defence system that would save the country in 1940 were already well developed. At first, the defences which were established were ad hoc and had a divided and confusing command structure. Gradually, as each successive threat was encountered, the defences grew in size and sophistication.[1]

The First Battle of Britain

Despite the first raid, in 1914 no conventional aircraft could bomb London and return safely to German airbases located in occupied Belgium. Instead, the only way to carry out this mission was to use Zeppelin dirigible airships, over 500 feet long and filled with highly combustible hydrogen gas. The effectiveness of the airships for strategic bombing was demonstrated by an attack on Antwerp on 26 August 1914. The defenders rose to the challenge posed by Zeppelin attacks in 1915 and 1916. The huge, lumbering, and highly inflammable airships proved relatively easy to destroy, particularly after the aircraft were fitted with forward mounted machine guns firing new incendiary and explosive bullets.

The development of a comprehensive air defence system can be traced to the first two highly successful daytime attacks on London on 13 June and 7 July 1917 by the Germans new Gotha twin-engine bombers. In the first of these raids a formation of fourteen Gothas arrived over the great city unmolested by the defenders. They released seventy-two bombs, killing 162 and injuring a further 400 people. All of the German planes managed to return to their base despite facing a series of uncoordinated attacks by a large number of defending aircraft.[2] In the second raid twenty-two aircraft dropped seventy-five bombs, killing fifty-seven and wounding nearly 200 people. Only one of the raiders was shot down by defending aircraft, although one battle damaged Gotha was forced to make an emergency landing on the beach at Ostend.[3]

Public reaction to the second raid was violent. There was rioting in London, angry mobs attacked stores they believed were owned by Germans. The repercussions in the newspapers and Parliament were almost equally fierce. On 11 July Prime Minister, David Lloyd-George, appointed the Prime Minister's Committee on Air Organization and Home Defence Against Air Raids, better known as the Smuts Committee after its principal member Lieutenant-General Jan Christian Smuts. The committee was charged to examine 'the air organization generally and [the] direction of aerial operations.'[4]

Within thirty days Smuts prepared two reports. Smuts' second report, which recommended the formation of an independent air force, has received the most attention from historians. Yet, it was his first paper, written in just eight days, which was instrumental in reforming the air defences. Smuts recommended that all of the air defences of London, including aircraft of both the Royal Flying Corps and the Royal Naval Air Service, anti-aircraft artillery and searchlights, be placed under the command of one man.

On 8 August Brigadier-General E. B. 'Splash' Ashmore was appointed to head the new London Air Defence Area (LADA) command, which was given the responsibility to defend the capital and all of south-east England. An Old Etonian, Ashmore entered the Royal Artillery in 1891 and learned to fly in 1912.[5] In 1917, he was on the western front in command of the artillery of the 29th division. Ashmore accepted the new appointment willingly since his dug-out north of Ypres 'was drenched with gas on most nights' and the members of the 'Expeditionary Force were inclined to look at the troubles of London somewhat light-heartedly.' He later confessed: 'The fact that I was exchanging the comparative safety of the Front for the probability of being hanged in London did not worry me.'[6]

Dealing with the immediate threat of daylight Gotha raids proved to be comparatively easy, and the Germans never seriously challenged Ashmore's rapidly improving defences. The German bombers staged three major daylight raids in August. The first raid, on 12 August by eleven Gothas against Southend, was met with only a sporadic series of attacks by defending fighters. Only one of the raiders was shot down, although three others, some perhaps with combat damage, crashed on landing. The second attack, on 18 August, proved to be an utter fiasco. Twenty-eight bombers took off, against meteorological advice, and promptly ran into severe head winds, which pushed them far off course. The entire formation was forced to turn back before crossing the English coastline and eight or nine of the bombers either crashed in the sea, were shot down after drifting into Dutch airspace or were forced to crash-land in German-occupied Belgium.

The third raid on 22 August involved fifteen aircraft, all that could be mustered after the disaster of the 18th. The plan was to bomb Sheerness, Southend or Chatham, and Dover. Five of the attackers were forced to turn back before reaching England, and the tiny force that remained was sighted by Kentish Knock lightship moored some distance off the coast, giving the defenders more than thirty minutes warning of the attack. While Ashmore mustered his defences for what he mistakenly believed was another attack on London, at the coast Royal Naval Air Service fighters and batteries of anti-aircraft guns were ready for the Germans. Three of the attackers were shot down and the rest did little damage to Dover, Margate and Ramsgate. It had been a disastrous defeat for the Gothas and they would never again launch a major attack against Britain during daylight.[7] It is important to note that the defences were so effective in the latter attack because of the early warning message received from the lightship.

As with the Zeppelins a year earlier, the Germans decided to switch their bomber force to launching night-time attacks. A series of night raids was launched, culminating in a raid on London by five aircraft on the night of 4/5 September. In these first series of evening attacks, a total of 152 people were killed and 177 injured. When the War Cabinet asked Smuts to comment on the latest developments, he could offer no easy solutions. He lamented that, 'Our aeroplanes afford no means of defence at night as they find it impossible to see the enemy machines even at a couple of hundred yards. They might just as well have remained on the ground.'[8]

Ashmore's command, however, would find the means to make the defences more effective. Key to air defence system was the rapid and accurate gathering and dissemination of information through a central command structure. The first step of the air defences was the initial detection of a raid. Against Zeppelins indications of attacks were frequently made by radio intelligence. The Germans were very lax about radio security and made frequent use of wireless sets both on the ground and in the air. Several of the German codes had been broken and, even if the messages could not be deciphered, a great deal of information was provided by signals traffic analysis and radio direction finding. Bombers, however, proved to be much more difficult to track by wireless. Radio intelligence could often detect the take-off of a raid, but 'their position and direction in the air could rarely be determined.'[9] It was the absence of accurate long-range early warning which posed the most critical weak link in the defences.

The defences around London consisted of three rings. The outer defensive ring of searchlights and guns was placed about 10 miles from the heavily populated areas of the city. It was designed to harass and break up bomber formations, and, if possible, attract the attention of patrolling fighters. Friendly aircraft were excluded from flying over the outer defensive zone, and the gunners were instructed to treat every aircraft flying overhead as hostile.

The middle ring, was reserved exclusively for fighters, which were aided by a large number of searchlights to illuminate the German bombers. Upon receiving the 'readiness' order, fighter squadrons would get the planes and pilots of the first patrol ready for immediate take off. When the approach of the enemy was confirmed, usually by observers on the coast, the fighters were ordered to take off and assume a predetermined patrol area. During the day squadrons were trained to fight as a group to overcome the defensive fire of any formation of Gothas. At night individual fighters flew standing patrols at predetermined altitudes hoping to spot an elusive attacker. By the summer of 1918 there were more than 200 modern fighter assigned to air defence duties.

The inner ring, located in London itself, consisted of more anti-aircraft guns and searchlights, as well a number of balloon apron barrages. The aprons consisted of three balloons, which supported an array of cables dangling beneath them. In total, over 20,000 personnel, 266 guns and 353 searchlights were required to man the ground defences.[10]

It took considerable time to make the LADA effective. By April 1918 a number of aircraft were fitted with radio telephones to allow pilots to receive the latest intelligence of the enemies' movements. It was not until nearly four months after the last major German air raid, on 19 May 1918, however, that the 'highly centralized intelligence and command system' was fully established. On 12 September a dedicated telephone communication scheme became operational. This allowed every component of LADA to be linked directly to twenty-five regional sub-control rooms. The sub-control rooms would plot spotting reports from their district with counters onto a large scale map. These plots were then read off by a 'teller' to a plotter in the central control room at the Horse Guards in London.

The central control room was the archetype of Fighter Command's headquarters used in 1940. It 'consisted essentially of a large squared map fixed on a table, round which sat ten operators (plotters) provided with headphones; each being connected to two or three sub-controls.' Information from the sub-controls was plotted on the map using 'an ingenious system of coloured counters.' The counters were removed at intervals to prevent the map from becoming overcrowded. The central control room assessed the information, and passed on the approach of raids back to the sub-control rooms. Ashmore surveyed the map from a raised gallery behind the plotters, and he could at a flick of a switch talk directly to any of the sub-controls.[11]

The new system provided tremendous possibilities. Previously, Ashmore and his staff could do little to influence the course of operations once a raid was underway. Now that LADA had the ability to be kept constantly informed of enemy movements, 'the power to manoeuver defending formations by day, and to concentrate patrols at will at night.' Ashmore calculated that this increased the chance of a successful interception 'about fourfold.'[12]

Despite Ashmore's claims, however, the effectiveness of the air defences during the First Battle of Britain remains a subject for historical debate. The last and largest raid carried out by the Germans occurred on 19 May. It consisted of forty-one aircraft, of which only thirty-one were able to launch an attack, and just thirteen bombed London. Six of the raiders were shot down, three by fighters and three by anti-aircraft guns. A further three bombers crashed in landing accidents. The German decision to terminate large-scale bombing, however, had little to do with the improvements in the defences. Instead, it came about because the German high command determined that they were getting more useful results in the night bombing of targets immediately behind the Allies front lines in France.[13]

While the LADA defensive structure would form the nucleus of the air defence system used by Fighter Command in 1940, its sophistication and effectiveness should not be overstated. The defences benefited from the technological limitations of the Gotha and the later Giant bombers, which operated at the very limit of their capabilities. A very high percentage of the bombers never reached their targets because of mechanical failures or were destroyed in landing accidents. The numbers of bombers were relatively small and the defenders always outnumbered them substantially. Major raids were usually at least several weeks apart. Since no fighter was capable of accompanying the bombers, they had to rely on their own defensive armament. No attack was launched directly on the defences themselves.

Yet German and British technological achievements were remarkable given the state of technology and the non-existence of doctrine prior to the war. To Ashmore and his predecessors during the war, the problems in providing an effective air defence were both organizational and technological. The rapid change in technology and tactics used by the Germans required equally speedy improvements to the defences. Ashmore, an excellent staff officer, could solve the organizational deficiencies of the defences, but he had little control over the technology. In large measure the growing effectiveness of the air defences was not the work of the traditional military establishment, but the direct result of a remarkable coupling of science to the war effort.

Science in the First World War

Air warfare was just one of many new forms of combat first encountered in the First World War. It was in that conflict that scientists began to have a central role in the development of new ways to manufacture raw materials and in the design and development of revolutionary new weapon systems.

It was not long before the major belligerents grasped a central truth: only by harnessing the intellectual skills of their scientists could they hope to stay ahead in the technological war. In Britain, the Admiralty established the Board of Invention and Research, while the Department of Munitions formed the Munitions Inventions Department. One of the first areas subjected to scientific research was aeronautics and related technologies. Much of the equipment used by LADA was specifically developed for the war effort. Radio telephony (R/T), which allowed reasonably reliable two way air-to-ground communication by the autumn of 1918 was a crucial component of Ashmore's defensive system.[14]

The largest research group devoted specifically to air defence questions was the Anti-Aircraft Experimental Section of the Munitions Inventions Department. It was created in January 1916 to investigate ways to improve anti-aircraft gunnery. Archibald Vivian Hill, better known simply as A. V. to his friends, was chosen to head the unit. Hill was a twenty-eight year old Cambridge physiologist who was just starting a flourishing research career when his Territorial Army unit, the Cambridgeshire Regiment, was mobilized for active service in 1914. Like most young scientists, Hill's professional talents were not recognized for the first two years of the war and only his expertise at rifle instruction kept him from sharing the same fate of many of his contemporaries in the trenches. Hill was selected to head the new research team in 1916 because his skills at applying rigorous mathematical analysis and in constructing scientific apparatus was seen as directly applicable to the problems of air defence.

Hill recruited an impressive team of 'young mathematicians and physicists who became known in the Service as "Hill's Brigands".'[15] The section used statistical analysis and experimental science to improve the accuracy of anti-aircraft gunnery by examining such things as the effect of high altitude winds and barometric pressure on shells and fuses and the accuracy of range tables. Hill would later claim that the experimental section created the science of operational research in the process.[16] One of Hill's research team turned its attention to developing devices for the short-range detection of aircraft, in order to direct searchlights onto attacking aircraft at night. These devices consisted of two or three conical horns or trumpets installed close together. The observer listened for aircraft using a stethoscope, training the device by attempting to equalize the sound detected in each ear. Bearings were then transmitted to searchlights which would illuminate the aircraft or airship for nearby anti-aircraft guns. The first two hundred Sound Locators No. 1 were ordered at the end of October, 1917.[17]

The Origins of Early Warning of Aircraft

While Hill's group focused its attention on improving anti-aircraft gunnery, searchlight and developing short-range sound detectors, one of the most pressing problems remained unsolved during the war. Ashmore's entire defensive organization was based on the fact that no accurate information on the direction, height and numbers of bombers was available until just before they crossed the coast. As a result, Ashmore resorted to organizing the defences into three rings in which fighter aircraft, the most effective weapon, operated only in the second circle. Aircraft defending London had to operate well back from the coast so that they would have the time to climb to reach patrol heights before the bombers passed overhead. This left a rather small area available for fighters to operate, no further than 10 miles in width. Fortunately for Ashmore this weakness in the defences was not altogether crucial, because of the comparative slow speed of the bombers. The fastest of these, the Gothas, normally operated at a top speed of only 80 mph when fully loaded.

The importance of early warning systems did not go unnoticed during the war, and research into this problem in Great Britain can be traced

back to early 1915 when physicist Frederick Lindemann began to conduct research on early warning systems at the Royal Aircraft Factory at
Farnborough. Lindemann was the son of a successful French-Alsatian
engineer who had immigrated to Britain at the age of twenty. His father
became a very wealthy businessman and married an American woman
whose father was born in Britain. His mother was taking the baths at
Baden Baden in Germany when Frederick was born somewhat prematurely on 5 April 1886. The family's country home in Devonshire was
provided with a well equipped private laboratory and observatory. The
young Frederick spent considerable time there learning basic science and
astronomy in this ideal setting. His father then sent him to high school
in Germany where Lindemann revealed himself to be a brilliant mathematician. Nernst accepted him as him for post-graduate studies at the
Physikalisch Chemisches Instut at Berlin. He received his PhD there in
1910; his thesis examined the applicability of then current theories of low
temperature physics to metals and metalloids.[18]

In 1914, Lindemann was denied entry into the armed forces as a result
of his place of birth and his family's roots in German-occupied Alsace.
Lindemann was determined to show that he was more loyal than any
native-born son and, if he could not do so in France, he would use his scientific talents in the service of the king.[19] By April 1915 Lindemann joined
the civilian staff of the Royal Aircraft Factory at Farnborough.

He began working on means to detect aircraft at a distance. He tried
several possible avenues of research; the most promising was using 'long
wave radiation.' It is unclear what type radiation he thought he could
detect; his later actions suggest it was either infrared emissions from an
engine or radio interference produced by aircraft magnetos. Nothing
apparently developed from his efforts to invent a device useful for the
long-range detection of aircraft.

Lindemann also explored as a possible means of providing distant
early warning the detection of sound made by an approaching aircraft.
He concluded, however, that sound was not a suitable medium for further research since atmospheric scattering meant that it could never be a
completely reliable means to detect aircraft. Lindemann was never one to
doubt his own findings. A decade later these early conclusions would be
his basis for challenging another research programme that would attempt
to use sound as a means for early warning.[20]

By the autumn of 1915, Lindemann would turn his attention away from
early warning to what he believed was the far more promising field of
aeronautical research. Here he would make a name for himself as one of
the pioneers in the scientific analysis of aircraft flight dynamics. His most
notable success was in developing methods that pilots could use when

their aircraft entered a deadly spin. He became well known for his decision to actually test out his theoretical concepts in the air. In order to do this he first had to learn how to fly.

While Lindemann ceased his research into aircraft early warning devices, other scientists continued to pursue the matter. They disagreed with Lindemann's conclusion that the detection of aircraft noise was useless. In June 1915, Professor T. Mather of the City and Guild (Engineering) College began experiments on detecting the sound of approaching aircraft. The source of Mather's scientific research is unclear, but he may have based his ideas on French technology that was being developed to detect sound behind the German trenches. The French devices were dubbed by the British scientist 'sound mirrors' apparently because they appeared to magnify and focus sound waves in a way analogous to how a telescope's mirror acted on light. Instead of the highly polished surface for reflecting and focusing light in a telescope, acoustical mirrors relied on the smooth surface of a very dense material like concrete or chalk to provide the best magnification and focusing of sound waves. Like a reflecting telescope, the sound mirror had a focal point located in front of the mirror.

Mather first built and tested three small concrete sound mirrors. Listening was done by a stethoscope attached to the focal point. In his

Acoustic Mirror carved in chalk, 1915.

laboratory Mather found that the mirrors increased by between five or six times the range in which he could hear the ticking of a clock.

After experiments in his laboratory proved the superiority of the parabolic design, Mather took his best mirror to the Royal Aircraft Factory for further tests. Various experiments by the factory's staff showed that a much larger mirror, at least 16 feet in diameter, was needed to magnify effectively the sound of an aircraft engine. That summer Mather continued his experiments at Bimbury Manor, near Maidstone. To produce a prototype quickly, Mather's team carved a 16-foot diameter mirror out of a chalk cliff. A small wooden listening platform was built 10 feet in front of the mirror for listening at the focal point. A series of test flights over the front of the mirror were undertaken, but only three successfully flew within the listening cone of the fixed mirror. An aeroplane powered by a 70 horsepower Renault engine flying at 4,500 feet was detected at 10½ miles away, at least four times the distance of an unaided ear. On 12 August they were able to detect a Zeppelin 12 miles away. Mather concluded that in service acoustical mirrors should be built of reinforced concrete, be 20 to 30 feet in diameter for 'a safe margin of power' and be mounted on gimbals so that they could be trained on approaching aircraft. Two or three of these mirrors working from a known base would form an accurate early warning system for aircraft.[21] Mather's enthusiastic reports to the War Office led to further tests in September by the Royal Flying Corps at Uphaven aerodrome. Inexplicably, they appear to have used the 4-foot mirror tested at Farnborough earlier in the year. The commander at Uphaven also declined the help of Mather's assistant. Not surprisingly, given these two handicaps, the acoustical mirrors were found wanting. On 6 October the War Office informed Mather's 'that the tests have shown that the invention does not live to the standard necessary to ensure its adoption by HM government.'[22] Despite several pleas by Mather asking the War Office to reconsider the mirrors, no further work was undertaken on long-range aircraft detection until late 1917.[23]

In November 1917, Captain W. S. Tucker, Royal Engineers, submitted a proposal to the Air Inventions Board of the Military Inventions Department concerning the development of an electronic long-range aircraft detection apparatus. Tucker, destined to lead British scientific research on aircraft detection until the discovery of radar in 1935, was a physicist with more than a year of experience in military technical research by 1917. He had obtained his degree in physics from Imperial College in 1902. In 1909, he had become an occasional student of Professor H. L. Callendar, receiving a DSc. from the University of London in 1915. Callendar, practised 'old fashioned physics,' which emphasized tediously exact experimentation over theory, an approach that was more closely related to engineering than modern physics. Tucker was profoundly influenced by his mentor's approach to scientific problem solving.[25]

He began his military scientific career in April 1916 working with Major (Dr) W. L. Bragg. In June he invented the gun microphone, the key component in gun sound ranging systems, which used sound to direct counterbattery fire against German artillery. In August 1916, he was appointed the research officer for gun sound ranging. In December he was selected to found and equip the experimental sound ranging section at Salisbury Plain. While there he established the accuracy of gun sound ranging using the Tucker microphone. The system was adopted throughout the British Army, and was also used extensively by the French and the Americans. In January 1917 he received a mention in dispatches for this work.

Tucker then adapted his microphones for use in the calibration of artillery muzzle velocities. He successfully tested this device in France in May 1917. As a result, gun calibration sections were established in France and England by the late summer.

With gun calibration and sound ranging established, General van Straubenzee suggested to Tucker in the autumn that he consider 'the possible application of electrical methods to aeroplane detection.'[26] Van Straubenzee probably made this suggestion because of the increasing aerial assault on Great Britain by German Gotha bombers. This led to Tucker's fateful submission to the Air Inventions Board.

Tucker told the board that he believed that he could develop a device for the long-range detection of aircraft using another adaptation of his microphones. The Tucker microphones could detect even the faintest sounds, and could be tuned to any given note so as to exclude extraneous sound such as those from guns, traffic, and friendly planes. He proposed constructing 'a large number of microphones in a series on a circular frame' of about 20 feet in diameter. The sound detected from the microphone would be detected optically, using a galvanometer. The galvanometer measured small changes in electrical current from the microphone by the movement of a magnetic needle.[27]

Within days the Air Inventions Board approved the establishment of a new research group, the Acoustical Research Section, with Tucker as head, located at the Military Inventions Board's experimental grounds at Imber Court, Thames Ditton. The focus of Tucker's work was to be the development of electronic long-range detectors of aircraft at night. By February 1918 Tucker had established the Acoustical Research Section laboratory. Personnel consisted of Tucker, designated as Superintendent of Acoustical Research, one or two other officers, and three Sappers. His second in Command was Lt E. Tabor Paris, who had worked with him on gun sound ranging.

By the beginning of 1918, Tucker had surveyed all devices tested or proposed for the detection of aircraft. Of particular interest to him was Mather's sound mirror. Early research had 'found that a microphone mounted in front of a rigid wall' which acted as a reflector of sound, was much more effective than one supported in mid-air. Mirrors not only magnified the sound, but focused it as well. As early as 11 January 1918, in one of his first progress reports, Tucker sketched out how the mirror could be combined with the microphone. The microphone replaced the human observer using a stethoscope at the focal point of the mirror, with accurate readings of the intensity of the sound being transmitted to an attached galvanometer or by listening with a telephone.[28]

For the remainder of the war, Tucker's research group made slow methodical experimental progress. Research included the fundamental analysis of the way sound moves through the atmosphere and identifying the specific tones generated by aircraft. Much work was also undertaken developing and improving equipment needed for aircraft detection. Particular emphasis was placed on tuning microphones to respond only to the sound of Gotha bombers, and developing amplifiers to magnify any sound detected.[29]

In June a field station was established at Joss Gap on the Isle of Thanet. By early September a prototype of the 20-foot disk of microphones was completed and tested with a variety of aircraft. The disk was mounted so that it could be traversed to follow an aircraft. The results seemed most promising, with aircraft being followed up to 8 miles out to sea. On 12 October a demonstration of the disk was given to Major General Ashmore and other officers. However, heavy winds and large rolling waves breaking on the nearby beach forced Tucker to reduce the sensitivity of the microphone. This decreased the maximum range an aircraft could be detected to between 3 or 4 miles. It was a disappointing result, but Tucker argued that the adverse winds, which had hampered the disk, also greatly limited hearing an aircraft with the unaided ear and would have likely prevented a Gotha raid.[30] By the armistice Tucker was a long way from perfecting a long-range early warning device and it remained uncertain if his work would continue.

2

Tucker's Acoustical Mirrors

At the end of the First World War, it was clear that air power would play an important, if not decisive role, in any future conflicts. Two main inter-related questions about air power, however, remained. The first was which organization or organizations were to control military air power. The very existence of the Royal Air Force was in some doubt, since the service was created at least as much to appease domestic political concerns as to fill any actual military requirements. Secondly, the issue of what combat role aircraft would play would remain the subject of debate throughout the interwar period. It was uncertain if their primary function was to provide support to the army and navy, or if an air force should emphasize independent strategic roles? If strategic roles took priority, the question of the amount of resources to devote to air defence or to offence bombing had to be addressed. The answers to these questions did not come quickly or easily.

In the first five years after the armistice the most pressing issue was whether or not there would be any air force at all. Throughout the 1920s the British armed forces laboured under the imposition of the 'Ten-Year Rule.' The guiding principle of the rule was that funding for the armed forces was based on the assumption that there would be 'no great war during the next ten years.' First proposed by the War Cabinet on 15 August 1919, the rule was gradually expanded and reinforced until the early 1930s. It greatly limited funding for defence, which was a popular policy with the war-weary electorate.[1]

Rapid disarmament left Great Britain with virtually no military aircraft and saw the almost complete dismantlement of Ashmore's air defence system. By 1920 not a single anti-aircraft gun or searchlight remained operational, the eleven fighter squadrons 'melted away until almost nothing remained', and Ashmore's efforts to retain the communication systems and control rooms were unsuccessful.[2]

The Formation of the Air Defence Experimental Establishment

As a reflection of the important role played by science during the war, one of the few elements of the air defence establishment retained was its research arm. This was not accomplished without some difficulties. Most of the scientists who undertook military research were anxious to return to their civilian occupations as soon as the war concluded. For instance, all three of the scientists who had developed short-range sound locators for the Anti-Aircraft Experimental Section had government or university careers which they returned to as soon as possible.[3]

Sound location of aircraft was one area in which a continuation of research was deemed particularly important. In early January 1919, a War Office sub-committee examined the state of sound detection research. It concluded:

> It would appear that the AA defences should to some extent be kept permanently mobilized, and a good lookout system maintained. This look-out must depend on some acoustical method, and should require as small a personnel as possible. The matter is still in its infancy and much experiment and experience is necessary before the best method can be decided on. If, as appears probable, flying develops commercially in the future, the detection and location of aircraft by sound will become of general and not only of military importance. This is further reason for continuing research on the subject.

The committee 'strongly recommended' that immediate steps be taken to retain at least some of the wartime personnel already familiar with aural means of aircraft detection.[4]

The War Office concurred and in early 1919 ordered the creation of the Acoustical Section of the new Signals Experimental Establishment (SEE) based at Woolich. The SEE was renamed the Air Defence Experimental Establishment in 1925, and the Acoustical Section was moved to Biggin Hill aerodrome in October 1929. (To avoid confusion the name ADEE will used throughout this book.) The acoustical section was to be given responsibility for all War Office acoustical research, including long and short-range aircraft detection and gun sound ranging.[5]

The only applicant to head the new group was Tucker. Tucker's qualifications were impressive and he was immediately hired. His second-in-command was Dr E. T. Paris.[6] All told between fifteen and twenty research personnel were on the establishment of ADEE at any given time between 1920 and 1935, two-thirds of whom were engaged in aircraft detection work. By the standards of the day it was among the largest military scientific programmes. It was also the only research centre examining air defence problems, although it remained under the control of the War Office.

In the immediate postwar period it was unclear if the RAF or the War Office would be given the responsibility for any future air defence organization. Simply by default, the War Office retained Tucker's group and put it under the supervision of the Board of the Royal Engineers. Although the technical arm of the War Office, the RE Board was not interested in directing or critically examining ADEE research except for doling out limited annual funding.[7] ADEE's research was classified and was virtually unknown outside of the defence establishment. Thus, until 1925 no one critically examined ADEE's research programme.

A better scientist than Tucker might have used the comparative independence offered him by this institutional indifference to make a significant technical breakthrough. As we shall see in the next chapters, the solution to the long-range aircraft detection problem dangled right in front of the eyes of many physicists and radio-engineers in the early twenties, but few made the intuitive leap to appreciate that a commonly observable phenomenon could be turned into a practical device. Those few who appreciated that the reflection of radio waves by aircraft might lead to the development of radar lacked Tucker's independence to pursue radically new avenues of research.

Tucker was no theoretician, but was instead a doggedly determined experimental scientist who could see but one solution to the long-range detection problem: his own. Nor would Tucker consider related problems and their solutions. In 1918, Tucker was in close contact with physicists working on the detecting of submerged submarines.[8] As in aircraft detection, early submarine detection research focused on developing passive listening devices – hydrophones. By 1918 extensive work was underway on active sonar systems, or asdic submarine detection systems, which relied on transmitting high frequency sound waves through the water and listening for their reflection from submerged objects. No one at ADEE ever considered how one might develop an active air detection system, because sound did not travel effectively through the atmosphere. If they had done so, a solid theoretical examination of the problem might have led them to radar.

Early Research on Long-Range Detection at ADEE's Acoustical Section

While the focus of this study is on long-range aircraft detection, between 1919 and 1939 the ADEE Acoustical Section was responsible for all aspects of military atmospheric acoustic research and development. Much effort was devoted to improving equipment already in service, such as gun sound ranging systems and short-range horn aircraft locators. The latter were the devices Hill's team had developed during the war for use with searchlight and anti-aircraft guns. Extensive fundamental research into the properties of sound was conducted, including examinations of the way sound travelled through the atmosphere in different meteorological conditions. Direct work on a long-range aircraft detection system was, therefore, only one part of ADEE's work. It was, however, crucial to maintaining and improving ADEE's standing within the military establishment.

By early 1920, significant developments had occurred in the research programme. At the Joss Gap Station the free disk system of detection had been abandoned in favour of mirrors. It appears that the disk system, which does not use a mirror but instead relied on the direct detection of sound by Tucker's microphones, was prone to jamming from the wind and other extraneous sounds. Tucker instead decided to copy Mather's earlier work by making an artificial cutting in a cliff face near Kingsgate Castle. The mirror had a diameter of 15 feet across, and was tilted at an angle of 15 degrees to the vertical.

At the front of the mirror an iron pillar supported a pivot which was at the centre of the curvature of the mirror. Attached to the pivot was a steel lattice arm, free to rotate in altitude and azimuth. At the extremity of the arm were placed either trumpets for listening with a stethoscope or microphones for electronic listening. The distance of the arms from the mirror was 'the focus position for sounds produced by a distant source.' The operator stood on a small platform erected in front of the mirror about 2 feet from its lower edge.[9] When using a stethoscope, two trumpets were mounted, each 1 foot long, 5 inches wide at the mouth, tapering down to 1 inch at the stethoscope attachment. This allowed for stereo listening which enabled the operator to bracket sound sources to achieve an intended accuracy of a few tenths of a degree.[10]

The 15-foot mirror suffered from a variety of significant technical limitations, but it gave valuable service in testing the basic theory of mirror technology. Aircraft were followed up to 10 or 11 miles out to sea using both microphones and sound trumpets. In late 1920, a seaplane was followed for 15 miles, but in less favourable conditions aircraft could only be made out at about the same time as they were heard with the unaided ear.[11]

In early 1921, the testing station was moved near the Hythe Musketry Training School. A new larger free standing mirror was constructed. The 20-foot diameter mirror was erected against a steep embankment abutting on to the Romney Marshes and overlooking the Channel. The site was carefully chosen to be near to the route of cross-channel civilian air-

Abbotscliffe 20-foot mirror.

Denge 30-foot mirror.

craft, which it was hoped would prove an alternative to arranging costly RAF overflights. The mirror was built of concrete with an umbrella of reinforcing iron rods. The listening arrangements were similar to those of the Joss Gap Mirror, except that initially no provision was made for electronic listening. Continuing problems with the susceptibility of the Tucker microphone to jamming meant that, for the time being while improvements were made, all listening would be done by trumpets attached to a stethoscope.[12]

The first full-scale testing of the new mirror occurred in September 1923. In a series of day and evening flights by Vickers Vimy aircraft, the mirror was put through its paces. The key test of the mirror took place over three evenings starting on 26 September. These flights reproduced 'as closely as possible' the conditions of an actual raid in which the listener had no previous knowledge, either of the time at which the aeroplane would arrive or of its bearing.

On average, the mirror detected aircraft for non-directive listening at 18 miles. Accurate bearings to the target, within a few degrees, occurred at around 10 miles or eight minutes of warning. In his official report of the September tests, Tucker painted a rosy picture of the mirrors accomplishments and its future potential 'to give early warning of [the] presence of aeroplanes in a time which would enable anti-aircraft defence units a few miles back to get into effective operation, without requiring them to be continuously on alert.'[13]

Testing of the mirror continued on a variety of aircraft in the spring and summer of 1924. Results were similar to 1923.[14] By early 1925, experience with the mirror, and the results from a long series of experiments on the type of sound made by aircraft and the way sound travelled through the atmosphere, led Tucker to propose a radically new mirror design. Experiments had shown that very long wavelength sound generated by aircraft seemed to travel the longest distance, but that the 20-foot mirror was ineffective at picking up this type of sound. In January Tucker proposed the building of a huge 200-foot strip mirror. By early May, a preliminary design was ready, with the costs of its construction estimated to be just over £3,000, a relatively small amount for such an important device.[15]

The Anti-Aircraft Research Sub-Committee

At virtually the same time Tucker made his request to construct the 200-foot mirror, ADEE's autonomy came to an end. Serious postwar planning on the provision of Home or Metropolitan air force and an air defence scheme began in 1922 when the government approved a scheme to provide twenty-three RAF squadrons for home defence. This number was increased on 20 June 1923 to fifty-two squadrons because of growing tensions with France over German reparations and the French occupation of the Rhur. This force of 394 bombers and 204 fighters was to be completed in five years, or by 1928-29. There was little enthusiasm for the plan outside the Air Ministry, it being seen as a 'melancholy alternative' to international disarmament.[16]

The politicians felt compelled to agree to these schemes because of an irrational fear of air attack that was fuelled, in part, by the propaganda of the air force and its supporters. As one senior RAF officer told the Naval Staff College in 1924:

> It is no exaggeration to assert that inadequately opposed air attack on the part of a first-class air power, such as France, would:
>   Cause immense casualties
>   Cause incalculable material damage
>   Sever our food supply communications within the country
>   Render Government in London impossible
>   Immobilise home force of the Army and Navy
>   Reduce production in factories to negligible quantity.

Nor, unfortunately, would the effect be confined within even those limits. It is probable that under the fearful strain of unremitting air attack, with its concurrent disorganization of food supplies, revolutionary elements might gain the upper hand with consequences which it would be difficult to foretell.[17]

While the RAF exaggerated the effect of strategic bombing, it made a reasonable assessment of the technical and tactical realities of the time when determining the ratio of bombers to fighters. The offensive strike forces were, at least on paper, given priority, but there was never any serious consideration given to abandoning strategic air defence. Planning for the revival of an air defence scheme began in 1922 with the appointment of a joint Air Ministry-War Office committee consisting of Air Commodore J. M. Steel and Colonel W. H. Bartholomew.

In February 1923, Steel and Bartholomew put forward a plan that became the basis for the home defence prior to the development of radar. The Steel-Bartholomew plan followed closely the scheme of Asmore's LADA. It consisted of an outer belt of advanced observer posts, an outer artillery zone, an air fighting zone and an inner artillery zone. The air fighting zone was set 35 miles back from the coast, except to the east of London where there was not enough room, to provide the minimum amount of time required for fighters to climb to 14,000 feet. Each fighter squadron was assigned a specific patrol area in which to await bombers either approaching or returning from their targets.

The Steel-Bartholomew plan was somewhat amended in 1924 by a committee headed by Major-General C. F. Romer, and with Ashmore among its members. Among the Romer Committee's recommendations was the formation of an Observer Corps to track aircraft and of a new organization, called Air Defence of Great Britain, to command all home air defensive and offensive forces.[18]

The formation of Air Defence of Great Britain marked the cessation of the postwar apathy towards air defence organization. ADGB was made responsible for the defence of Great Britain from aerial attack, and included War Office ground troops, bombers and maritime patrol aircraft within its remit. Its first commander was Air Marshall Sir John Salmond.[19]

Shortly after the formation of ADGB, the first serious effort to find new ways to improve air defences commenced. In the spring of 1925 the Committee for Imperial Defence established the Anti-aircraft Research Sub-Committee (ARC) 'with the object of examining whether scientific developments had brought any new conditions into existence which effected the efficiency of aircraft.' The chairman was Lord Haldane. Other

The author at Acoustic Mirrors.

members included Lord Chatfield, the third Sea Lord; Air Vice Marshall Sir W. G. H. (Geoffrey) Salmond; Lieutenant-General J. F. Noel Birch, the Master General of the Ordnance; and Sir John Anderson, the principal secretary of the Home Office. Scientific representatives on the committee were the geneticist J. B. S. Haldane, and the physicist Professor Frederick Lindemann.[20]

Lindemann would take the lead in evaluating ADEE's research. After the war, like most military researchers, he had left behind aeronautical research at Farnborough and returned to academic life. In 1919, he was appointed to the chair of Experimental Philosophy at Oxford, and headed the Clarendon Laboratory. He was a good, although not brilliant, physicist whose productive research years lay behind him. Lindemann spent his time unsuccessfully attempting to turn the Clarendon Laboratory into a worthy rival for Cambridge's Cavendish Laboratory. He was a fierce and stubborn, but often ineffectual, critic who took an instant dislike to acoustical mirrors, in large measure because in 1915 he had concluded that sound could never be an effective medium for locating aircraft at a distance because of atmospheric scattering. However, his analysis of what was wrong with the technology and his recommendations to replace it were usually as ill-founded as what he condemned.

Long-range detection of aircraft was a key component of the committee's investigation. In all of the planning documents up to 1925, it was recognized that early warning of the approach of distant aircraft was crucial. The Steel-Bartholomew plan anticipated sound locators making the initial detection of approaching enemy bombers. Increasing the performance of early warning was considered vital in order to 'move the fighting zone forward', providing more opportunity for fighters to intercept attacking aircraft.[21]

The ARC conducted a wide-ranging investigation into realistic technical solutions, such as improvements to fighter aircraft, barrage balloons and anti-aircraft guns, as well as utterly fanciful notions, like the development of floating aerial minefields and death rays. Acoustical mirrors fell somewhere between the two extremes.[22]

The RE board presented a position paper on the current state of sound location to the committee at its second meeting. We need not re-examine this paper in detail, but it was based on some remarkable assumptions, not the least of which was that attacking bombers would only attack at speeds of 80 mph! The paper urged approval of the £3,000 required to build the 200-foot mirror, and outlined for the first time a system of mirrors to defend London. To cover 200 miles of coast line 10 mirrors would be required. Total cost would be £250,000, and 175 personnel would be required to man them.[23]

Front view of a 200-foot mirror.

Discussion of long-range aircraft detection began on 24 June 1925 at the ARC's second meeting. General Birch, who had recently attended a demonstration of sound mirrors at Hythe, told the committee that progress 'appeared to be satisfactory.' However, he hesitated to place 'any great measure of reliance in this method' in its current stage of development. The science of acoustics was 'very complicated' and among problems still being worked out was the production of an instrument that could be truly selective. Also, it might prove to be very expensive. It was generally agreed, however, that sound location 'would be most helpful,' although some doubts were expressed by Lindemann.[24]

At the next meeting on 8 July a full discussion was held on long-range aircraft location with Colonel Waite, head of the RE Board, and Tucker present. Waite began by explaining that the mirror at Hythe gave, on average, five minutes earlier warning than the unaided ear. Direction could be computed with up to 6 degrees of accuracy. It was hoped that the new larger mirror would detect aircraft at twice the range of the Hythe device.

Lindemann was unimpressed and began his first of many attacks on acoustical detection. He pointed out that neither the accuracy nor the warning time was sufficient to justify continuing the project. He doubted that the new mirror design would prove effective because, while a sixty horsepower engine could 'theoretically' be heard 200 miles away, there was 'a very great difference between theory and practice ... due to damping of wind eddies, etc.', which caused atmospheric scattering of sound. The whole scheme could be replaced by posting observers in small ships off the coast.

Unfortunately, the only account of this confrontation between Tucker and Lindemann is contained in the reserved and unemotional official minutes. However, the minutes can not hide the fact that Tucker systematically demolished Lindemann's arguments and showed that the committee's scientific advisor knew very little about atmospheric acoustics. Tucker explained that test results from the 20-foot mirror could not be extrap-

View of the forecourt of a 200-foot mirror and disposition of microphones.

olated to work out the anticipated performance of the 200-foot mirror, because they detected sounds at different wavelengths. He added that atmospheric scattering was not an important factor in limiting the range of the mirrors, and was almost insignificant at night, the time when the RAF most feared a surprise attack by enemy bombers.

Much to Lindemann's chagrin, Tucker won this opening debate, gaining the support of Salmond. The air marshal told the committee that an increase in the range of detection of from 10 to 15 miles 'would be of considerable value in the air defence of this country owing to the very short margin of time which was available to get our defending aircraft up to the requisite height to engage the enemy.'

Despite some further discussion led by Lindemann concerning the jamming of the acoustical system, the committee concluded that there should be no delay in continuing with ADEE's experiments and that funds should be secured to build the 200-foot mirror. However, funding for the mirror was contingent on there being no further scientific objections to the device, and Lindemann was to be given full access to ADEE research reports.[23]

Lindemann's reaction to what he perceived as a humiliating defeat was consistent with his behaviour in his better known scientific disagreements on defence issues. For the next fifteen months he undertook a fierce attack on acoustical detection which effectively paralysed Tucker's work and eventually the ARC itself. Lindemann forced a thorough scientific re-evaluation of acoustical mirrors, while promoting investigation into other means of aircraft detection. Fortunately for Tucker, Lindemann was not a particularly astute scientist, nor did he have the type of political influence he possessed from 1934 onwards when he was supported by his great friend, Winston Churchill.

Lindemann's criticism foundered on the complete failure to find any other method of aircraft detection, and his inability fully to understand the intricacies of atmospheric acoustics. Lindemann promoted the development of apparatus that detected either infrared or radiant heat from aircraft engines or electro-magnetic radiation given off by the aircraft magnetos. The former was beyond the capabilities of science before the

Second World War. The latter was a technological dead end because air-craft magnetos were carefully shielded in aircraft to avoid interfering with onboard radios. Low-technology alternatives that were also promoted by Lindemann, such as observation from small naval craft or kite bal-loons, also proved unworkable due to sound interference.[26] Remarkably, Lindemann failed either to read or comprehend the significance of one submission to the committee which, as we shall see, outlined the possibil-ity of using reflected radio waves as means to detect aircraft.

Lindemann's scientific criticisms of acoustical detection were somewhat more effective. On 28 September he provided a report to the ARC highly critical of the 'comparatively disappointing' results of mirrors. Lindemann presented a detailed mathematical analysis that indicated that even under ideal conditions no mirror could be effective at more than 15 miles range, nor could they hear aircraft more than 4 miles further than an unaided ear. The latter was a remarkable statement given that ADEE had achieved far better results with the mirror at Hythe.

Yet Lindemann had found a key problem with acoustical mirrors that had been carefully downplayed in Tucker's research reports – their incon-sistent performance depending on atmospheric conditions. While Tucker had emphasized in the main text of his reports the successful tests of the mirrors, there were numerous trials where aircraft had not been detected to any significant degree before they were heard by the unaided ear.

Lindemann believed that there were two insurmountable causes of this inconsistency. The first involved scattering of sound by the atmosphere. This was the continuation of Lindemann's earlier argument, and, despite a detailed mathematical analysis, it had no more validity than in July. The second was the existence of the so-called acoustical horizon. This was a legitimate criticism, and one of the two fundamental flaws in acoustical detection that would ultimately see it replaced by radar. The other, per-haps not as obvious in 1925 as it would be by the mid-1930s, was that the increase of aircraft speed significantly eroded the warning time provided by any mirror in any atmospheric conditions.[27]

The acoustical horizon was a complex phenomenon which Tucker did not fully explain in any of his reports prior to Lindemann's investigation. Only in February 1926 would he define the term for the ARC:

> On certain occasions an aeroplane approaching an observer can be seen for some miles before it is heard, and at the moment of first hearing the sound comes in suddenly and strongly ... We have here an analogy to the Sunrise for a visual horizon and we may speak of the source of sound 'rising' over the 'Acoustical Horizon', hence the use of the word 'horizon.'[28]

The acoustical horizon was a variable depending on temperature gradients in the atmosphere and the direction and strength of the wind in relation to the observer. Refraction of sound decreased in direct relation to a decrease in air temperature. Therefore, usually the higher the aircraft, the lower the temperature and the greater the distance of the acoustical horizon. Wind blowing toward the aircraft decreased the acoustical horizon; similarly wind blowing towards the observer increased it.

Lindemann also reported to the ARC an important flaw in ADEE's programme; that neither ADEE or anyone else knew what percentage of the year conditions were likely to be acoustically favourable. Without such evidence, and given its limitations, he argued that approval be withheld from further major expenditures on the mirror programme.[29] Lindemann's analysis was persuasive, and the ARC determined that no further work on acoustical mirrors should be authorized until some of the key questions were addressed.[30]

These studies were not completed until the end of April 1926, too late for inclusion of the 200-foot mirror project into the 1926-27 fiscal estimates.[31] Lindemman had delayed the mirror programme, but he could not kill it. The reports indicated that Lindemann's findings were far too pessimistic. From very flimsy meteorological data, it was concluded that on 51 per cent of the days wind conditions would be favourable for sound detection, giving a mean acoustical range of 16.7 miles at 10,000 feet, 20.6 miles at 15,000, and 24.5 miles at 20,000.[32]

Lindemann was not satisfied with these reports, no doubt because they were in large measure the work of Tucker and used data supplied by ADEE. Throughout the rest of 1926 and 1927 Lindemann managed to paralyse the ARC and forced ADEE to undertake a new round of more detailed experimentation on the way sound travelled through the atmosphere. Beginning in June 1926 the ARC began to wind down, meetings became far less frequent and most of its work was shunted to a variety of scientific sub-committees.

In December 1927, the ARC took the unusual measure of appointing a scientific sub-committee to write what would become its final report. This sub-committee consisted of the Directors of Scientific Research at the Admiralty and Air Ministry, and the President of the Royal Engineer Board.[33] Noticeably absent were the ARC's scientific advisors. The scientific sub-committee took only two months to produce a detailed analysis of the present state and possible future direction of anti-aircraft research. They gave a lukewarm endorsement to continuing with acoustical detection because it was 'the only system which can be applied at present with any hope of success.' The sub-committee recommended that construction begin as soon as possible on the 200-foot mirror and additional 20-foot mirrors. However, the sub-committee warned:

The scientific principles underlying the use of acoustic instrumentation are sound, but that the method is limited with respect to long-range detection by meteorological conditions. How far these limitations can be minimized appears to us to depend upon the results of the fundamental investigation of acoustics [currently underway].[34]

With one important caution, the ARC endorsed its sub-committee's recommendations and passed them to the Committee of Imperial Defence in March 1928. While the ARC agreed that ADEE's acoustical program should continue, they warned that its limitations meant that it was desirable 'that the possibility of the development of other methods should always be kept in mind.'[35]

The 200-foot Mirror

Despite the lack of enthusiasm of its endorsement, the ARC's final report must have been a great relief to Tucker and the acoustical section. The mirror programme had survived the most scathing criticism. While the ARC delayed mirror construction, important developments as a result of the committee's investigation had paradoxically improved the chances of full-scale development.

Most importantly, it was through the ARC that ADEE research won the enthusiastic support of the Royal Air Force. By December 1926, the Air Council declared that acoustical detection was 'fundamental to the scheme of defence.'[36] In April 1927, the council for the first time laid out a clear military requirement for mirrors. 'Acoustical apparatus' were required to fulfill two main objectives:

1. The detection of hostile aircraft at a distance of 25 miles from the coast, with an approximate indication of their bearing from the observer.
2. The approximate notation of their height, speed, course and number at a distance of 10 miles; and

The Air Council urged the War Office to consider the building of a line of 20-foot acoustical mirrors, to provide a five minute early warning of approaching aircraft. As a preliminary step, they suggested that three new 20-foot mirrors 'be erected for experimental purposes in conjunction with the Hythe mirror.'[37]

The Air Council's requirements were a complete endorsement of ADEE. The first requirement could only be met by 200-foot mirrors, the second by the use of mirrors similar to the Hythe apparatus, and the third by the

Observer Corps. In July 1927, the War Office authorized £1,300 for the building of two 20-foot mirrors, one on the beach of Dungenness near Denge and the second at Abbotscliffe between Dover and Folkstone. The mirrors were each approximately 8 miles from the Hythe mirror. They were to be used in experiments in accurate sound location by the use of triangulation.

However, the War Office refused to sanction the building of the 200-foot mirror, despite being advised in August by the Air Council that the instrument was 'fundamental to the system of detection' upon which they would rely on in time of war. The Air Ministry felt so strongly about the need to proceed immediately that a commitment was made to cover one-third of the cost of the mirror, the first time major funding for acoustical detection had come from outside the War Office.[38]

While wrangling over the funding of the 200-foot mirror continued through 1928, the three 20-foot mirrors were tested together for the first time. The trial was a failure, because the mirrors were found poorly situated to work together, subject to sound interference, and unable to determine altitude close to the coast. Despite this, Air Defence of Great Britain command determined that 'a line of acoustical mirrors' would eventually be required along the south-east coast.[39]

In 1929, the 20-foot mirrors at Hythe and Denge were replaced by a new 30-foot mirror design. Shaped like a giant bowl, the mirrors were tilted 45 degrees from the horizon to allow for accurate determination of elevation. The 20-foot mirror at Abbotscliffe was retained because its position on a high cliff mitigated many of the inherent design faults.

However, key to Tucker's scheme was the 200-foot mirror. Work on the prototype at Denge was finally approved in 1929 and basic construction complete in early 1930. By 1931 the mirror was ready for preliminary testing. In examining the design of the new mirror it looks like one of the crazy fanciful machines of Rube Goldburg or Heath Robinson. However, it was based on ten years of research in the way sound travelled through the atmosphere, and on the methods to detect this sound effectively. ADEE's research revealed that 'the most penetrating sounds for long distance transmission are the lowest pitched sound with the greatest wavelength.' While the smaller mirrors were effective in picking up sounds with a wavelength of from 2 to 3 feet, low pitched sound had wavelengths of from 15 to 18 feet, and, thus, required a mirror ten times the size.

Also, ADEE researchers had realized that elevation angles given by this long-wave low-pitched sound, for aircraft height resolution, would be too small to be determined at any great range. They therefore decided to build mirrors that would determine direction alone. This had the advantage of allowing the designers to abandon a symmetrical round layout in favour of a long strip of gently curved concrete 200 feet in length, but only 26 feet high.

The mirror was built of reinforced concrete. In front of the mirror was a concrete forecourt which gently sloped down to a flat walkway. Bordering the platform was a low concrete wall, placed at the mirror's 'focal surface', some 75 feet from the base. The wall was concentric with the mirror's surface.

Initially, listening was done manually by one or two 'sentries' patrolling the walkway next to the mirror's focal point. However, Tucker had maintained his belief in the superiority of electronic means of identification through the use of microphones. Tucker conducted research on improving microphones throughout the 1920s. For the mirror he devised a sophisticated new type of hot wire microphone that was specially tuned to respond to very low frequency sounds generated by aircraft. The microphones consisted of a 'double resonating cavity made by screwing the neck of a cylindrical bottle through the base of a cylindrical drum.' The drum had a circular hole cut in its lower end and the sound entering the drum was magnified by resonance, the sound surging between the drum and the interior of the bottle which created a very strong disturbance in the neck of the bottle. It was here at the neck that the hot wire was placed, and when sound was received the wire was cooled and the electrical resistance changed.[40] This change in electrical resistance was amplified electronically and transmitted to a loud speaker at the listening station.

Two persons were required to man the microphone listening station. One operator manipulated a series of switches which would isolate the microphone receiving the maximum sound from the target, while the second reported findings to a central control station by telephone. Initially, the listening station was located in a temporary hut to one side of the mirror. In 1934, the listening station moved to a sound-proof chamber in a structure attached to the centre of the rear of the mirror. The chamber had a small window cut through the mirror to provide a clear view of the entire forecourt area.[41] The 200-foot mirror was the most complex and sophisticated project ever undertaken by ADEE and it is not surprising that numerous minor technical problems had to be sorted out before full-scale testing of the device could begin in 1933.

Operational Testing and the Thames Estuary Scheme

Experimental testing on the 200-foot mirror began in the late spring of 1930. Much of this early work was undertaken to test the mirror's magnification powers and to determine the best design of the forecourt and positioning for the microphones. The 30-foot mirrors at Dungenness and Hythe were also tried for the first time. A series of flights by RAF Vickers

Virginia aircraft during nineteen nights in June showed that in the right acoustical conditions the 200-foot mirror could detect aircraft over 20 miles from the coast. The 30-foot mirrors were able to track targets up to 15 miles out, but in poor acoustical conditions this often shrank to as little as 6 miles.[42]

A similar program of tests on both types of mirrors was carried out in June and July 1931. Details of this series of tests have not survived; the only known results are of four nights testing of the 200-foot mirror by low flying (under 5,200 feet) aircraft. On three of these nights the mirror performance was as Tucker had promised; aircraft were detected from between 20 to 24 miles off the coast or between two-and-a-half to three times the distance of detection by the unaided ear. But on the fourth night 'very gusty' winds reduced the range to 9 miles, only 4 miles more than an unaided listener. Among the major innovations introduced in 1931 was the establishment of a central plotting room at Hythe where plots from all the working mirrors were recorded, with intersections of tracks being made on a plotting board to provide an accurate position of approaching aircraft. ADEE intended this to be the model for a much larger installation in a national defence scheme built around mirrors.[43]

In 1932, testing of the mirror system was expanded to include operations by Royal Air Force personnel. The air force decided that the 1932 tests would determine whether mirrors had evolved sufficiently to 'warrant adoption and heavy expenditures involved in their production in quantity.'[44] The operational testing took place between 6 June and 21 July, with a total of twenty nights of flying provided by an RAF flying boat. The 30-foot mirrors at Denge and Hythe and the 20-foot mirror at Abbotscliffe were manned by military personnel, while operation of the 200-foot mirror was undertaken by ADEE scientists. The control room at Hythe was run by two RAF officers. The RAF personnel took about four weeks to master the basics of listening. In the last two weeks of the programme they accurately detected aircraft at just slightly less than 10 miles. This was less than the ranges achieved by ADEE personnel in 1930, but the difference was ascribed to the lower elevations flown by the flying boat in 1932.

The maximum range of the 200-foot mirror was not tested during this exercise, but it was integrated into the programme by providing early warning of the approaching aircraft. Once an aircraft was detected by two of the smaller mirrors, the control room plotted the position of the aircraft by intersections of the bearings received. Aircraft were then followed, with new bearings being received at thirty second intervals. The 1932 program proved that acoustical detection worked on most nights and that regular air force personnel could easily be trained in its use.

Tucker claimed that the adverse results occurred when aircraft were forced to fly at very low altitudes owing to low clouds.[45] However, Tucker failed to mention in his official report the inability to detect the target aircraft on the night of 19 July. This failure was caused by interference from other RAF aircraft participating in the annual ADGB exercise. The mirrors were jammed by patrolling aircraft overhead, which indicated that they could not be used in conjunction with fighters. This omission could only have been a deliberate deception on Tucker's part, to hide one of the fundamental weaknesses of the mirror system.

Before this deception was caught, discussion began on the next stage of the mirror program. On 22 October, Geoffrey Salmond, now an air marshal and commander-in-chief of ADGB, sent a memorandum to the Air Ministry on mirror defences. Salmond endorsed the construction of a chain of 200-foot mirrors along the coast. Salmond commented: 'I regard this as of the utmost importance in the solution of the problem of warning where time is so vital a factor.' However, Salmond mentioned that he was concerned with the mirrors being susceptible to interference. Salmond had been informed by RAF mirror personnel that heavy rain, wind and ground noises such as motorcycles could hamper mirror operations, yet he was still unaware of the more serious problem of aircraft interference. As a result, all that he would recommend for 1933 was the continuation of the 1932 programme of training RAF personnel to use the 200-foot mirrors.[46]

In December Tucker proposed a bolder scheme. He too called for construction of an entire chain of mirrors to protect the South East coast and London. By building twenty-one 200-foot mirrors 'the coast line would be pushed out twenty miles.' The 30-foot mirrors would be deployed at 're-entrant parts of the coast to supplement the 200 foot mirrors.' Tucker identified the approaches to Portsmouth and the Thames Estuary as suitable sites for the smaller mirrors, a total of fifteen in all. The total cost of the project, minus the four mirrors already built, would be £67,200. In order to test the system, Tucker proposed the construction of the Thames Estuary defences be started immediately. This would involve the building of two of the large and nine of the small mirrors.[47]

Tucker's wish to build the Thames Estuary portion of the mirror defence system was not well received at ADGB. On 31 December Salmond asked ADEE to answer a short questionnaire on mirrors before he would agree to proceed further. In this, two questions struck at the heart of the limitations of this technology. First, Salmond asked if the 200-foot mirrors could determine the height of approaching aircraft. The second question revealed that Tucker's deception had failed:

It is noted that on the 18th and 19th July 1932, interference from aircraft participating in the Air Defence of Great Britain Exercises prevented any reliable data being obtained from the experiment then being conducted with the mirror at Hythe. This fact is of considerable interest as the possibility of employing patrolling aircraft to reconnoitre for enemy aircraft in the vicinity of the coast is contemplated as part of the air defences.

Salmond insisted on knowing if this type of interference could be eliminated with practice.[48]

Tucker could not provide positive answers to either question. The 200-foot mirror could neither determine height nor screen out the interference from patrolling aircraft. Tucker felt that standing patrols were simply 'preventable local jamming,' and he suggested that 'no useful purpose would be served by mirror and patrols functioning in the same area and the same time.' Instead mirrors and patrolling aircraft should complement each other, with the aircraft patrolling only during the day in clear skies, and the mirror would function at night and in thick weather.[49]

On 7 February 1933, a meeting of all senior personnel concerned with mirror research and operational deployment was held to discuss these problems. As for the problem of interference with aircraft, the minutes of the meeting record a remarkable conclusion: 'It is quite clear that patrolling aircraft and mirrors could not function simultaneously. Although they could not be made complementary in the simultaneous sense, they could be complementary in the consecutive sense.'[50]

This technical uncertainty left enough doubts so that no new mirrors would be authorized in fiscal year 1933-34. Instead, testing and training would continue with the existing single 200-foot mirror during the summer. The ADGB summer exercises would determine the fate of Tucker's invention. R. E. Peirse, the deputy director of Operations and Intelligence at ADGB, with ADEE's assistance, drew up plans for realistic operational testing during the summer exercises.[51]

Work with the large mirror began in mid-June with ADEE personnel. For the first time the mirror was tested at extreme ranges, a maximum of 22.1 miles was recorded with a mean of all tests of 17.8 miles. All testing was carried out in favourable conditions.[52] On 26 June a team of one officer and fifteen other ranks arrived at ADEE for three weeks of training. Although the RAF personnel were not fully trained, due to poor weather on the last week of the programme, when the ADGB exercise began on the evening of 17 July, ADEE staff withdrew. In the next 60 hours the air force team detected and reported five day raids by formations and seven night raids by individual aircraft; all the aircraft sent by ADGB at the mirror defences. The maximum warning time achieved was a sensational

29 minutes, but this was on a lone night raider flying at 12,000 feet. Yet, first detection frequently took place at less than 10 miles range. This provided an early warning time of only five minutes; the same time it took for the information to be calculated at the control centre and transmitted to ADGB headquarters.

In his official report of the exercise, Tucker played down the unsatisfactory results. He argued they were caused by the inadequate training time and methods, as well as easily correctable technical glitches. Tucker explained that the training had generally been undertaken at night on single aircraft, and that he was unaware until the exercise that the RAF expected the mirror to be operational during the day detecting multiple aircraft attacks. The problem of sound interference with other aircraft had occurred again, but it now appeared feasible that listeners could be trained to screen out much of this extraneous noise. The technical problems centred on the fact that he had not expected the mirrors would be used against day bombers, and that, as a result, the microphones were not tuned to the right frequencies.

However, Tucker could not avoid mentioning the one limitation of the technology that could not be easily rectified. Tucker asked that the performance of the mirror not be 'judged from their performance on one particular occasion, because of the wide fluctuations in range caused by weather conditions.'[53]

Despite the less than satisfactory results, ADGB were optimistic about the mirrors. In his official report on the 1933 exercise, Air Marshal Sir Robert Brooke-Popham, the new C-in-C of ADGB, stated that the information provided by the mirrors was of 'definite value.'[54] In December, the Air Ministry approved the construction of the Thames Estuary system, which would serve as a full scale model for a national mirror early warning system. It would consist of two 200-foot and eight (later reduced to seven) 30-foot mirrors. The larger mirrors would provide early warning, while the smaller ones would accurately track any aircraft attempting to approach London by flying down the Thames Estuary. A central control room would be built to plot all the data received by the mirrors and transmit the computed location of the aircraft to ADGB headquarters.[55]

The approval of the Thames Estuary system marks the high point of the research programme of William Tucker and his staff at ADEE's acoustical section. For sixteen years Tucker's team had experimented, developed and tested acoustical mirrors and it seemed at last that they had triumphed. Within just a few months Tucker's work was propelled from comparative obscurity to the centre of attention.

Although the mirrors remained top secret, soon the press was reporting on Britain's new weapon. In *The Sunday Pictorial* of 10 June 1935 a headline read: 'Air Ministry's Defence of London Secret/Acoustic Ears – That will make Surprise Raids Impossible – Watch the Coast – New System will be the Ears of London.' The story read as follows:

> I understand that the new system involves the use of a new invention by means of which sound is made visible. In other words, it will be possible to see the approach of hostile aircraft long before they are in the line of vision. An expert in acoustics has perfected the device, further details of which must not be revealed, and it will subject to exhaustive tests in the neighbourhood of the Thames estuary before being adopted generally for coastal defence. Acoustic ears will act with equal efficiency in daylight or darkness, and by the vibrations set up it will be possible to tell the type and, therefore, the nationality of aircraft. Surprise attacks, should the invention come up to expectations, would thus be rendered almost impossible.[56]

Late 1933 and 1934 saw further validation of ADEE's work. Brooke-Popham reported that on the 1934 air exercises: 'The mirrors were transmitting a wealth of information which actually proved an embarrassment by reason of its volume and the difficulty in sorting out the steady flow of reports including as they did movements of aircraft on the civil routes and the homing formations.' He foresaw that the Thames Estuary system would 'necessitate the institution of special measures to deal with [this information] and perhaps to ultimately even the establishment of a special intelligence section.'[57]

Despite Brooke-Popham's endorsement, throughout 1934 and the first half of 1935 the Thames Estuary system languished. Initially, this was not the result any lack of zeal on the part of the RAF, but instead was caused by tortuously slow negotiations for land needed to build the new stations. Finally, by June 1935 all the land was purchased, clearing began on several sites, and the first equipment for the mirrors was ordered. However, on 14 August the Air Ministry, without explanation, suddenly suspended the project. When the War Office asked for justification of the decision, the Air Council laconically replied: 'That pending further results of the experiments with radio detection [the Air Council] desire to suspend <u>temporarily</u> the construction of the acoustical mirror system in the Thames estuary.'[58]

This sudden stoppage came as a complete surprise to Tucker and his superiors at the RE Board. Shocked and outraged by this decision, Colonel F. J. C. Wyatt, the President of the RE Board, complained bitterly about Air Ministry developing in secret a competing early warning system which they claimed was far superior to acoustical detection.[59]

Tucker was stunned by the sudden ending of his project. At the moment of success everything that he had worked for was suddenly rendered obsolete. At first he could not understand what had happened. As the true nature of the challenge to the mirrors emerged, Tucker could only fall back on desperate measures. He began by belittling the new technology. Claiming that: 'The [new] system appears to offer no useful substitute for sound. It will, however, give early warning to count the number of aircraft threatening the system. It can only be used to put troops on the alert.'[60] All his efforts were in vain. Each month interest in the mirror program diminished. On 26 August 1937 the Air Council announced that they had 'finally decided to abandon work on the acoustical mirror system.'[61] One year later the War Office ordered ADEE to suspend all work on long-range acoustical detection and dismantle and place in storage all equipment from the mirror sites.

3

The Scientific, Political and Social Background to Radar

Tucker's programme was destroyed by the development of radar, a far superior technology because it relied on the detection of reflected radio signals. Of all of the events that will be discussed in this book none has been told more often than that of the British discovery of radar. Although it has been portrayed as a rather simple tale, the radar program actually occurred as the result of a complex and remarkable coming together of political, diplomatic, military, economic, technological and scientific developments. As Robert Watson-Watt, the father of British radar, would comment in one of his first public articles on the war-winning technology:

> The most important thing about the British development of radio-location as an instrument of war is that it happened at the right time. The essential difference between the British effort and the most nearly corresponding effort in most other countries is to be sought in those intangible factors which assured to us, at each stage of development, an adequate (though often a no more than adequate) margin of time for meeting successive crises of the war.[1]

Before considering the events that led Watson-Watt to propose the development of radar for aircraft location, therefore, we must examine why early 1935 was the 'right time' and also consider those 'intangible factors' which made the British programme unique.

The Scientific Background

The state of technological and scientific knowledge about radio waves meant that the discovery of an electromagnetic means of detecting objects at a distance was almost inevitable in the mid-1930s. Radar, in some form or another, was under development in at least eight different countries before the Second World War began. These projects all had their separate origins, because most work on radar was undertaken as classified military projects, but all built on scientific research that can be traced back to the late 1880s.

In 1887-88, Heinrich Hertz demonstrated that shortwave radio signals could be refracted or reflected. Hertz made these observations during his famous series of experiments which proved James Maxwell's theory of the existence of electromagnetic radiation at a lower frequency than light. He used as a transmitter a spark gap device, in which electricity jumped a small gap between two metal plates; this happened suddenly, and a visible spark could be seen. Maxwell had predicted that other unseen waves moving at the speed of light would be created by the oscillations of the electrons jumping the gap. Hertz constructed the first receiver of these radio waves, a wire loop with a small gap; there were brass knobs on each end of the loop at the gap. When the electromagnetic waves generated by the transmitter reached the loop, a spark was observed between the gap similar to the spark from the transmitter. Hertz concluded that the second spark was created by the passage of these unseen electromagnetic waves. Once detected, Hertz then demonstrated that these were electromagnetic waves because, like light, they could be deflected and refracted. Hertz's experimented with radio signals just 66 centimetres in wavelength.

This aspect of Hertz's experiments was largely forgotten by the end of the nineteenth century when attention shifted to the study of very long-wave radio waves, between 600 and 200 metres wavelength, which Marconi had demonstrated could be used for long-range communication. This switch to long wave occurred because radio signals at these frequencies appeared to follow the curvature of the earth and there existed very powerful transmitters in this frequency band. Shorter wavelengths could not be generated with much power and were thought to travel in a straight line. This appeared to limit the maximum effective range of radio signals of less than 200 metres wavelength. It had the unfortunate consequence, however, of limiting research into radio waves that were prone to being deflected, refracted, and, most importantly, reflected.[2]

Modern radar developed as a logical progression from improvements to radio and, later, television technology. During the first two decades of the twentieth century major developments were taking place in radio. Radio was transformed from being a scientific curiosity to being a vital part of international telecommunications. Among the many technological developments was the radio valve or vacuum tube. Based on the research of British engineer J. A. Flemming and the American Lee DeForest, radio valves in various forms were becoming the basic component of radios by

1920. Valves could convert DC to AC, act as a voltage or current amplifier or perform as a frequency changer, and, using the heterodyne principle, separate audio and radio frequencies or act as a carrier, placing audio sounds onto radio waves. This jack-of-all-trades was rapidly developed during the 1920s, with a vast variety of new tubes being introduced on a regular basis. Most importantly, for our story, is that the radio valves became available to amateur or ham radio operators soon after the First World War.

After 1918, wartime restrictions on the use of radio by amateur enthusiasts were gradually relaxed. However, access to commercially valuable long and medium wavelengths was severely restricted. Instead, radio hams were given access to a variety of shortwave radio bands. These enthusiasts soon discovered that short wave radio was not restricted to line of sight transmissions. In November 1923, two American amateurs established two-way radio contact with a French ham operator in Nice. These results were quickly duplicated by hams throughout the world. Although the cause of this phenomenon was not understood until 1925, these long ranges were made possible by the shortwave radio signals being reflected by the ionosphere, a previously little understood layer of the upper atmosphere. The reflected waves could carry round the globe, with the expenditure of much less power than was required for long wave signals. This discovery sparked immediate interest in short and ultra short (also known as Ultra High Frequency – UHF) wave signals. In the late 1920s and early 1930s a variety of new high power short wave receiver and transmitter radio valves, antenna systems and related devices were introduced. Particularly important was the perfection by 1930 of a low voltage cathode ray tube (CRT) which could graphically display received signals. UHF radio proved particularly applicable to short-range voice radio-telephony, a technique, which we shall see, would become an integral part of the air defence system.

As radio researchers turned their attention to shorter wavelengths, they began again to observe the reflective properties of these signals. In the autumn of 1922, Albert Hoyt Taylor and Leo Clifford Young of the US Naval Aircraft Radio Laboratory began research on 'the properties and propagation characteristics' of radio waves of 5-metre wavelength. One property on which they reported was the ability of a solid object, like a ship, to reflect these waves. Their report to the Bureau of Engineering stated: 'Possibly an arrangement could be worked out whereby destroyers located on a line a number of miles apart could be immediately aware of the passage of an enemy vessel between any two destroyers of the line, irrespective of fog, darkness or smoke screen.' They proposed to continue work on this idea but the navy did not make funds available, and the proj-

ect ceased, not to be revived for another eight years. Young would publish
an outline of his work in a radio magazine in early 1924.[3]

The initial applications of radio reflection were in the measurement of
distances. In late 1924 and early 1925, scientists in both the United States
and Great Britain used this phenomenon to demonstrate 'the existence of
a reflecting ionospheric region in the upper atmosphere.' A large number
of scientific research projects using radio waves to probe the extent and
composition of the upper atmosphere continued until the Second World
War. Radio reflection was also utilized to develop devices for geodetic sur-
veying and in aircraft altimeters.[4]

By the beginning of the 1930s, therefore, it would only be a matter of
time before it was realized that the reflection of radio waves could be
used to detect aircraft at long range. Several radio researchers published
papers where in they commented on the interference caused by aircraft. In
December 1931, British Post Office radio engineers reported that during
tests of shortwave radio telephone equipment in Scotland they observed
'a fluttering of signals whenever a plane passed nearby.' In 1933, two
American researchers at Bell Telephone Laboratories working with ultra
short waves reported interference from aircraft even when the aircraft was
out of sight.[5]

The importance of these observations did not go unnoticed for long.
In 1930, radar research began at the United States Naval Research
Laboratory. On 13 June 1933, the United States Navy filed a patent application in the name of the NRL researchers for a 'system for detecting
objects by radio.'[6]

In 1933, Dr Rudolph Kühnold, Chief of the German Navy's Signals
Research division began work on radar detection system. The origins of
this work stemmed from Kühnold's earlier development of an underwater
acoustical device similar to modern sonar. This shows that it was possible
to make the transformation from acoustical to radio detection systems.[7]
Tests on radiolocation devices began in France in January 1934 and in the
Soviet Union in July of that same year.[8]

It is more surprising, therefore, that it took until 1935 before a programme to develop radar began in Great Britain. Sean Swords, in his
technical history on early radar development, managed to trace at least
three separate suggestions or formal proposals for the development of
similar devices dating from as early as 1926. The first of these suggestions
was made on 4 August 1926 by O. F. Brown, a scientist with Department
of Scientific and Industrial Research, to the Anti-aircraft Research
Committee. He suggested to the committee a variety of ways that a cathode-ray oscillograph, a precursor to the CRT, could be used to enhance
existing acoustical mirror technology. He added that it was 'possible that

a method of location in azimuth could be based on the use of the cathode-ray direction-finder or the short wave radiation excited in the metal of aircraft by magnetos or by *secondary excitation in a strong field emitted from a ground transmitter.*[9] Unfortunately for Lindemann, he either did not see Brown's paper or missed the significance of this particular suggestion. Lindemann, in his quest to find an alternative to acoustical mirrors, certainly pursued research into the detection of radiation from aircraft magnetos.

In 1928, L. S. B. Alder of the Royal Navy's Signal School proposed the development of a device 'for the employment of reflection, scattering, or re-radiation of wireless waves by objects as a means for detecting the presence of such objects.' The device could be used as an aid to navigation or as a means to detect 'approaching ships or aircraft.' There is no record that officials at the Signal School took any notice of Alder's proposal.

The most extensive research by British defence scientists into using the reflection of radio waves for long-range detection undertaken prior to 1935 occurred at the Signals Experimental Establishment in 1930-31, by W. A. S. Butement and P. E. Pollard. Ironically, Pollard was under secondment to SEE from Tucker's Air Defence Experimental Establishment. In late 1930, the two conducted a series of experiments using a shortwave radio source and receiver. Based on the results of this work, on 26 January 1931, they proposed the building of an 'apparatus to locate ships from the coast or other ships, under any condition of visibility, or weather.' The 'apparatus depends on the reflection of Ultra Short Wave Radio Waves by conducting objects, e.g. ships.' SEE allowed them to conduct a limited series of experiments and they were able to detect a mast at about 100 yards away. However, here too the significance of this work was missed by both the War Office and Admiralty and the research program was soon terminated.[10]

The Social Position of Science in British Society and Government

The existence of these separate proposals from government scientists working at three different ministries illustrates the extensive employment of scientists by the British Government. It also shows that government researchers often had limited freedom to pursue independent lines of inquiry, although, as we have already seen in the case of William Tucker, some government scientists enjoyed too much freedom from outside scrutiny. Yet, despite this limitation, one of the unique factors which allowed the radar program to succeed was the extensive government, academic and, to a lesser degree, industrial scientific community and the social status it enjoyed. It would be the way that these scientific resources were mobilized by the state and the ability of scientists ultimately to be able to direct important policymaking that played a vital part in the success of the radar defence system.

In 1915, the Munitions Inventions Department and the Board of Invention and Research were established to supervise military research programs. In the postwar period, as we have seen, efforts were made to retain much of the wartime scientific research capabilities. A variety of research centres such as ADEE were established from wartime pro-grammes, while other preexisting research facilities, including the Royal Aircraft Factory, renamed the Royal Aircraft Establishment (RAE) in 1921, and the National Physical Laboratory, were greatly expanded from their prewar establishment.

The administrative functions of the wartime Munitions Inventions Department and the Board of Invention and Research were taken over by permanent scientific departments in the navy and air force, while the War Office failed to create a centralized administration for research. The new Royal Air Force created a Directorate of Scientific Research in 1924. The directorate was established to 'control all the Air Ministry activities in regard to scientific research within the Ministry, at the RAE and outside.'[11] However, the directorate was not given control over research into aircraft armament and wireless. Remarkably, in 1924 the Air Ministry believed that it would need to conduct little research into radio and related technol-ogies. The director reported to the Air Member for Supply and Research.

The first Director of Scientific Research (DSR) at the Air Ministry was Henry Edgerton Wimperis, previously in charge of the Air Ministry's only research laboratory located at Imperial College. Born in 1883, Wimperis was a Cambridge-trained mechanical engineer when he joined the Royal Naval Air Service in 1915. He was made the head of the RNAS (later the Air Ministry's) Laboratory at the Imperial College of Science. Wimperis used his prewar experience working on fire directors to aim long-range artillery, to design a drift bomb sight, essentially a mechanical computer, which took into account an aircraft's speed and height and the wind's velocity and direction. Called the Wimperis Course Setting Bombsight when introduced into service in late 1917, it was claimed to be the best device of its kind in existence. It was widely used by British and Allied forces both during and after the war. One Wimperis sight was used by the bombers of American army officer Billy Mitchell to sink the German battleship *Ostfriesland* during his famous series of tests in the early 1920s to demonstrate the power of aircraft.[12]

In addition to this elaborate government research organization, close links were forged between the state, industrial and academic science through a variety of boards, research associations and committees. The two most important of these were the Radio Research Board (RRB) and the Aeronautical Research Committee (ARC). Through the ARC and RRB the armed forces had close contact with civilian scientists. The ARC had broad powers to advise both government and industry on aeronautical problems, to supervise Air Ministry and other government departments' aviation research, and to promote aeronautical education. The RRB provided funding and supervised a variety of radio research projects, particularly in short and ultra short wavelengths. Among them were early ionospheric radio experiments. Both committees provided excellent opportunities for defence scientists to meet other researchers and to be made aware of their current work. One such link was formed on the RRB, where Wimperis, the Air Ministry's representative, came to learn of the research of Watson-Watt, a radio scientist working at the National Physical Laboratory. The two men became close friends.[13]

Despite the sheer size and complexity of the British Government's scientific resources, much of the research was constrained by narrowly defined missions, a shortage of funds, and control by managers who were often neither scientists nor engineers. As J. D. Bernal commented on the National Physical Laboratory in 1939: 'The impression produced is that work of routine examination occupies too dominating a place, and sterilizes, so to speak, the rest of its activities. Examination of materials or processes by a national laboratory seems to be limited to attempts to correct defects as they occur in practice.'[14]

The same could be said of most military research facilities, which had several additional problems until the mid-1930s. As A. P. Rowe, Wimperis' assistant at the Air Ministry, recollected:

> Very few scientists could be attracted in time of peace to work on armament problems and, even had they been available, the money for their salaries and their experiments would be very difficult to find. In those days, armament research work was almost wholly controlled by competent and conscientious RAF officers who concentrated upon improving the weapons familiar to them. Thus there was little scope for civilian scientists in armament work. Perhaps above all there was the curse of secrecy.[15]

According to a 1927 parliamentary report on science in the civil service, 'the curse of secrecy' doomed many military scientists to 'scientific obscurity.' Scientists 'were unable to publish in the scientific journals advances in knowledge gained by them during their work.'[16]

In some areas, such as anti-submarine warfare and aeronautics, slow but steady progress was made on a clearly defined set of research problems. In other areas, such as acoustical mirrors, mediocre scientists, a lack of clearly defined research goals and outside supervision led to failure. The best research talent and managers were lured away by higher salaries in industry and universities, the attraction of pure research or the potential to profit from industrial applied research, and a general antipathy towards all things military. In 1925, one university mathematician wrote in a scientific newsletter that military research was 'essentially second-rate and dull.' No matter how 'public-spirited and praiseworthy', this type of science was not for 'a first-rate man with proper personal ambitions.'[17]

There was certainly some validity to these views. The armed forces tended to see its scientists as being the producers of gadgets only. The specifications of new equipment and the development of new doctrine to utilize any new technology was left strictly in the hand of those in uniform. Analysis of what was required was based not on any system of scientific or statistical analysis, but on often faulty military intuition and expertise.

While state science languished, academic science in Britain flourished in the inter-war years. This was perhaps the last golden age of British science, not yet eclipsed by the vast resources of the United States. It was the era of Ernest Rutherford and his extraordinary talented group of followers at the world famous Cavendish Laboratory at Cambridge University.[18] Rutherford's followers were at the forefront of unlocking the secrets of the atom. British science prospered in almost all areas of research, whether it were in the physical, biological, medical or chemical sciences.

University science and scientists were generally highly regarded in society and wielded tremendous influence. There was great debate amongst scientists as to the degree to which they should use their influence and become involved in the political process. A large number of the younger scientists, like their university colleagues in the humanities, became involved in left wing politics, from socialism to communism. They advocated for British science to pursue an activist agenda to improve the human condition. These scientists were usually vocally pacifist. Other scientists, generally older, more established and politically conservative, were less enthusiastic about bringing politics into the unsullied world of pure science. Professor A. V. Hill, in his Huxley Memorial Lecture of 1933, warned that the Charter of the Royal Society did not allow its members to 'meddle with politics.' This had the advantage of fostering tolerance of scientific pursuits. Scientists, Hill continued, 'should remain aloof and detached, not from any sense of superiority, not from any indifference to the common welfare, but as a condition of complete intellectual honesty.'[19]

While Hill disdained involvement in politics, he did not imply that scientists should remain aloof when basic freedoms on which they relied to pursue science with 'complete intellectual honesty' were threatened. In this same speech Hill made one of the first public condemnations of Nazism's 'violation of freedom of thought and research.' British scientists were among the first groups to come face to face with the evils of the Nazi regime. No other group of British professionals were as closely linked to their German counterparts; science was, after all, an international activity and Germany the other great centre of European science. Jewish scientists were among the first victims of the Nazi regime. Hill became one of the first members of the Academic Assistance Council, which was formed in 1933 to aide scholars who were already fleeing persecution.[20]

A year later, physicist P. M. S. Blackett, a left-wing Fabian socialist, came to similar conclusions. He expressed his deep concern of the extent that 'science was being used for war preparations, and that many eminent scientists had left Germany, science was being used for anti-working class activities, and scientific fact was being deliberately distorted to accord with Nazi teachings.'[21] Within days of Blackett writing this, he joined with the conservative Hill, united in their desire to utilize science to protect Britain from the growing threat from Nazi Germany. By December 1934 the British government once again asked for scientific assistance in solving the air defence question, but this time there was a sense of urgency that had not existed since the First World War.

The Political and Diplomatic Situation

Political, diplomatic and military developments in 1933 and 1934 created an ideal climate for the development of a new type of air defence technology by early 1935. A lack of political commitment to the 1923 fifty-two squadron plan for home air defence, based on the Ten Year Rule and economic concerns, continually pushed backed its completion from 1928, until by 1929 the estimated completion date was set by the government as 1938.[22]

Doubts about the validity of the Ten Year Rule were only expressed in political circles in the early 1930s. On 22 March 1932, as a result of the Japanese attack into Manchuria, the Committee of Imperial Defence (CID), the highest military advisory body, recommended the abolition of the rule. Two days later the Cabinet agreed, but the economic crisis of the Great Depression precluded any additional funds for defence. To meet the immediate fiscal crisis, overall spending on the armed forces actually was to continue to drop until the spring of 1933. Instead, the National Government of Ramsay MacDonald chose to pin their faith on disarmament talks at Geneva.[23]

These talks opened in February 1932. They immediately became polar-
ized between Germany's demands that equity be achieved by other powers
disarming to match the size of its Versailles Treaty limited armed forces,
and France's equally unyielding stipulation of her rights to maintain sufficient forces to guarantee her security. Initially, British sympathies lay
with the Germans. Adolph Hitler's assumption of the Chancellorship on
January 1933, however, brought about an immediate change in Germany's
position. When the conference resumed sitting in October 1933, Hitler
announced Germany's rejection of any disarmament formula, and its
withdrawal from the conference and the League of Nations.

In response to Germany's actions and continuing tension in Asia, the
CID established the Defence Requirements Sub-Committee (DRC). By
January 1934, at least one member of the committee, Sir Warren Fisher,
had branded Germany as the 'ultimate potential enemy against whom our
"long view" defence policy would have to be directed.' In late February,
after lengthy and often acrimonious discussions, the committee recommended that a significant expansion of the armed forces be undertaken
during the next five years at the cost of £76 million. This included an
increase of the strength of the RAF Home Force to 77 squadrons and provision of a five division expeditionary force for the continent.

While the government was willing to accept the DRC's analysis that
Germany was now the ultimate enemy, it was unwilling to accede to all
of the large expenditures recommended to make Britain ready for war by
1939. The attack on the DRC's proposals was led by the Chancellor of
the Exchequer, Neville Chamberlain. Chamberlain believed that Britain
simply could not afford an-across-the-board increase in military expenditures and argued for focusing on the air force expansion alone.[24]

The rationale behind focusing on building up the RAF at the expense of
the other two services was the result of perceptions of the growing military threat of modern aircraft. Behind the shield of the Royal Navy, the
only short-term threat to Great Britain came from the air. Germany would
take many years to create a creditable navy, but an air force could be built
up relatively quickly. Moreover, rapid technological progress in aviation
assured Germany a virtually level playing field.

More important was the popular belief that bombers could rain terror
and destruction onto Britain's cities. Memories of the First World War
bomber offensive were still fresh. As we have seen, although they caused a
political crisis, Gotha bomber attacks proved to be indecisive in determining the outcome of the First World War. Despite the dramatic improvement
in aeronautical design since 1918, in the 1930s no air force had the technological capability to destroy a city or force the civilian population's morale
to break. Technological reality, however, was irrelevant. Public pronounce-

ments of doom from the air by politicians were common in this period. The most famous was made by former Prime Minister Stanley Baldwin in the House of Commons on 10 November 1932:

> I think it is well also for the man in the street to realize that there is no power on earth that can protect him from being bombed. Whatever people may tell him the bomber will always get through. *The only defence is offence*, which means that you have to kill more women and children more quickly than the enemy if you want to save yourselves.

Such political statements were supported by the proliferation of newspaper and magazine articles, popular books, and, ultimately, movies on the subject of future war. Many of these accounts of the next conflict started off with an unstoppable aerial assault on London and other major cities.[26] Both sides of the rearmament debate were willing to believe in the efficacy of the bomber.

It was in this environment that during the spring and summer of 1934 the Cabinet debated the DRC's recommendations. On 30 April Lord Londonderry, the Secretary of State for Air, told the Cabinet: 'There is a strong public pressure to increase our air force, and it is time the Cabinet made up its mind definitely and told the country the truth.' The Prime Minster Ramsay Macdonald, who had been strongly committed to disarmament and pacifism, replied that 'Such proposals were not going to be popular.'[27] It was a difficult transition for those who had put their faith in the League of Nations and disarmament and no quick decision could be reached.

Unfortunately for the government, the rearmament debate was not confined to the corridors of Whitehall for very long. Among those who ensured the debate became public was Winston Churchill, now a backbench MP, but one with impressive connections to those in power and the media. As early as 1921, Churchill had written an article for Nash's *Pall Mall* magazine, titled 'Shall We All Commit Suicide', in which he expressed his deep concern about the extraordinary destructive capabilities of current and soon to be invented weapon systems. Churchill wrote that a 'German' informed him that the next war would be fought with 'electrical rays which could paralyze the engines of motor cars, could claw down aeroplanes from the sky and conceivably be made destructive of human life or vision.' He speculated that new forms of explosives would soon be invented in which a bomb 'no bigger than an orange' could 'blast a township at a stroke.' Equally frightening were further development of poison gases and the spreading of plagues like anthrax. In 1921, Churchill believed the only answer to avoid disaster was the avoidance of war.[28]

However, by 1934 Churchill was convinced that only rearmament could deal with the threat posed by Nazi Germany. In the spring, when he learned that Hitler had authorized the creation of an illegal and secret air force, Churchill began a campaign to force the government's hand. On 7 June, Churchill spoke at Wanstead in his constituency about the slow pace of air rearmament. He agreed that the government had promised plans were being made but 'All that ought to have been done long ago. We ought to have a large vote of credit to double our Air Force; we ought to have it now, and a larger vote of credit as soon as possible to redouble the Air Force.' There soon followed an article in the *Daily Mail* with similar criticisms of the government. It is difficult to assess Churchill's role in forcing the pace of government policy, but certainly he did much to publicize the threat and to win over a doubtful electorate.

His persuasiveness was enhanced by the accuracy of his information. Churchill's sources included Desmond Morton, the head of the Government's Industrial Intelligence Centre, who briefed him on many top secret intelligence estimates of Germany's rearmament programme. His willingness to use classified information in public debate made him a dangerous thorn in the side of all governments until the war began. Churchill became the leading spokesperson for those advocating pushing the pace of not only increases in aircraft production but also of air defence measures.

On 18 July, the Macdonald Government finally agreed to accept Chamberlain's amendments to the DRC's programme. A £50 million package was approved with the lion's share of the funding going to the RAF. This became known as expansion scheme A. As one army officer lamented, the decision was the direct result of the 'air panic' fomented by Churchill and the media, and accepted by a gullible general public.[29]

The cabinet's decision to approve Scheme A did not appease Churchill and his supporters. Adding his voice to the growing chorus demanding more and speedier action was Professor Lindemann. Churchill and Lindemann had forged a strong friendship since 1921, when 'the Prof', had acted as Churchill's source for his article, 'Shall We All Commit Suicide.' The friendship between the austere, vegetarian, teetotalling physicist and the all encompassing extrovert politician and author is one of the more remarkable relationships in modern British history. While on the surface they appeared to have little in common, the men shared several important character traits. They had both shown great personal bravery, Lindemann as a fearless test pilot in the First World War, and Churchill in the Sudan, the Boer War and on the Western Front. Lindemann always craved acceptance by British society, which he felt had ostracized him unfairly because of his German-sounding Alsatian name and his accidental German birth.

Churchill's consuming ambition was to achieve the premiership which his father had never obtained.

By the early 1930s, both saw their careers fall apart. When the Conservatives were defeated in 1929, Churchill was cast out of government. He then proceeded to alienate himself from Stanley Baldwin, the leader of the party, over his stand on India. In 1931, when the first coalition government was formed, Baldwin ensured that Churchill remained on the back benches. During the same period it was becoming increasingly obvious that Lindemann's goal of revitalizing Oxford's Clarendon Laboratory as a serious rival to Cambridge's Cavendish would not be achieved in his lifetime. Lindemann dabbled in too many different scientific areas, without achieving notable success in any. While a new breakthrough in atomic physics was announced almost monthly by one of Rutherford's students, Oxford remained very much in the scientific backwater.

Lindemann considered himself to be one of the nation's leading experts on the applications of science to war. He entered the debate about the growing threat of air attack in a letter to *The Times* which appeared on 8 August 1934 under the title 'Science and Air Bombing.' Lindemann argued that the debate in the House of Commons on air defence issues had been based on the assumption 'that there is, and can be, no defence against bombing aeroplanes and that we must rely entirely upon counterattacks and reprisals.' Lindemann feared that this situation might allow 'gangster Governments' to jeopardize the whole future of Western Civilization. He warned:

> To adopt a defeatist attitude in the face of such a threat is inexcusable until it has definitely been shown that all the resources of science and invention have been exhausted. The problem is far too important and too urgent to be left to the casual endeavours of individuals or departments. The whole weight and influence of the Government should be thrown into the scale of endeavour to find a solution. All decent men and all honourable Governments are equally concerned to obtain security against attacks from the air, and to achieve it no effort and no sacrifice is too great.[30]

Later that same month Lindemann went on vacation with Churchill to the continent. Being on vacation, however, did not stop them from pursuing their quest. In early September, they decided to go to Aix-en-Provence and interrupt the vacation of Baldwin, now the Lord President of the Council and second most important man in the government, to press their case. Baldwin considered his holiday sacrosanct time in which he carefully avoided the outside world. This did not stop Lindemann and Churchill

Lindemann and Churchill.

who urged Baldwin to increase the pace of rearmament and research into air defence. The only result of this meeting was to impress upon Baldwin that these troublemakers were not going to be easily appeased.[31]

In late November, the House held another debate on the subject of German rearmament and air defence. Churchill once again led the attack, and gave an emotional speech, the content of which had certainly been heavily influenced by Lindemann. Churchill warned that German air rearmament was the greatest danger to British security, since air attacks might lead to from thirty to forty thousand people being killed in London

Gov't under pressure to invest in radar

A Background

and a further three to four million being driven into the countryside. He
cautioned the government not to neglect 'the scientific side of air defence
against aircraft attack – the purely defensive attitude against air attack.'
He then stated: 'I hope that there will be no danger of service routine or
prejudice, or anything like that, preventing new ideas from being studied,
and that they will not be hampered by long delays such as we suffered in
the case of tanks and other new ideas during the Great War.' Churchill also
spoke of the need to be able to inflict damage on the enemy as much as he
could inflict on Britain. In order to do this he demanded that Britain maintain a superior air force to Germany. Churchill received almost a standing
ovation when he sat down.[32] Baldwin responded by attacking Churchill's
figures and stated that he was exaggerating the threat, but pledged that
the government was determined to have a superior air force to that of
Germany.[33]

Churchill refused to accept the government's pledge for good reason.
The government's acceptance of air force expansion did not by any means
complete the endorsement of a new arms race or of the inevitability of
a future war beginning with a devastating bomber offensive. The British
Government's appeasement policy is well known, but the government
also pursued a less well known policy of trying to limit, through international agreement, the use of aircraft in a future war. Although these efforts
were no more successful than disarmament or appeasement, attempts to
establish rules for air warfare demonstrated politicians' desire to limit
the destructive capabilities of bombers.[34] Yet, while there remained great
reluctance to build up an offensive bomber force the political will and
money was there to support research into air defence by the end of 1934.

The Royal Air Force and Air Defence

The RAF was equally willing to accept new means of air defence. A
remarkable amount has been written on RAF planning and strategy
during the inter-war era. Most historians have focused on the RAF's twin
policy of promoting the use of aircraft as an inexpensive means for imperial policing and the development at home of a bombing force for strategic
interception.[35] By 1928, Malcolm Smith explains, the Air Staff considered
this strategy meant that the 'enemy air force should be ignored, *at least
beyond the close defence of the most vital targets* and the tactical defence
of the bombers themselves. The Air Staff now claimed that while the Army
and Navy had to destroy or neutralize the enemy army and navy, the task
of the RAF was to paralyze the enemy's productive centres of all sorts and
to stop all communications and transport.'[36]

Smith and others have correctly stressed the primacy given in air planning to the bomber, but this has not been balanced sufficiently by analysis of senior RAF officers' support for improvements in defensive technologies. The impression left is that senior officers of the service were opposed to the development of air defences that might hinder the creation of a strike force.

In large measure the air force historians have allowed themselves to be led astray by the Air Staff planning documents, which focus on offensive, as opposed to defensive tactics. However, this was the result of a general consensus that air defence, if technologically feasible, should be vigorously pursued. There was much more debate on the way which bombers should be utilized, thus there are many more planning documents on strategic bombing in the official papers.

Since its formation in 1925, much of the work of ADGB was directed towards improving the air defences, particularly around London. Even before the appointment of ADGB's first commander, it was decided also to create a sub-command, known as Fighting Area Headquarters, to control fighters and other defence forces in Air Fighting Zone. While the 1923 fifty-two squadron home defence scheme languished, a very high propor-

Sector Control Room.

tion of what money was available went to fighter aircraft and building up the defensive infrastructure.[37]

Work was also undertaken to recreate and improve the command and control arrangement of Ashmore's London Air Defence Area. The air fighting zone was divided into a number of sectors, each one being assigned to one or more fighter squadrons, as well as searchlights and observers. Each was provided with a sector control room. The sector control rooms were in turn connected to the Fighting Area Headquarters Operations Room.

The operations room served 'to provide an intelligence centre upon which are focused all sources of intelligence which may furnish information of value to the Defence Commander' and to direct and coordinate 'the activities of all units of the defending force, both air and ground.' The room was organized very much along the same lines as Ashmore's command centre in 1918. At the centre of the room was a table on which was placed a large scale map of south-eastern England. Around the table were fifteen plotters' keyboards, each of which was connected by direct telephone link to two or three fighter sectors, observer centres or other operational headquarters. Directly behind the plotters on a platform raised 2 feet above the plotting table was the Tellers' platform for 'reporting the information appearing on the plotting table to other interested Headquarters.' Directly behind the tellers was the Control Dias, raised some 6 feet above the floor of the room. Here operations officers would transmit orders to and receive information from sector controls, advanced fighter aerodromes, and ground defences. Behind the Control Dias was the office of the Air Officer Commanding Fighting Area.[38] Sector control rooms were smaller versions of the main operations room.

All of the Fighting Areas control and operations rooms were designed to process and disseminate information as quickly and as accurately as possible. Counters used to plot information on the plotting maps were 'coloured red, blue or orange, as were also the successive 5-minute segments of the operations room clock.' The colour of the counter matched the 5-minute time interval in which the information was received and there were never more than two colours of counters allowed on the table. This ensured that all information was no more than ten minutes old. Information on height and numbers was displayed by other counters. Other information required by the controllers – the state of readiness of aircraft, weather, etc. – was marked on chalk boards located along the far wall.[39] *maybe include ?*

Research also continued on improving air-to-ground and air-to-air radio telephony (R/T). This was necessary if the most current information on enemy air raids was to be passed to patrolling fighters. In 1932, a new model R/T, the TR 9, was introduced. It had an air-to-ground range of 35 miles and 5 miles when operating between aircraft. The fitting of the TR 9

in all aircraft allowed for the first time the direct transmission of information from the sector commander and the fighter aircraft.

Testing of the air defence system took place in a series of annual air
exercises. The results of these exercises were very controversial, as they
seemed to suggest that the bomber might get through, but that it would
rarely returned home unscathed. From the statistical information alone it
appears that the defences were highly effective. In 1928, fully 84 per cent
of raids that crossed the coast were intercepted. Similar high interception
rates were obtained in all of the air exercises of between 1931 and 1935,
yet, few senior air force officers were sanguine about these results. It was
recognized that the annual air exercises were conducted under increasingly unrealistic conditions which enhanced the defenders' performance.
For instance, night attacks were greatly assisted by the fact that peacetime
safety regulations required all bombers to fly with their navigation lights
on. As a result, searchlights were usually not needed by defending fighters,
which could see their targets up to 4 miles away. This was well within the
normal minimum sighting range during the day! As well, most raids operated only under the most ideal weather conditions and the bombers were
often recalled when it became foggy or misty, even though the bombers
'could have easily carried on to their targets.'[40]

Even more importantly, many senior officers were aware that aircraft
performance was on the verge of radical improvement. The close research
relationship between government, the military, science and industry in
Britain and in other countries was fuelling a revolution in aeronautics.
New technologies, such as all-metal internal framed monoplane construction, streamlining, and far more powerful engines, had the potential of
rendering the entire defensive system obsolete. British bomber design in
the early 1930s was well behind the cutting edge advances already being
introduced into civilian and military service elsewhere.

The report of the 1935 air exercise pointed out that bombers would
soon be operating at 'the modest figure of 280 mph,' while Hart bombers, the fastest then in service, flew at just 140 mph. As a result 'the Furies
[fighters] had just double the time to bring off their interception that
our fighters will have in the next war.'[41] In 1933, Air Chief Marshal Sir
Robert Brooke-Popham, the Commander-in-Chief of ADGB, outlined the
dilemma that the air defences would face in the near future:

> Change in conditions has definitely been brought about by the increasing
> speed of the aircraft in the last 10 years. It is not merely a problem of
> relative speed, but of absolute factors, some of which are invariable. The
> speed of the bomber is steadily increasing; on the other hand two other
> factors remain constant: firstly, the distance of London from the coast

and secondly, the time that must elapse between the aircraft being seen by observers and the defending aircraft leaving the ground.[42]

Brooke-Popham argued that while in the past fighters were intended to intercept the bombers at the front of the air fighting zone and attack them through the entire width of the zone, they would soon only be able to meet the enemy at the rear of the zone. Fighters would, therefore, only be able to launch a single brief attack or follow the bombers over the target. This would result in the fighters interfering with anti-aircraft guns in the inner artillery zone. The only other option was to have fighters flying regular standing patrols. This was certainly feasible given that the aircraft could be vectored into attack position by R/T, but standing patrols would demand far more fighter resources and put tremendous strain on men and material.[43]

Speed, however, was only one of the factors which made the results of air exercises rather dubious; another was the altitude at which bombers operated at during the tests. Between 1933 and 1935, the average altitude of simulated attacks during the day increased from 5,700 to 11,500 feet, and at night from 5,600 to 9,500 feet. Although effort was being made to operate at more realistic altitudes, even as late as 1935 few raids operated at even 20,000 feet, the minimum expected for any future daytime attacks. At any altitude over 10,000 feet, with even limited cloud cover, the observer corps could often not track raids visually, and their short-range sound detectors were so unreliable that they 'did not give the controller sufficient information on which to order up his patrols with any confidence.'[44]

In March 1940, Air Marshal Hugh Dowding, then Commander-in-Chief of Fighter Command, looking back at the air exercises of the late 1920s and early 1930s commented:

In peace time we obtained flattering results partly because all interceptions were made over land, when the Observer Corps were generally able to give some approximate estimate of height, and partly because no really high flying raids were ever encountered.

Most of the raids came over at 7,000 or 8,000 feet, and the Bomber Command [*sic*] were generally unable to stage high flying raids owing to a lack of oxygen and warming arrangements for the crew. It was rare that a raid as high as 12,000 feet was staged.

Yet one other problem placed the entire defensive scheme into doubt. Not only were aircraft flying faster and higher, but they also were increasingly able to fly longer distances. By 1935 it was accepted that not just London and coastal centres were vulnerable to attack. Virtually the entire country, including industrial cities in the Midlands, Yorkshire and Lancashire, was within range of bomber attacks. This in turn made the idea of flying standing patrols more ludicrous, given the far larger area that would have to be patrolled.[45]

Throughout this period, 'the importance of extending the period of warning allowed [was] fully realized and [was] constantly borne in mind' by senior air force officers.[46] As we have seen, in all of the exercises from 1931 acoustical mirrors were tested because they offered the only possible solution to extending ground based early warning. Yet, even as the Thames Estuary system was authorized, it was obvious that it was at best a limited improvement, which was also being rendered obsolete by the technological revolution in aeronautics. Other schemes to improve early warning, including flying long-range reconnaissance patrols and placing sound detectors on ships off the coast, were tested and found wanting.[47]

The realization that Germany was building up a secret air force, the growing political demand for action, and the appreciation of the growing threat of obsolescence of the air defences spurred the Committee of Imperial Defence (CID) into action in 1934. On 2 August the Home Defence Committee of the CID established a sub-committee on the Re-orientation of the Air Defence System of Great Britain. Brooke-Popham was made the chairman. Also serving on the committee were senior RAF and War Office officers involved in air defence and military intelligence.

In November Brooke-Popham's committee received a report on future early warning requirements prepared by the staff of ADGB command. Assuming the attacking aircraft to be approaching at 250 mph and allowing five minutes for the transmission of intelligence information and eleven minutes for the fighters to scramble, the paper stated that a fighter squadron would need sixteen minutes to get into an position to intercept an aircraft at 20,000 feet – which figure translated into a warning distance of at least 70 miles.[48] This distance was more than three times the range of the 200-foot mirror operating on days free of any atmospheric disturbance. Something radically different was needed if the air defences were to remain effective.

So by the end of 1934 there was the scientific, political, military and economic will to develop more effective air defence technologies. It was clear that Tucker's mirrors were not the answer. What was needed was a radically new approach to the way science was utilized by the armed forces, which in turn might lead to the creation of fundamentally novel technologies to combat the bomber menace. The answer to the bomber dilemma was lingering in the air, given the atmosphere of the time all it would take would take would be for someone to ask the right question of the right person. The only question, however, would be if this would occur in sufficient time to make Britain secure before she had to face the growing might of the German air force.

4

The Discovery of Radar

Less than a fortnight after receiving the ADGB report on future interception requirements, the Brooke-Popham Committee met with Frederick Lindemann. Lindemann had been invited to appear on 14 November, but he had first refused, because he believed that the committee held the attitude that there was no possible defence against air attack other than a counter-attack by bombers. In early November 1934, Lindemann began to demand that the government appoint a powerful and independent committee to investigate new means of air defence chaired by a highly respected statesmen and consisting of scientific and service experts. He personally lobbied several government ministers, including Lord Londonderry, the Secretary of State for Air. He only reluctantly accepted Brooke-Popham's invitation on 27 November.

At this meeting Lindemann held himself up as the leading scientific expert on air defence, disingenuously blaming the chairman of the ARC, Lord Haldane, who had died in the summer of 1928, for the failure of the earlier committee. After politely listening to Lindemann's criticisms of both his own committee and of the Air Ministry, Brooke-Popham asked Lindemann what 'particular new means of air defence' he had in mind.

Lindemann outlined several possible avenues of research that he had first suggested to the ARC, including the detection of infrared emissions from aircraft. He used the opportunity to get back at another old nemesis by attacking Tucker's sound mirrors. Lindemann concluded by briefly mentioning one possible new line of inquiry. He was 'sure that there were other means of detecting aircraft. For example, he was confident that the reflection of wireless waves might be applied to aircraft detection.'[1]

This tantalizing reference to radar marks the first time that a civilian scientist had brought recent developments in radio to the attention of the government. Yet Lindemann's suggestion was made almost as an afterthought; his primary purpose was to get the Brooke-Popham Committee to support his notion of a new and independent scientific inquiry into air defence technology. Lindemann succeeded in his goal; the CID Committee agreed to include his idea in its final report.

Lord Londonderry was not as open to appeasing the politically trouble-some physicist. The day after Lindemann's meeting with the CID Committee, the Air Minister informed him that he was 'afraid that it would hardly be useful to consider the setting up of a body of this character with a roving commission over such a wide field.' Londonderry contradicted the assertion that the Ministry was not open to new specific suggestions or was not actively examining ways of improving the air defences. Unfortunately, Lindemann could not be informed of the latest research since it was 'impossible to make public in any way details of the Air Ministry scheme for defence against air attack.' Londonderry withheld from Lindemann that action was already underway, within Air Ministry's Directorate of Scientific Research, which would sideline his concept of an independent scientific committee and lead to another scientist, quite separately and more successfully, proposing the use of radio reflection as a solution to the air defence problem.[2]

Death Rays and Scientific Committees

The growth of political and military concern about German rearmament and home air defence that occurred throughout 1934 did not go unno-ticed by the Air Ministry's small team of scientists. In June, Albert Percival Rowe 'had a slack week and in a weak moment' began an informal, but 'thorough' investigation of all Air Ministry's files on the problem of air defence. Rowe could not realize that this informal investigation would dramatically change his life and the history of the world.

Rowe was born in Launceston, Cornwall in 1898 and attended one of the Royal Naval Dockyard schools. In 1918, he entered Imperial College, London, where he completed a first-class honours degree in physics in 1921. One year later he completed a postgraduate diploma in air naviga-tion. He was one first scientific officer hired by the Air Ministry's new Directorate of Scientific Research in 1924. By 1934, Rowe was the assist-ant to the Director of Scientific Research, Henry Wimperis.

In his memoirs, Rowe states that he found fifty-three Air Ministry files related to air defence which clearly showed that 'the Air Staff had given conscientious thought and effort to the design of fighter aircraft, to meth-ods of using them without early warning and to balloon defences.' It was clear that little or no effort had been made to call on science to find a way out. Rowe wrote a memorandum to Wimperis, which summarized the con-tents of the Air Ministry files and advised him to inform the Secretary of

Dr A. P. Rowe (seated with glasses) with the senior staff of the Telecommunications Research Establishment in 1945.

State for Air 'that unless science evolved some new methods of aiding air defence, we are likely to lose the next war if it is started within ten years.'[3]

The actual memorandum written by Rowe does not survive, so there is some conjecture as to its contents. It appears from later developments that Rowe found that one of the recurring proposals for a new anti-aircraft weapon was for the use of high energy radio transmissions to shoot down enemy bombers by incapacitating the crew, melting the airframe or damaging an aircraft's engines or electrical system. Only two of these proposals survive today, neither in Air Ministry records, but they are an indication of the type of crackpot schemes which Rowe wondered might contain some kernel of truth.

The first is from the First World War. In December 1917, R. Russell Clarke, who worked at the Admiralty's top secret cryptographic department, the famous Room 40, approached Winston Churchill, then the Minister for Munitions, to outline his idea for constructing a 'Heat Ray' to be used for offensive purposes. Clarke began his proposal with an ironic and likely unintentional pun. He wrote that, 'The idea of a heat ray for offensive purposes is not novel. Mr. Wells suggested it in his "War of the

Worlds", but he gave no idea of how it was to be accomplished.' Clarke, however, did know how such a device might work. He suggested that 'a generator could be discovered for producing electrical oscillations of very short wave length, and of sufficient power, these very short waves could be reflected by a carefully laminated reflector composed, say, of a number of wires arranged on a parabolic surface like the ribs of an umbrella. This parabolic surface would concentrate the oscillations, creating a beam which would generate intense eddy currents in it, and if the effect could be made powerful enough, the body would be melted.' Clarke stated that he had built a prototype of the generator before the war, although it was too small and primitive to generate short wave radio signals.[4]

Churchill told Clarke to send his proposal to the Ministry of Munitions immediately. The minister's office forwarded Clarke's memorandum to the Military Inventions Department (MID). Written on the memorandum was a marginal note ordering MID to conduct an 'urgent' investigation of Clarke's scheme. The marginal note is unsigned, but given his persistent, unscientific and indiscriminate promotion of gadgetry, it is highly probable that Churchill himself wrote it.

The MID produced a detailed analysis of Clarke's proposal on 10 January 1918. It found every aspect of the scheme flawed. No such scheme could work, the MID investigator reported, because 'the image of the ray could not be made intense – if the target is 1,000 times as far from the source as the mirror the heat per square surface that is possible to develop is one two millionth. The mirror can reflect only by itself absorbing energy; so that the intensity at the target is further reduced. The target will itself reflect, otherwise it cannot absorb; so the effect on it will be infinitesimal; and the damage nothing.'[5]

Despite this devastating assessment, inventors continued to send heat ray schemes to all the service ministries during the interwar period. The second of these proposals that has survived stems from a press report of 9 February 1933 on the invention of a heat ray device by Mr Coxhead of Maidenhead. The inventor told the newspaper:

> I have invented a new electrical heat ray; its force or power is derived from the electrical subterranean energy of the earth. On [31 January] I transmitted the minimum of power which easily reached the firmament in the form of an illuminant clearly depicted upon the sky; and, although the ray itself is invisible in transit through space, it showed the ray had travelled a great height.[6]

This publicity garnered Coxhead a meeting with two sceptical Royal Engineer officers on 17 February. The officer's reported:

[Coxhead] claimed to have destroyed a beech tree with it from a distance of about 150 yards. We visited the site with him and saw the transmitter of his apparatus set up, but it gave no clue as to its components. The tree appeared to be the victim of a normal tree disease and it is difficult to believe that anything Mr. Coxhead may have done with his "ray" had done anything to the tree at all.

The officers could not get Coxhead to explain how his device worked and they quickly surmised he had little technical knowledge. The engineers concluded that it would be 'useless' to pursue this invention any further unless Coxhead provided some technical details.[7] Needless to say, no such details were ever forthcoming.

Rowe explained in his memoirs that in the 1920s the Air Ministry was inundated with similar claims, usually by inventors who alleged that they were able to kill rabbits at short distances. Like Coxhead, these inventors would not reveal any technical information, at least not until the ministry provided some money to continue their research. The Air Ministry managed to deter these proposals by making it known that it would offer a £1,000 prize for any inventor 'who could demonstrate the killing of a sheep at a range of 100 yards, the secret to remain with the owner. The mortality rate of sheep', Rowe reported, 'was not affected by the offer.'[8] Still, as Watson-Watt would later recall, in 1934 'the newspapers were full of [stories about] rabbits being killed at four or five hundred yards, and motor bicycles being stopped, cars in Germany being stopped by some invisible means, so that it was all up in the air, and I'm afraid it was *only* in the air.'[9]

Rowe was likely influenced by much more than these bizarre and unsubstantiated proposals and reports. As we have seen, radio technology had dramatically improved since the First World War. The power and reliability of short wave radio systems was growing exponentially, and certainly there was some mathematical possibility of death rays being built. Whatever argument was contained in Rowe's memorandum, it was persuasive enough to have Wimperis continue to investigate.

Since the Air Ministry contained no scientific radio experts of its own, a product of the 1924 decision to forego research in this area, Wimperis had no choice but to continue his investigation with outsiders. On 15 October Wimperis lunched with A. V. Hill because he wanted to have 'a good talk with him on radiant energy as a means of AA defence.' By 1934, Hill was Britain's most respected physiologist; in 1922 he had shared the Nobel Prize in Physiology and Medicine with German scientist Otto Meyerhof 'for his discovery relating to the production of heat in muscles.' Since 1926 he had held one of the Foulerton Research Chairs funded by the Royal Society and conducted his research at University College, London.

The combination of his knowledge of human physiology and air defence technology made Hill a logical choice for Wimperis to pursue his investigation into death rays. While the content of their discussion is unknown, Wimperis left the meeting convinced that he had to make further inquiries on death rays.[10]

On 12 November, Wimperis composed one of the two crucial documents for the development of the air defence system that would save Britain in the summer of 1940. It was a brief one and one-half page note outlining the results of his discussion with Hill for Air Marshal Sir Hugh Dowding, the Air Member for Supply and Research; Air Marshal Sir Edward Ellington, the Chief of Air Staff; and Lord Londonderry. Wimperis began by reminding them that if one looked at 'the stupendous technical advances of the last fifty years one cannot but wonder what equally striking advances can possibly lie ahead of us in the next equal period of years.' Wimperis was 'confident that one of the coming things will be the transmission by radiation of large amounts of electric energy along clearly directed channels. If this is correct the use of such transmissions for the purpose of war is inevitable, and welcome in that it offers the prospect of defence methods at last overtaking the attack.' As a result of improvements in aviation technology there was a pressing need 'to intensify our research for defensive measures and no avenue, however, seemingly fantastic, must be left unexplored.' There was possibility of 'radiation affecting the human body and, perhaps, the metal fuselage and wings.'

These developments needed 'careful watching' and in order to pursue new technical developments in air defence technology Wimperis proposed the formation of a committee of 'two or three scientific men' whose 'findings may sometimes prove visionary, but one cannot afford to ignore even the remotest chance of success: and at the worst a report that at the moment "defence was hopeless" would enable the government to realize the situation and know that retaliation was the sole remedy – if such it can be called.' The terms of reference for the committee should be 'to consider how far recent advances in scientific and technical knowledge can be used to strengthen the present methods of defence against hostile aircraft.' He also recommended that the new committee be responsible to the Committee for the Imperial Defence.

Wimperis recommended the air defence inquiry compose of four members. He would serve as the Air Ministry's representative and Hill as one of the scientists. Joining them would be P. M. S. (Patrick) Blackett, Professor of Physics at Birkbeck College, London and the chair would be Henry Tizard, the rector of Imperial College, London.[11]

By 12 December the Air Ministry approved of Wimperis' proposal, directly adopting his broad terms of reference for the new Committee for the Scientific Survey of Air Defence (CSSAD) and accepting his nominations for membership. Rowe was added to the committee to act as its secretary. The only other alteration to Wimperis' proposal was that the new organization remained under the auspices of the Air Ministry; no effort was made to expand its authority by making it part of the CID. This alteration was made by Air Marshal Hugh Dowding, Wimperis' immediate superior, who was noted for never allowing outside powers to interfere with his authority.[12]

Wimperis had made well considered choices for membership in what he frequently referred to in his diary as 'my new committee.' He had undoubtedly crossed paths with Hill during the war, when both where engaged in aeronautical research. Hill was an obvious choice. Less obvious was his decision to include Patrick Blackett.

Born in 1898, Blackett was the youngest member of the CSSAD. He started training to become a naval officer, at the age of thirteen, training at the Royal Naval Colleges at Osborne and Dartmouth. He finished at Dartmouth in August 1914, academically first in his class. Even before he had written his final exam, Blackett was mobilized to begin service as the First World War began. He was at both the Battle of the Falklands in December, and the Battle of Jutland in May 1916.[13]

In 1919, Blackett was among 400 young naval officers sent to Cambridge for six months to finish their education. Already discontented with life in the navy, it took just three weeks at Cambridge to convince him to resign his commission. He flourished at Cambridge, immersed in his studies and in a burgeoning social and intellectual world. He became heavily involved in various socialist circles. His excellent naval education served him well at university. Blackett was fascinated with physics and this brought him to the Cavendish Laboratory in the last days of the professorship of J. J. Thomson. Here he learned of the latest developments in the study of the atom, from many of the people who had made these discoveries in the years before the war. In 1921, Blackett received a coveted First in Natural Sciences (Physics).

He then received a fellowship to begin postgraduate research and teaching at the Cavendish, which was now being run by Ernest Rutherford. By the late 1920s Blackett was recognized as being among the best experimental physicists of his generation. In the early 1930s, he developed with 'Bebbe' Occhialini, an Italian research student, an ingenious system, in which two Geiger counters linked by a coincidence circuit, allowed the automatic photographing of cosmic ray tracks, including some of the first and the very best of a positron. After the Second World War he would be awarded a Nobel prize for this work.[14]

In October 1933, he was appointed to the chair of physics at Birkbeck College and in the same year elected a member of the Royal Society. In

1934, Blackett began his involvement in defence related research when he was appointed as a scientific member on the Aeronautical Research Committee (ARC).

The choice of Tizard, a second-rate experimental scientist, as chairman of the CSSAD might appear unusual on the surface. He had not published a single scientific paper for seven years and even earlier had stopped pursuing active research. Tizard later admitted that he had ended

Henry Tizard.

his research career because: '[I] would never be outstanding as a pure scientist ... younger men were coming on ... all with greater ability in that respect.'[15] Outwardly nothing indicated that he had the natural qualities of leadership required to head one of the most important committees ever established by a British government. Of medium height, medium build, and looking very much like the stereotypical English bourgeois bureaucrat, Tizard did not have a particularly commanding presence. At times, according to his good friend C. P. Snow, 'he looked like a highly sensitive and intelligent frog.'[16]

Tizard, however, transcended the limitations of his pure scientific capabilities; his ordinary exterior hid one of the great scientific leaders of this century. In fact, Tizard was one of the best of a type of scientist new to this century ... the scientific manager and administrator. Scientific administrators have proven to be as important to the development of science in the twentieth century as inspirational theoreticians such as Einstein or brilliant experimenters like Rutherford. This century has seen the development of 'big science', where some of the most important work is done by huge research programs involving hundreds and sometimes thousands of individuals working collectively towards a common goal.

Henry Tizard was born in Kent on 23 August 1885. His father was a captain in the Royal Navy at the time, soon to become the Assistant Hydrographer of the Admiralty and an elected Fellow of the Royal Society. His mother was the daughter of a prominent civil engineer. If not for poor eyesight, Tizard would have followed his father into the navy. With this avenue closed to him he won a scholarship to Westminster, one of the more prestigious of the British public schools. He soon showed excellent abilities in mathematics and science. In 1904, he went to Magdalen College, Oxford, where he had been awarded yet another scholarship, to begin his university education. He graduated in 1908 with a coveted first-class degree in mathematics and chemistry. That autumn he was admitted to the University of Berlin, as a postgraduate student in chemistry with Professor Walther Nernst. Due to financial limitations he was forced to leave Berlin after one year, having undertaken no publishable research. This was mainly the fault of Nernst, who instructed him to undertake several unproductive research projects.

After a year as a researcher at the Davy Faraday Laboratory at the Royal Institution in London, Tizard was offered a fellowship to be a resident tutor at Oriel College, Oxford in 1911 and a demonstrator and lecturer at the university's Electrical Laboratory. However, the First World War interrupted his career, changing his life forever by removing him from the serenity of Oxford. For the first time he became aware of the need for scientists to play a role in military research. [17]

Tizard's military career began in the army in 1914. Unlike other scientists, however, he never risked his life in the trenches. Soon after enlisting Tizard was persuaded to volunteer for the Royal Flying Corps and joined the Experimental Flight of the Central Flying School at Upavon aerodrome. As aviation was becoming increasingly important to the military, there was growing need to speed technical advancement. Tizard, along with a handful of other scientists, was assigned by the Royal Flying Corps to develop scientific methods of analysis for aviation. He soon became one of the pioneers of the new science of aeronautics.

Soon after his arrival at Upavon in 1915, Tizard decided that the only way he could understand aircraft was to learn to fly them. This he did at the age of thirty, despite intense opposition from the station commander, who only agreed to allow a scientist flight time if it was confined to days when the weather was too bad for combat pilot training. Similar bureaucratic intransigence limited scientific work until a separate airfield for scientific testing was established in early 1917 at Martlesham Heath near Woodbridge. Tizard was appointed the scientific officer in charge.

On 7 July, while flying a Sopwith Camel, Tizard attacked a Gotha when it crossed the coast. Diving down from 17,000 feet on the German bomber, he managed to get into an attack position just 100 yards away from his prey. Both his machine guns soon jammed. He was able to clear one gun to make a second attack, but it again failed to function properly. As a good scientist he carefully noted the cause of each gun failure, a notoriously common problem at that time.[18]

By the end of 1917 he had established methods of accurately measuring aircraft performance in various weather and service conditions, and had developed a variety of new flying techniques. It was here that Tizard developed his particular expertise of getting operational personnel to work closely with scientists. So successful was he that when the Royal Air Force was established in 1918, he was appointed Assistant Controller, Research and Experiments, at the new Air Ministry. He became the acting Controller in the last three months of the war, after his superior died in an air crash.

Although Tizard returned to Oxford briefly at the end of the war, his future career was focused on government scientific administration. In 1920, he accepted the position of assistant secretary to three boards being established at the Department of Scientific and Industrial Research to coordinate defence research of all three services. Tizard's administrative abilities led him quickly to the senior-most positions in DSIR; by 1927 he was appointed the Permanent Secretary and, therefore, the leading scientific administrator in the British Government. In 1924, Tizard turned down the Air Ministry's request for him to become the Director of Scientific Research. Instead, he urged the RAF to offer Wimperis the position.

Financial worries caused Tizard to leave the DSIR in 1929 and accept the position of rector at the Imperial College of Science and Technology. However, he remained heavily involved in government research. He had been a member of the Aeronautical Research Committee since 1920 and in 1933 he was appointed chairman, a post that he would hold for the next ten years. Through this body Tizard remained a leading authority on aviation research.

During his lengthy public career Tizard developed an extraordinary ability to communicate to any audience. Whether he was talking to senior civil servants, air marshals, politicians, scientists, fighter pilots or lowly office messengers, he was able to bridge the often huge gap in language and understanding when he discussed complex technical problems. 'He was a master of the searching question', able to analyse a complicated problem, break it down into its simplest components, and then find the right query to make to the right expert.[19] Tizard was a team player, running his committees by consensus. He listened carefully to all sides before making a decision. His honesty and clear thinking inspired immense loyalty among most who worked with him. If he had a serious flaw, it was his inability to understand those who did not meet with his own high standards.

Wimperis knew him well, both professionally and personally. They were the two most important members of the ARC and frequently socialized together. This included engaging in regular games of golf, belonging to the same club committee and frequently dining together. Yet there was no favouritism shown in choosing Tizard; Wimperis knew that he was the ideal choice to chair the committee. It would prove to be a remarkably astute decision.[20] From the time of this appointment until the summer of 1940, Tizard was to dominate scientists in the determination of air defence policy. The CSSAD became known as the Tizard Committee.

Prior to the Tizard Committee's first meeting, Wimperis prepared 'A Note on the Problem of Air Defence' to brief the scientists. It is a remarkable document because it shows the very firm grasp that the Air Ministry had on the growing technological problems involved in preventing 'enemy aircraft from reaching industrial areas and ports' and the desperate measures being considered to solve them. Interestingly, the main issue discussed was not how to provide immediate solutions, but for the situation as it would exist in 1940, when bombers could fly at over 250 mph and up to 30,000 feet. The scientists would have to concern themselves with developing countermeasures to a whole range of potential attack scenarios, including attacks at day or night 'at any attitude from "hedge-hopping" to 30,000 feet,' and in a variety of weather conditions.

Destruction of aircraft could only be achieved by improved anti-aircraft fire, fighter aircraft and balloon barrages, but none of these offered any real assurance of success. Various other technological solutions had been investigated. Consideration had been given to having RAF bombers drop bombs onto enemy aircraft. The use of poison gas or of mysterious vapours to stop aircraft engines was discounted as being equally impracticable. Other possibilities included the development of missiles or pilotless aircraft to ram approaching bombers.

Death ray proposals were 'attractive', but had not been seriously investigated since the prospect of success seemed remote. Yet at least one army research group at the Signals School at Woolwich had conducted an unsuccessful investigation into using radio waves to detonate aircraft bombs, and the possibility of physiological effects on crews 'justified a constant watch on the position' of this technology.

Wimperis understood that the key to air defence depended on knowledge of approaching aircraft's position. A general idea of the altitude, speed and direction of approach of a bomber would greatly assist intercepting fighters, while more accurate short-range detection would significantly increase the effectiveness of anti-aircraft fire. Existing visual and acoustics methods were being eclipsed by improvements in aeronautical technology. There were other means of detection under investigation, but all had their distinct limitations. Infrared radiation from aircraft engines was absorbed by atmospheric water vapour and masked by "Black body" radiation from the earth and from cloud. Electro-magnetic radiation from an aircraft could be easily shielded. Even the detection of radio waves was mentioned, but Wimperis was not optimistic stating that, 'there seems little hope of a solution on these lines.'[21]

Radio Reflection

While the Tizard Committee was being organized, Wimperis remained active in pursuing his inquiries into death rays. In mid-January 1935, Wimperis made the fateful decision to ask Robert Watson-Watt, the head of the National Physical Laboratory's Radio Research Station and the government's leading expert on radio physics, to come to his office at the Air Ministry for a discussion.

Robert Watson-Watt was born on 13 April 1892 in Brechin, Aberdeenshire, Scotland. His father was a self-employed joiner and carpenter, Presbyterian Sunday School teacher and active temperance leader. His mother was the daughter of a Dundee millwright and also a keen believer in the evils of alcoholic beverage. He won a scholarship to attend the local high school and a Carnegie bursary so that he could continue his education at the University of St Andrews' University College at Dundee. Being from a practical family,

Watson-Watt chose to study engineering, but he soon discovered a passionate interest in the more esoteric world of physics. In 1912, he graduated at the top of his class and was immediately offered the position of assistant to the Professor of Natural Philosophy (physics). In the next two years he turned his attention to learning about wireless telegraphy.

During the First World War, Watson-Watt joined the Meteorological Office where he lead investigations into methods of using radio to locate thunderstorms. The work was only partially successful, but it did lead Watson-Watt into the bourgeoning scientific civil service. This position gave him access to cutting edge technology in radio direction finding, as used by the Admiralty to track German naval wireless communications. He also learned of the development in the United States of the first primitive cathode ray tube, and quickly grasped that the device could be used to provide a simple and fast visual indication of a radio signal. In 1922, Watson-Watt was able to secure access to one of the first two CRT oscilloscope to arrive in Great Britain. The other oscilloscope went to E. V. Appleton, a Cambridge trained physicist, who was investigating the newly observed phenomena of atmospheric radio reflection. Appleton soon began collaborating with Watson-Watt's group at Slough, although Watson-Watt did not participate in Appleton's Nobel Prize winning experiments which proved the existence of a reflective layer in the upper atmosphere. Watson-Watt's team, however, were the first British researchers to adopt a superior American developed experimental technique to use radio pulses to measure atmospheric reflection. Watson-Watt coined the term ionosphere for the ionized layer which reflected radio transmissions.

In 1927, Watson-Watt was named to head the NRL Radio Research Station. In effect, outside of the Post Office, Watson-Watt became the senior radio scientist in government service and a permanent member of the Radio Research Board. Through the RRB he met Wimperis.[22]

Watson-Watt's personality, with the possible exception of that of Frederick Lindemann, is the most complex of any of the prominent figures in the story of British radar. He was an extremely able, if not erratically brilliant, hands-on experimental scientist who could command great loyalty from his laboratory co-workers. While he prided himself on his ancestral connection to the eighteenth century steam engine innovator James Watt, Watson-Watt had little experience with the methods of adapting research devices to mass production. He also shared with many scientists of the period an antipathy towards industrial researchers and technical experts. At best, he was an indifferent administrator, who had the potential of being downright destructive.

This inconsistency was the result of his erratic, almost manic personality. He could be shy and reserved with strangers, but when in the company of close friends or when a situation aroused his passions, he could be extraordinarily verbose. He spoke quickly, sometimes incomprehensibly, never using one word where a thousand would do. As one radar scientist recollected: 'He used a sort of verbal karate, in which you paralyse your opponent with a triple negative and then knock him out while he is trying to work out what it was you meant.'[23] At times he seemed to have a brilliant intuitive grasp of a situation, at other times he was either oblivious to impending crisis or provided answers that were so off base that they appeared idiotic.

Fortunately for Britain, Watson-Watt was on his best form on 18 January when he arrived at the Air Ministry.[24] Wimperis had not told Watson-Watt the reason for the meeting. Watson-Watt had been acting as a radio consultant to the Air Ministry for some time and had met with Wimperis several times in the previous six months, but up to this point it had always involved providing straightforward advice on improving communications. As the radio scientist sat down Wimperis exclaimed:

> Watson-Watt I've asked you to come and see me, not in my official capacity nor in yours, but because you are a friend, a personal friend, for whose discretion and judgement I have considerable respect. We are not to talk officially, but I wish you to tell me what you think about the possibility of a ray of damaging radiation which might be used in defence against air attack.

Watson-Watt thought to himself that 'Harry means a death ray.' He responded to this bizarre request by saying: 'I really don't think the physics are at all promising; unless we have an absolutely startling break-through in physics, in new knowledge, I would think that it is no bet at all. But I'll go back to my desk and do some arithmetic, and I'll give you a reasoned answer.'[25]

Watson-Watt returned to his office and being a 'lazy person' he left a note for one of his 'brightest young men', Arnold Frederick 'Skip' Wilkins, since 1931 a junior scientific officer at the radio research station. He requested that Wilkins 'calculate the amount of energy which, when radiated in the form of high frequency oscillations, would be required to raise the temperature of 75 kilograms of water by 2°C' at a distance of 10 kilometres. Wilkins 'deduced at once that the question of the practicability of the "Death Ray" had again arisen' since 75 kilograms of water was approximately the amount of water in an adult male body. He concluded that he was being asked to calculate the energy required to induce a fever among aircrew. Watson-Watt confirmed his suspicions and urged Wilkins to provide a definitive answer. Wilkins calculations proved convincingly and 'as expected, that, even using an array of the largest aperture commercially tested, the transmitter power' required was 'astronomical and the project quite impracticable.'[26]

After he informed Watson-Watt of his results, Wilkins was asked if he 'had any suggestions to make as to how the presence of aircraft approaching the coast could be detected.' It was a leading question since Watson-Watt already knew the answer and was testing his junior staff member. Wilkins easily passed the test. He had collaborated in the 1931 research in which Post Office engineers had observed that aircraft interfered with ultra short wave radio transmissions. He suggested to Watson-Watt that this 're-radiated' energy might be used to detect aircraft.

Watson-Watt had been suggesting the same thing since perhaps as early as 1932. In his memoirs he states that around the summer of that year he had privately told William Tucker and other ADEE scientists: 'You are listening up the wrong tree, irrespective of barking dog interferences. You are working far too near the stops of your technique, you should be thinking now of radio methods.'[27] Tucker ignored his advice. Watson-Watt decided to try one more time to promote the use of radio reflection as a means to detect aircraft at long range. In the concluding paragraph of a five page memorandum which outlined the impossibility of building a death ray, Watson-Watt informed Wimperis that 'meanwhile attention is being turned to the still difficult but less unpromising problem of radio-detection as opposed to radio-destruction, and numerical considerations on the method of detection by reflected radio waves will be submitted when required.'[28]

This was the second key document that would transform the air defences of Britain. To Wimperis, Watson-Watt's new direction must have seemed like a godsend. Wimperis had already been subjected to some good natured leg pulling about his investigations from his immediate superior, Air Marshall Sir Hugh Dowding, the Air Member for Research and Development.[29] Both men, however, well understood the seriousness of the situation.

Daventry

Watson-Watt's memorandum had only just been received at the Air Ministry when, on 28 January 1935, the Tizard Committee met for the first time. There was great interest in Watson-Watt's proposal and he was asked to submit a follow up report on radio reflection as soon as possible. Watson-Watt already had Wilkins hard at work preparing a second paper examining the question of radio reflection in detail, which was completed around 6 February. The two scientists had concluded that their 'best chances of detecting an aircraft were, other things being equal, by using a wavelength for which the wing-span would approximate to a half-wavelength aerial;' this meant a wavelength of some 50 metres.³⁰ The original draft of this memorandum does not survive, and it is difficult to reconstruct it from the revised version of 28 February. Yet it is known that it was sufficiently compelling to convince Wimperis and Tizard that Watson-Watt should immediately receive £10,000 in funding to begin developing a prototype aircraft detection device as soon as possible. However, Hugh 'Stuffy' Dowding was not as easily impressed by calculations on paper. He told Wimperis, 'These scientific fellows can prove anything on paper, I would like to have a demonstration.'³¹

Dowding, in his position as Air Member for Supply and Research (after 14 January 1935 Air Member for Research and Development) and in his subsequent appointment as Commander-in-Chief of Fighter Command, which he held from 14 July 1936 to 25 November 1940, was the RAF officer most closely connected with the development of radar and its integration into the air defence system. Born on 24 April 1882, Hugh Dowding joined the army in 1899 when he entered the Royal Military Academy at Woolwich. From 1900 until 1910, he served the Royal Artillery in a variety of colonial postings. In January 1912, he received a coveted position to the Royal Army Staff College at Camberley. While there he acquired the nickname 'Stuffy', which likely referred to his somewhat aloof personality. Dowding also became fascinated with the military potential of aeroplanes. Since the army did not provide basic flight training, he arranged to take private lessons, which would be paid for when he joined the Royal Flying Corps (RFC). He received his 'ticket' to fly after a grand total of just one hour and forty minutes in the air, which included time as a passenger. After finishing the staff college course, Dowding received advance flight training from the RFC, and then in the spring of 1914 he asked to be transferred back to the Royal Artillery.

His return to the more traditional arm of the army was short-lived; officers with staff college training and flying experience were a rare commodity when the First World War began, and he soon found himself transferred back to the RFC. During the war he held a variety of positions, including aerial observer, running an experimental program involving fitting wireless into aircraft, front-line squadron commander and as a supervisor of aircrew training in England. He willingly transferred to the new RAF in 1919 with the rank of group commander. For the next decade he served as a staff officer in various home commands, and in RAF imperial policing operations in Iraq and Palestine. He was appointed to be the Air Member for Supply and Research on 1 September 1930. By 1935, he was one of the senior-most officers of the RAF, and he saw himself as being the heir apparent to the Chief of Air Staff. While his imperious nature and his refusal to delegate authority made him unpopular with many other senior RAF officers, his firm grasp of emerging technologies and his enthusiastic support for them endeared him to the scientists who worked on the air defence system. He developed a particularly close working relationship with Tizard in the years ahead.[32]

Dowding, always stubborn when he made up his mind, could not be dissuaded by a letter from Wimperis that a preliminary test was unnecessary nor was he persuaded that radio detection of aircraft was the key to air defences. Dowding informed Wimperis that his 'only hesitation' in approving the £10,000 was that the project was 'purely defensive' and that if they went '*all out*' for it immediately then 'some offensive device which might be in the offing, may have to take a back seat.' Dowding, however, could see that radio detection might do more than simply replace the acoustic mirrors, but might also be developed to allow anti-aircraft guns 'to be directed against invisible aircraft.' Still he asked that the Tizard Committee consider research that would improve '*offensive* methods' before rushing in to embrace Watson-Watt's proposal. All he was asking for, he told Wimperis, was a brief delay before committing the limited resources of the NPL radio researchers to this one project.[33]

Watson-Watt could not fathom Dowding's hesitation and demanded an opportunity to discuss the manner with him. They met at Farnborough, but Dowding refused to reconsider his position. Watson-Watt reluctantly agreed with the demand for a preliminary experiment using existing ionospheric equipment. Watson-Watt later claimed that he bluntly told the Air Marshal something like the following:

> If we don't get the indications we expect, it will not be because we are wrong in our theory or seriously wrong in our rough figuring. It will be because we will have "lashed up" a rough equipment, made of parts meant for other purposes ... or have done one of a hundred things that should be avoided in a crucial demonstration, but can't all be avoided in a hurried demonstration.[34]

The experiment as devised by Watson-Watt, with assistance from Wilkins, was a brilliant piece of scientific improvisation. For a transmitter they opted to use the British Broadcasting Corporation's short-wave radio station located at Daventry. It operated at around 50-metres wavelength, and had one of the most powerful signals available. As a receiver they employed an existing apparatus that had been developed to investigate the transmission of long distance, short wave radio signals. This device con-

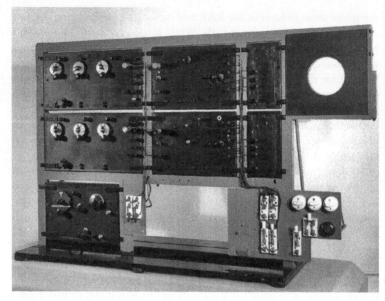

Daventry Receiver (front).

Daventry Receiver (rear).

sisted of 'two aerials spaced about twenty feet apart in a line running in the direction' of the transmitter to the receiver. Lines connected the antennae to separate receivers. One of the lines first went through a variable phase changer, which was tuned to produce a signal that was anti-phased, but equal in amplitude, to the other. The two signals being in exactly opposite phase, but of equal amplitude would cancel each other out when mixed together before being sent for display on a cathode ray tube. 'Should a ray from the aircraft be produced, this condition of balance would not hold and a deflection would be produced on the cathode ray oscillograph.'

On the night of 25 February in a biscuit-coloured, caravan-bodied Morris truck, Wilkins, accompanied only by a Mr Dyer, the radio section's driver, arrived in a grassland field near the village of Weedon, not far from Daventry. Wilkins, with some assistance from Dyer, began to assemble the receiver. This task was complicated by the failure of lighting in the interior of the truck, which forced Wilkins to fumble about in the dark 'during the process of connecting up' all of the lines and equipment. Wilkins was just able to make sure the system was in balance before the Daventry transmitter shutdown for the night. They then found that they had to hike to a nearby hotel. The Morris had sunk into mud, which froze while they assembled the set, locking it firmly into place.

Watson-Watt arrived the next morning in his Daimler accompanied by his twenty-three year-old nephew Patrick, and Rowe. Being a secret experiment, Patrick was left sitting on the grass at the gate into the field. As an added security precaution, Dyer was ordered to move away from the equipment while the test was in progress. As had been arranged, at 9:45 a.m. a Heyford bomber, piloted by Squadron-Leader R. S. Bucke flew over the Daventry masts. Bucke was not told the reason for this flight; he was simply instructed to follow a compass course to carry him over the transmitter, and then he was to follow a heading 20 miles south towards London. He was to repeat this trip back and forth until 10:30.

Rowe, Watson-Watt and Wilkins huddled around the CRT as the Heyford flew overhead. On the first pass the aircraft did not come close enough to the direction of reception of the primitive antenna array and nothing was observed. On the second pass, however, as soon as the aircraft headed south they saw 'beats of very good amplitude' on the display screen for a little over two minutes. During the third run signals were detected for 4 minutes and for the fourth, for 4.3 minutes. A quick calculation, based on the bomber being requested to fly at 100 mph, indicated that it was tracked for a maximum distance of just under 8 miles. Given the primitive nature of the equipment and the haste with which it had been assembled, all present were aware that the results were outstanding. Rowe remarked that it had been 'the finest demonstration he had ever witnessed.' One can

Condense this → radar has been shown work

only wonder if any of these three men realized that their entire lives were about to change as a result of this remarkable morning. Watson-Watt was certainly overcome by the excitement. It was only when he reached the outskirts of London that he suddenly remembered that he had left Patrick back at the test site. Fortunately, when he returned Patrick was still peacefully enjoying the unusual February sunshine, unaware that his uncle had almost left him stranded.[35]

Rowe reported to Wimperis:

> It was demonstrated beyond doubt that electro magnetic energy is reflected from the components of an aircraft's structure and that it can be detected. Whether aircraft can be accurately located remains to be shown. No one seeing the demonstration could fail to be hopeful of detecting the existence and approximate bearing of aircraft, approaching the coast, at ranges far in excess of those given by the 200 ft. mirror.[36]

Wimperis' relief that the test was successful can be seen in a rare emotional outburst in his diary. He simply wrote: 'Glad indeed to have a letter from Rowe (I was at Folkestone) saying that yesterday's radio test had been successful! Good Watson-Watt!' On 28 February, he informed Dowding of the results. Dowding was equally pleased. He told Wimperis that he could now have all the money he wanted, within reason.[37]

On the same day, Watson-Watt completed his revised memorandum on 'the Detection and Location of Aircraft by Radio Means.' He outlined many of the technical approaches that would be used and problems that would need to be overcome to make an effective air defence early warning system. The most important suggestion was that research be focussed on a technique developed for measuring the ionosphere. This involved the transmission of very brief radio pulses, equally spaced in time, which would allow for the distance between craft and sender to be measured directly by observation on a CRT screen which would be calibrated with a linear distance scale. By sending a thousand pulses a second and superimposing the successive images, the CRT screen would show an 'easily visible sustained image permitting close measurement and even showing the advance of the craft.' The direction of approaching aircraft would be determined by simple triangulation of the results from three different receivers. The height of the aircraft could be determined by measuring the 'angle of elevation of a descending radio wave.'

Watson-Watt also discussed which frequency would be used by the system. He saw 50-metre wavelength as having the advantage that the radio equipment already existed, but it had the disadvantage of being subject to disruption by ionospheric reflection, which would cause interference with short wave radio communications and the transmissions would be readily detectable in foreign countries. However, none of these objections were sufficient to force delays in preliminary testing and all could be dealt with. Watson-Watt proposed that an elaborate cover story be created to shield the research from foreign eyes. This involved having the Air Ministry make an offer to the Department of Scientific and Industrial Research 'of facilities for ionospheric investigation and other work for the Radio Research Board at a convenient flat and isolated site at Orfordness, [which was at a] suitable distance from Slough for special experiments.'[38]

On 1 March, Wimperis and Watson-Watt took the train to view the Orfordness site. Wimperis believed that the station would 'do excellently for our new radio research.' He was keenly aware of the importance of the work that was going to be undertaken and confided to his diary: 'I am most hopeful of the new system going almost all the way to make air attack on this country a very doubtful success.'[39]

The Tizard Committee had been handed the one vital piece of technology with which it would transform the air defence system of Britain and at the same time recast the role of the scientist in the military establishment. Yet the success of radar was by no means predetermined. Wimperis' belief that the new technology would succeed was not misplaced, but it was a matter of faith guided by scientific reason. In March 1935, radar was very much in its infancy; the experiments at Daventry were just the very first preliminary steps in a long serious of experiments before it would become Britain's shield.

5

Committees and Politicians

The discovery of radar and the formation of the Tizard Committee could not have come at a more opportune time. On 31 January 1935, just three days after the first meeting of the Tizard Committee, Brooke-Popham's committee completed its interim report. The report called for a sweeping reform of the nation's air defences. While the RAF continued to believe 'that the most important function of aircraft [was] offensive action by bombers,' a defensive system was 'also seen as a necessity.' The improvement in bomber performance necessitated a vast expansion of the area that needed protection. Rather than a series of defensive rings around London, they advocated 'a continuous defensive system from Portsmouth, round the eastward of London, to the Tees' which would extend some 26 miles in depth from the coast. The arrangement of the defences would remain similar to Ashmore's First World War London Air Defence Area. Anti-aircraft batteries and searchlight companies would still be placed in an outer artillery zone extending 6 miles from the coast. A fighter zone of 20 miles depth would lie behind the artillery zone in which fighter squadrons would operate, supported by searchlights for night operations. An inner artillery zone would protect London, and similar defensive areas would be established for important industrial and port cities such as Birmingham, Sheffield, Leeds and Manchester.

In order for this defensive scheme to become fully operational, a huge expansion of the existing defensive forces was necessary, well beyond what Cabinet had approved in the summer of 1934. For instance, the Air Ministry had estimated that it would require twenty-five fighter squadrons for Home Defence, but Brooke-Popham's committee considered this number 'must be regarded as a minimum' in which some squadrons would be forced to operate both during the day and at night. In the 'event of the German Air Force expanding more rapidly than anticipated when the estimate was made, an increase in fighter squadron' would become necessary. The War Office would have to provide 43,500 personnel to man the fifty-seven anti-aircraft batteries and ninety searchlight companies, more than double the manpower utilized in 1918. Total cost of the land defences alone would be £5½ million more than already approved. The whole scheme would be completed in three stages. The first stage could be completed by 31 March 1940, at the still considerable cost of £2¼ million for the ground units alone. Yet, until all stages of the defensive scheme were completed, no adequate measure of protection would be provided for any part of the country.

The committee's proposals were based on existing technologies, and provided a less complete understanding of the need to develop new means to deal with air attack than Wimperis' 'A Note on the Problem of Air Defence.' Even though the Air Ministry's scientific committee had been approved some six weeks earlier, Brooke-Popham was unaware of its formation. He kept his commitment to Lindemann by recommending the formation of a new CID scientific sub-committee on air defence.[1]

Just as radar was discovered and the work of the Tizard Committee was getting underway, the Brooke-Popham report began a major re-evaluation of the air defence system. The air defence system would become part of the ongoing debate on rearmament already taking place in Whitehall, Downing Street, Westminister, and Fleet Street. While this tumultuous debate was in progress the scientists of the Tizard Committee would begin their work which would transform the air defences and the way that science was utilized for military purposes. The scientists would try to avoid the dirty world of politics; however, right from the beginning they would find that leaving the safe confines of the academy placed them directly in the private and public political battles that would rage around the rearmament programme and diplomatic policy towards Hitler's Germany.

The Tizard Committee

The was a peculiarly British institution. It had none of the executive powers that Lindemann envisaged would be necessary to fulfill its mission; yet, paradoxically, it had far greater influence than any political creation. The source of the committee's authority was not readily apparent. Right from the start Tizard, Hill and Blackett had insisted that they retain their academic independence by not becoming reliant on the Air Ministry or any other government body for their livelihood. They remained academics. Blackett and Hill continued their successful research careers.

Tizard, on the other hand, had a very well paying but not terribly demanding academic administrative appointment. He remained the chairman of the Aeronautical Research Committee, but was soon heavily involved in running the Committee for the Scientific Survey of Air Defence. Tizard in effect became the one full-time scientific member of the committee, something he did voluntarily because he well understood the vital nature of the work. The administrative burden alone was heavy. The Air Ministry at first provided no office space and only sporadically was a typist available to assist with the paper work. Rowe helped all that he could, but he had other responsibilities within the Air Ministry and on CID committees. Tizard was forced to conduct confidential interviews in his flat in St James Court, and would eventually hire, at his own expense, a part-time confidential secretary to assist in the committee's work.

Unable to wrangle regular access even to the ministry typing pool, the Tizard Committee had to rely on other less tangible factors to be effective. Tizard's biographer says that instead of executive power the CSSAD utilized persuasion, the 'old boys net', and the goodwill of individual commanders.[2] There is no doubt that this assessment is partially correct. All members of the committee had some connections with serving officers. Among officers, there was a general consensus that the committee's work was important, something that could have only been reinforced by discovery of radar just as the work began. Moreover, the committee's approach to the problem of air defence met with widespread approval. The CSSAD did not begin its work with a desire to criticize, but instead it carefully learned from those most intimately connected with all aspects of the problem, before it made any recommendations on how to improve air defence technology.

The approaches taken by the Tizard Committee to win the trust of the armed forces can be seen in its first few months of operation. On 21 February 1935 the second meeting of the committee was held at the headquarters of Air Defence of Great Britain with Brooke-Popham, the Commander-in-Chief; Air Vice Marshal Joubert de la Ferté, the Air Officer Commanding, Fighting Area and other senior officers. Wimperis proposed that instead of a formal list of questions they have an 'impromptu' discussion 'since the committee might be thought to be going outside its terms of reference.' Rowe also pointed out that all of the questions 'would be covered by one question – "what should we do?"'[3] Tizard agreed and asked Rowe 'to stress' that the committee's purpose was simply to look for background against which they could assess the value of new angles of approach to the scientific problems involved.[4]

This impromptu, humble, and non-aggressive approach led to an extremely wide-ranging and free discussion at ADGB prior to a tour of the central control room. The committee learned about every aspect of the air defence system and the operational commanders' appreciation of the tactical and strategic problems they expected to face in any upcoming conflict. Joubert told them that 'early detection [of approaching aircraft] was the main problem' and that extending out early warning to at least 50 miles from the coast would 'revolutionize the present means of controlling aircraft.' Each zone of the defences had its own unique weaknesses. The outer gun zone, which existed merely because fighters could not get to altitude to meet bombers at the coast, was only 6 miles wide. This would give anti-aircraft guns just 1½ to 2 minutes firing time. Night interception in the fighting area was becoming more complex as bomber performance improved and no data was available on how effective searchlights would be in establishing contact and illuminating the next generation of aircraft. This situation was only to some extent eased by the provision of radio telephones in fighters, which made it much easier to relay information to pilots while in the air. Finally, Brooke-Popham frankly admitted that there was 'no adequate defence against low flying ("hedge-hopping") aircraft.' Any bomber flying below 5,000 feet 'would be a definite menace.'[5]

By early April the committee completed the first draft of its first interim statement. It was a sweeping overview of the existing state of air defence technology and future avenues of research. Tizard instructed Rowe, who compiled this draft, that the report was to be 'intentionally provocative and controversial.' The report outlines numerous faults within the air defences which would be exacerbated in the years ahead. Fighters, for instance, needed to have at least a 30 per cent excess in speed over bombers given the present state of early warning and be provided with a continuous stream of information on the location of their targets until these targets were spotted. The need for pilots to achieve visual detection of a bomber was such a severe limitation 'that fighter aircraft cannot alone prevent hostile aircraft from reaching their objectives.' Bombers would fly above clouds or in fog, a situation that would hamper interception during the day and render searchlights, vital for night detection, next to useless. A means, therefore, had to be found to direct anti-aircraft fire through clouds.

Anti-aircraft fire, however, had its own major technical limitations. Gun-laying predictors then under development, which calculated the correct firing position for the guns against a fast moving target, were unlikely to be able to deal with bombers taking evasive action. The existing 3-inch gun, which had a small bursting area and a maximum operational height of 18,000, required immediate replacement. They urged that the new 3.7-inch gun be rushed into service. The committee also suggested that steps be taken to test the effectiveness of balloon barrages.

Long-range early warning was considered to be vital. Acoustical mirrors were given a scathing review because 'the increased speed of aircraft and improved technique of silencing will more than outweigh technical advances in acoustic research.' Watson-Watt's research was held out as offering the best possible solution, in the short term, for distant early warning, and, in the long term, of locating aircraft 'with considerable accuracy in any weather conditions.'

In general the committee argued that the problems of air defence would be solved, if they could, by experimental methods. However, they found that 'the amount of experimentation hitherto conducted [was] totally inadequate' and that the Air Ministry had made a 'trifling' effort to determine the likely methods to be used by enemy air forces. Until this was done no one knew what proportion of effort should go into AA guns, fighter aircraft and balloon barrages.

The committee did offer conditional hope, however, that the situation could be dramatically improved. The report concluded:

> Surveying the situation generally, the Committee believes that if detection and approximate location at 60 miles from the coast is practicable; if accurate location can be provided for AA gunnery, in all weather conditions; if the number and efficacy of AA guns is largely increased; and if an effective low altitude balloon barrage is provided, then hostile aircraft will be so harassed after crossing the coast that piloted bombing aircraft will not achieve decisive effects on the conduct of war.[6]

There are some remarkable differences between the first and final draft of the interim report, which appeared on 16 May. Some of the changes can be explained by further investigations conducted by the Tizard Committee in the intervening six weeks. Watson-Watt provided the committee with his first detailed account of his research programme in the only meeting held during this period. Comments on the initial draft by officers at ADGB led to the inclusion of a section concerning the positive psychological effect on civilians of visible air defences such as anti-aircraft fire and balloons.

These additions, however, do not explain the general shift in tone that took place. On the whole the final draft was far more positive, less critical, and anything but 'intentionally provocative and controversial.' The report now summarized the committee's findings with this much more optimistic statement:

> If these means for defence can be made available, in conjunction with the counter-offensive and means for passive defence, it appears that the problem of air defence will present far less difficulty than it does at present. If the experiments proposed in this report are successful, defensive measures can be greatly increased. It is probable that nothing that

the Committee has considered will prevent a determined enemy from making a serious and partially successful raid on London; enemy losses should, however, be such that continuous repetition of such raids would be improbable.[7]

The change in tone was quite deliberate because the audience it was directed at went through a fundamental change soon after the first draft was completed. The alterations were designed so that the report would be 'suitably worded and wrapped up' because the CSSAD was being thrust into the centre of the continuing political debate over rearmament and air defences.[8] Harsh yet fair criticisms were tolerable as long as the statement remained an internal Air Ministry document, but for politicians, some of whom were highly critical of the government, this watering down was a political necessity.

The Politics of Air Defence

That the Tizard Committee could avoid being uncomfortably thrust into the political spotlight was a forlorn hope. Even before the committee had held its first meeting, Lindemann, ably supported by Churchill and Austen Chamberlain, guaranteed that air defence research would become a large part of the continuing public debate on rearmament.

In 1935, Chamberlain would act as Lindemann's principle conduit to the senior-most levels of the government at Question Time and in debates in the Commons. Chamberlain, like Churchill, was only a backbencher, though a senior one. He had last served in the Cabinet as First Lord of the Admiralty in 1931. However, with Churchill, he played a leading role in warning of the growing danger of Hitler and of the need for rapid rearmament, something that is often forgotten in accounts which emphasize his far more flamboyant friend.

By late 1934, Lindemann was so obsessed with promoting his views on air defence research that it became a holy crusade that would last until Churchill rejoined the Cabinet as First Lord of the Admiralty in September 1939. In response to Lord Londonderry's letter, which dismissed Lindemann's request for a CID committee on air defence, he stated that he was convinced that the problem of developing a defence against night bombers 'was more important than any other, far more urgent than the quest for a cure for cancer or tuberculosis or indeed any other form of research.' Lindemann believed that the forthcoming conflict would follow a pattern similar to the events of 1917-18, although the attacks would be far more devastating. Daylight raids would be dealt with by the existing defences, but attacks at night would destroy Britain's cities, unless science found a new way to defeat them. He continued to be highly suspicious of the Air Ministry, which he believed held the view that counterattacks by RAF bombers were the only possible defence. Both Lindemann's strategic and tactical view of the air defence problem were faulty, as was his evaluation of the Air Ministry. Like any good zealot, Lindemann could not be put off by an unsubstantiated claim that the ministry was investigating new defensives measures, something he made plain to Londonderry.[9]

For Londonderry this caused a political dilemma. Churchill and Lindemann had already shown their willingness to use the press and radio to attack government policy. Churchill and Chamberlain were willing to ask embarrassing questions in the Commons, some of which were difficult to answer because of security and foreign policy considerations. Moreover, Londonderry served in the House of Lords and the defence of the Air Ministry had to be left to his under-secretary, Sir Philip Sassoon, or other members of the government. Should air defence research and planning become public, efforts to improve the system might be compromised.

Since stonewalling did not work, on 20 December 1934 Londonderry tried to appease Lindemann by informing him of his decision to create the CSSAD and advising him to communicate directly with Tizard. Lindemann's mistrust of the Air Ministry and his dismay at having been excluded, however, combined to lead him to believe that the Tizard Committee could never be effective and that he now had to appeal over Londonderry's head.

He first turned to Lord Hailsham, the Secretary of State for War, to find out if the Cabinet supported Londonderry, and to ask advice on how he could get a 'proper' committee established. Hailsham was sympathetic and advised Lindemann that the only way to move the Cabinet in his favour was to start a press campaign. Lindemann was not yet ready to make a public appeal, however; he instead consulted Austen Chamberlain who advised that a preferable course of action was to make a direct appeal to the Prime Minister.[10] On 7 January 1935 Lindemann composed a letter for Austen Chamberlain and Churchill to sign and send to Prime Minister MacDonald. They requested that Macdonald review the correspondence between Londonderry and Lindemann and urged him to appoint a sub-committee of the CID to investigate air defence research. No mention was made of the Tizard Committee, since it was assumed that MacDonald was aware of it.[11]

MacDonald, however, was ill for much of his last year in office. This is why Baldwin took the lead in many of the parliamentary debates on rearmament from the autumn of 1934 and succeeded MacDonald as Prime Minister on 7 June 1935. MacDonald was not up to date on the latest initiatives at the Air Ministry when he received the letter from the three men on 9 January. He immediately replied: 'I quite agree with you that it should not be a departmental committee, and have already made up my mind that it is something that the CID ought to take in hand.'[12]

This ill-conceived response was corrected by MacDonald on 15 January when he asked Chamberlain to disregard his earlier erroneous correspondence and to await further developments. Chamberlain had already been informed by Lindemann of the 'secret' Tizard Committee. Chamberlain forwarded his copy of MacDonald's second letter to Lindemann, scrawling on the bottom: 'Most unsatisfactory. I have written at once to say that no Departmental Committee will satisfy us.'[13]

MacDonald scrambled to repair the political damage. Wimperis was summoned to Downing Street on 15 January to assist in drafting a proper response. The letter was dispatched to Chamberlain three days later. It outlined the composition of the CSSAD, assured them that the committee would report to the CID and that MacDonald would personally be informed of its activities. Lindemann was offered the opportunity to testify before Tizard's group.[14]

This about face further exacerbated the situation. Lindemann wrote to Chamberlain upon receiving MacDonald's letter:

> It is really deplorable the way he chops and changes. I gather the Departmental Committee has no power and can only serve to hold things up. In any event it could not possibly get vigorous work carried on in the intermediate regions between the various Defence services, which will probably prove the most important.

Lindemann continued by stating that he believed the majority of the Cabinet was behind them, but that the Air Ministry appeared to hold a veto. This was an 'absurd' situation. Churchill was even more blunt. He wrote to Lindemann: 'I quite agree [the situation] is most unsatisfactory. The man is a hopeless twister and the only thing now is to have a debate.'[15]

There was then a further exchange of letters between MacDonald and Chamberlain, The letters have not survived, but Lindemann's letter to Churchill on this exchange has. Here 'the Prof' revealed his innermost thoughts about the Tizard Committee and shows himself to be not only a narrow minded bigot, but a fount of misinformation and innuendo. Lindemann admitted that Tizard was, of course 'a good man', but mistakenly believed that both Tizard and Wimperis received salaries from the Air Ministry, and therefore were suspect. As for Hill and Blackett, neither 'ever had anything to do with aeroplanes.' What was worse was that Blackett held 'himself out as a communist.' What topped all these other concerns was the CSSAD's 'terms of reference and complete lack of status' which left it with 'no power to proceed with experiments.'[16]

On 28 January Austen Chamberlain wrote to MacDonald requesting a meeting with himself and Churchill on air defence research. Two days later MacDonald responded by trying to co-opt these troublemakers. He had the Air Ministry offer Lindemann full membership on the Tizard Committee. Chamberlain warned Lindemann that if he accepted it would effectively muzzle him. Lindemann would not be silenced and he replied to the offer: 'I thank you for your letter of the 30th January. I understand the whole question is soon to be debated in Parliament, and I think you will agree that in the circumstances it is best for me to defer giving a definite answer for the time being.'

The mention of bringing the issue up in a forthcoming parliamentary debate caused a great deal of consternation, since this might reveal secret research work and embarrass the government. In the next two months there would be several defence debates in the House. Final preparation was underway for the publication of the first post-ten year rule defence white paper in early March. The government would have to defend itself from both those who believed they had gone too far and other who believed they had not gone far enough. By the end of the fiscal year, next year's defence estimates would have to be approved. The last thing the government wanted was potentially embarrassing questions on the state of air defence research.

On 13 February Lindemann met with Londonderry for a full airing of the issue. Since the only account of this meeting is Lindemann's, we can only speculate as to what actually took place. According to 'the Prof' the Air Minister began by expressing his inability to understand why they were not satisfied with his departmental committee. Londonderry also expressed a reluctance to go outside his department since this reflected badly on it. Lindemann responded with his normal arguments for a CID committee, but expanded his ideas to suggest that this committee should be approved by the House of Commons. With parliamentary approval, the CID committee could be given its own budget to overcome 'any departmental inertia.'

Londonderry replied that he would be 'very much aggrieved' if this committee found anything his departmental experts had overlooked. Lindemann said that not only had they unquestionably overlooked new approaches, but that the Air Ministry's experts felt bound to try to prove that any new avenue of research would not work. According to Lindemann, at this point Londonderry capitulated and agreed to write to MacDonald advising to support the creation of a CID committee on air defence research. Lindemann left confident of success, unless on his return to the Air Ministry Londonderry was 'over-persuaded by his officials' to change his mind.[17]

Two days later Lindemann, Chamberlain and Churchill met with MacDonald. The Prime Minister informed them that they had convinced him of the need for a CID committee. Lindemann later claimed that MacDonald also agreed to wind up the Tizard Committee as soon as it completed a report on its investigations. When pressed as to when this would occur, MacDonald said that it would take place before the Air Estimates were introduced in the second half of March. MacDonald's account of the meeting makes no mention of this promise.

If such a promise had been made, MacDonald and Londonderry must have anticipated in return support for their rearmament policy, or at least a toning down of the public debate. If this was the case, then they were sadly mistaken. On 18 February, just three days after meeting with MacDonald, Lindemann gave a public talk in which he continued to sensationalise the air threat and revealed the existence of the Tizard Committee. He explained the reasons for not trusting the Air Ministry, not accepting their invitation to sit on the departmental research committee, and for his demands that it be replaced by a more powerful body responsible to Parliament.[19] On 22 February Churchill, speaking in his constituency, demanded that the government fulfill its commitment to keep the RAF at least as strong as the German air force.[20]

Whether these public pronouncements influenced the government's decision on forming a CID committee is unclear, but no immediate announcement on a new committee was forthcoming. On 27 February, during question period in the House, Capt. Harold Balfour, a conservative backbench MP and supporter of Churchill's views on rearmament, decided to prompt the government. He asked Sassoon 'whether consideration has been given to the advisability of an early investigation of the possibilities of countering air attacks by utilising the recent progress of scientific invention; and if so whether it is intended to enlist assistance outside the Government service?'

If Churchill's support anticipated the announcement of a new committee, they were sadly mistaken. Instead, Sassoon delivered a carefully prepared response. He officially revealed the existence of the Tizard Committee, provided a list of its members, and stated that this small group was 'best equipped to make rapid progress in this most important matter.' They only indication of Lindemann's views was contained in a promise 'to bring the report of this committee before the Committee of Imperial Defence.' Balfour followed up this response by asking if Lindemann would be invited to serve on the committee. Sassoon assured the House that 'an invitation was conveyed to this gentleman several weeks ago.'[21]

The public announcement of the Tizard Committee was noticed by the press. As opposed to the controversy surrounding rearmament, both right and left united to treat air defence research as being of crucial importance. *The Manchester Guardian Weekly*, which opposed increases in military spending and the new defence white paper, was solidly behind the government's new air defence initiative. But the paper was not sanguine about prospects for success. The newspaper reported:

> One cannot envy the committee of distinguished scientists who have been appointed by the government 'as a matter of urgency' to investigate new possibilities of countering air defence. People often say that an effective 'reply' to the bomber is bound to be discovered ... It may be so but there are no grounds for any immediate optimism ... We wish the eminent scientists well, but there is no denying they have taken on a full size job.[22]

Chamberlain continued to press the government to honour its commitment to establish a CID committee. On 11 March a major debate was scheduled on a Labour Party motion condemning the government's White Paper because it was 'completely at variance with the spirit in which the League of Nations was created.' During question period, just before the defence debate began, Austen Chamberlain asked Baldwin, the acting Prime Minister, when the Tizard Committee was expected to make its report; and whether he would appoint a CID committee 'with wider powers and sufficient funds for experiments to devise methods of defence against night-bombing?'

Baldwin indicated that the duration of the Tizard Committee's investigation depended 'on what emerges.' The committee might issue more than one report, even a whole series of interim reports could be issued before its work was completed. Baldwin then made a remarkable admission: 'They have already recommended that practical steps should be taken to follow up one very interesting proposal, and this is being done without delay.' This could only refer to Watson-Watt's experiments. Undoubtedly the discovery of radar had influenced the government's policy on air defence research.

One can only wonder what observers made of this cryptic statement. Fortunately, Baldwin was not pressed to give further details. Instead, Chamberlain continued by asking if this purely departmental committee could deal with an issue that 'inevitably' affected more than one department. Baldwin replied that, while established by the Air Ministry, the Tizard Committee, consisted of 'gentleman of first-rate scientific eminence who [were] not attached ... in any way to any department.'

The question of air defence was barely touched on in the following debate, which saw the Labour Party motion easily defeated. The government, however, knew that it would not be so fortunate in the debate on the air estimates, which was scheduled to commence on 19 March, unless it took some action to appease their most vocal critics on the right. The situation was exacerbated by Hitler, who, on 16 March, announced he was re-introducing compulsory military service in Germany. To compound matters the latest figures cast doubts on Baldwin assurances in November that the RAF would retain a considerable advantage in first-line strength.

Austen Chamberlain sent a private notice to MacDonald during this period, asking him to inform the House of any further steps the government would be taking to 'provide defences against aerial bombing.' Just before the debate began MacDonald replied in a statement to the House. The government finally gave in to Lindemann's principal demand, as MacDonald announced the formation of a CID subcommittee on air defence. The new committee would have 'the direction and control of the whole inquiry, and the necessary funds to carry out experiments and to make researches [sic] approved by this committee.' The Tizard Committee would also continue, since it had 'already made concrete proposals for a promising line of research for which the necessary funds [had] been made available and preliminary experiments already started.'[23] This was the second enigmatic reference to radar by a member of Cabinet in just eight days. The name of the new body would be the Air Defence Research Sub-Committee (ADRC) of the Committee for Imperial Defence.

The government hoped that this announcement would ease the criticism it would receive from Churchill and his followers in the air estimates debate. It appears to have worked. While Churchill still called on the government to do more, his criticisms were somewhat muted. Again little mention was made of air defence research.

The press reported positively on the announcement. *The Manchester Guardian Weekly* stated that the CID Committee indicated that the government 'further recognized' the importance of 'discovering some counter to the bombing aeroplane.' The announcement was viewed as an official repudiation of Baldwin's 'famous speech' that 'the bomber would always get through.' Although the magnitude of the problem was not discounted, finding a solution was viewed as being crucial since there was 'no denying that the threat of the aeroplane [was] the chief cause of the present European feeling of insecurity.'[24]

Lindemann read the press reports the next morning and immediately penned a note to Sir Maurice Hankey, the Secretary of the CID, expressing his delight and asking if he should postpone his Easter vacation to assist in setting up the committee. Hankey replied that he should by all means go on his vacation since a chairman of the committee had not yet been selected. There is no indication that Lindemann noticed that the Tizard Committee would continue to function.[25]

The Air Defence Research Committee

The formation of the ADRC marked the first significant step in a growing rift between Tizard and Lindemann that would evolve into one of the most fascinating and long lasting personal feuds in modern British history. The dispute between the two scientists is filled with bitter irony. The two men were once close friends. They had first met in Germany before the First World War when the both studied under Nernst. They dined and travelled together and Lindemann had even suggested they become roommates. Tizard refused this offer because he wanted to have the opportunity to learn German and feared that Lindemann, despite being fluent in the language, would not help him, but instead insist on speaking English. He was also concerned about sharing with someone who was much better off financially. Tizard also had experienced Lindemann's temper when the two had engaged in a sparring match at a local gymnasium. After Lindemann discovered that the much smaller Tizard was a far more skilled boxer 'he lost his temper, completely, so much so that [Tizard] refused to box with him again.'[26]

In the Cherwell papers there are several touching exchanges of correspondence between the two men dated from 1913 to 1927. In several of his letters Tizard refers to his friend affectionately as 'Lindy'. Lindemann began many of his letters, 'Dear H. T.' In 1915, Lindemann invited Tizard to join him at Farnborough when he was setting up his experimental flight. Tizard expressed great interest in the invitation because at the time he was 'bored' and wanted to take a more active part in the war. Lindemann was a guest at Tizard's wedding and godfather to his first son. In 1919, Tizard actively supported Lindemann's successful application to head the Clarendon Laboratory.[27]

The close friendship of the years before 1920 did not last. The two men gradually drifted apart as they followed very different professional and personal paths. Tizard was a powerful civil servant who associated with other government scientists like Wimperis, while maintaining close contact with his academic colleagues. Lindemann socialized with Churchill and his circle, rarely condescending to attend scientific conferences. His main battles were fought at Oxford, where he gradually began to transform the Clarendon Laboratory into a good, but not first-class, centre for research.

According to R. V. Jones, Lindemann's most important doctoral student, there was trouble between the two men in the 1920s. In 1926, at Tizard's instigation, Lindemann was appointed a member of the Council of the Department of Scientific and Industrial Research. Lindemann's caustic personality made him unpopular with other board members. Tizard also nominated his old friend for membership on the Aeronautical Research Council. When the ARC refused the nomination Lindemann blamed Tizard for this failure.[28]

It is uncertain how much Tizard knew about the events leading up to the government's decision to create the ADRC. He had met with Lindemann in late 1934 to exchange ideas on air defence research. While Tizard knew that Lindemann was at centre of the lobbying effort to supplant his committee, it cannot be ascertained if he was aware of the extent of the vitriolic criticisms directed at it. Tizard's reaction to the announcement of the formation of the ADRC was one of puzzlement and annoyance. He could not understand how his committee could operate given the similarity of the mandate of the new more politically powerful organization. At the very least, he wanted the invitation to Lindemann to join the CSSAD withdrawn.

On 21 March, Dowding wrote to Tizard to reassure him that his committee would continue to operate without outside interference and to express his gratitude 'for the good tempered manner in which [he had] accepted this irritating political intervention.' There were grave political difficulties in withdrawing the invitation to Lindemann before he provided a definite answer. Dowding concluded, however, that 'there will certainly be no question of our renewing our invitation, or stirring him up to reply – in fact we hope that the matter may lapse. From what I hear, also I think it extremely improbable that Baldwin would ask him to join the main committee if he is left to follow his own inclinations.'[29] Tizard, however, remained unconvinced, noting that the mandate of the new committee if 'interpreted literally', would take away 'all initiative' from the CSSAD.

In early April, however, Tizard was offered membership in the ADRC and reassured by Hankey that neither the Prime Minister, War Office or Air Ministry wished to interfere with his committee. Moreover, Dowding, an enthusiastic supporter of the CSSAD, would also be a member of the ADRC. Rowe consoled Tizard by reporting that 'there seems to be an impression that the new committee will sit once or twice and fade away.' Coordination between the two committees would be enhanced by Rowe's appointment as joint secretary of the ADRC.[30]

On 10 April the membership of the new committee was announced. Sir Philip Cunliffe-Lister (soon to be Lord Swinton), Secretary of State for the Colonies, was chair. Joining Tizard and Dowding as members were: W. Ormsby-Gore, MP and First Commissioner of Works; Lord Weir, the former Secretary of State for Air; Sir Warren Fisher, the Permanent Secretary to the Treasury; Sir Frank Smith, Secretary of the DSIR; Vice Adm. R. G. H. Henderson, 3rd Sea Lord and Controller of the Admiralty; and Lieutenant General Sir Hugh Ellis, the Master General of the Ordnance. This was the powerful ministerial level committee that Lindemann had demanded, but without him or any of his closest supporters as a member.

In the first two months of its existence the ADRC appeared to follow a rather lackadaisical approach to the question of air defence. The committee appeared to be political window dressing designed to quiet the government critics, meeting only twice before 7 June. It operated more formally than Tizard's. Rather than interviewing operational and technical experts, the ADRC ordered that a series of reports be produced on a variety of subjects related to air defence. Chief amongst these documents was the first interim report of the Tizard Committee. On the whole, the ADRC deferred scientific and technical investigations back to the Tizard Committee.[31]

If it was the government's intention to deter its critics by using the ADRC as a front, it was playing a dangerous political game. Evidence that government figures on German air strength were grossly in error began to mount in the spring of 1935. On 22 May, during the defence debate, Baldwin was forced to admit that the figures on the rate of German aircraft production he had provided the House in November were far too low. The government now proposed another accelerated aircraft procurement programme, Scheme C, to provide the RAF with 1,500 front-line aircraft by early 1937. Churchill and his supporters, well briefed by a growing cadre of officials on current intelligence information from Germany, argued that even this new programme was insufficient.

The slow pace of activity of the ADRC was not ignored. On 25 May, *The Daily Telegraph* published a letter from Lindemann which condemned the inaction of the government in pursuing development of new means of air defence. Lindemann expressed the hope that the ADRC was pursuing the matter with the necessary vigour. While he did not 'expect that they should publish what they are doing' he wanted 'some reassurance that the problem is at last being attacked on a scale commensurate with its importance.'[22]

On 4 June, Churchill asked the government how many times the ADRC had met. When Churchill was informed that the committee had held just two meetings, he announced that he would make air defence research part of the adjournment debate scheduled for 6 June. As a result, MacDonald was provided with a briefing note by Wimperis outlining the situation in regard to Lindemann's invitation to join the Tizard Committee and had answers carefully prepared for any contingency.

Churchill opened the debate by outlining the history of his, Lindemann's, and Chamberlain's efforts to steer the government towards energetically pursuing means to combat the bomber menace. It was a passionate speech, although not without some major inaccuracies. He claimed that the Tizard Committee existed as early as 1932, and that it had completely failed to find any new methods for improving air defence. Churchill compared the government's efforts to a 'slow-motion picture.' He added that 'if a really scientific Committee had been set to work and funds provided for experiments, twenty important experiments would be under way by now, any one of which might yield results decisive in the whole of our defence problem.' The purpose for raising this issue in the House was to stimulate and stimulate action on the committee. If a successful defence could be found, Churchill argued, it would benefit 'every single nation in the world' since 'once it was found that the bombing aeroplane was at the mercy of appliances erected on the earth, and that haunting fear and suspicion which are leading nations to the brink of another catastrophe would be abated by such a discovery.' For Britain a successful air defence was even more important, as it would restore 'the old security of our island.'

MacDonald gave a masterly rebuttal of Churchill's charges. He started by explaining that Churchill and his followers had misunderstood the purpose of the ADRC. This committee was not intended to pursue its own research, but, instead, to coordinate research activity and 'to see that investigations were being pushed ahead with all due expedition.' The committee had already accomplished this speeding up of research activity. Nor was the ADRC designed to replace existing committees, since 'it would have been foolish if it had scrapped for instance the Tizard committee' since it is 'composed of exactly the type of men with the type of experience and knowledge which such an investigation requires.' As for Professor Lindemann, he had neither accepted nor rejected the invitation to join the Tizard Committee. 'The fact of the matter [was] that this committee [was] working day by day on the most important questions involved in this investigation and that its progress [was] very marked indeed.'

The Prime Minister continued by advising that it would 'not be in the national interest' for him to provide specific details about progress already achieved. However, MacDonald assured the House that no policy or fiscal barriers were being allowed to stand in the way of finding a solution to the bomber menace. He concluded that he could report with pleasure that as a result of a recent interview with one of the principal investigators he felt 'able to take an optimistic view of the outcome of these researches.'

Austen Chamberlain was the first member to respond to MacDonald's speech. He was obviously impressed, and was now willing to believe that the government no longer accepted the dictum that 'the bomber would always get through.' He took responsibility for Lindemann's failure to accept the invitation to join the Tizard Committee until there was a debate in the House on air defence research. Given the tone of MacDonald's response to Churchill, Chamberlain now would advise Lindemann to accept the government's proposal.[33]

MacDonald had, for the time being, managed to appease the concerns of Churchill's group. It was among his last official acts as Prime Minister; that evening he resigned. Baldwin assumed the premiership and appointed a new cabinet. Cunliffe-Lister, the chair of the ADRC, succeeded Londonderry as the Secretary of State for Air. The government was determined to manage the politics of air defence better. This decision came at a very opportune time, since the details of the ongoing research program were beginning to be of interest to the press.

While the newspapers had not paid much attention to the earlier allusions to new means of defeating air attacks, the veiled references to radical new approaches was not missed this time. *The Manchester Guardian Weekly* wrote a lengthy editorial piece in which it tried to assess what possible means could have been developed which led MacDonald to have 'an optimistic view.' The newspaper assessed various possible air defence technologies, and since it could find none that held out any real promise, it wished that its readers could know the grounds for this optimism. It all seemed too convenient that a solution to bomber defence had been found. The editorial concluded: 'Altogether, while the work to obtain a positive counter to the bomber is necessary, we have not much hope that it will be successful. The world will be flung back again on the need to destroy military aviation when it has failed in trying to tame it.'[34]

The Daily Telegraph insisted that, despite MacDonald's concerns about security, 'a time must come when the House [would] be taken more fully into confidence.' The government had to press forward the investigations because 'the dread of attack by air is based on past experience of the absence of an effective counter.'[35]

The government followed two related policies to try to prevent further public disclosures. The first part of the government's effort to end the debate centred on making a decision about the future of the air defence system. The recommendations of the Brooke-Popham Committee had languished since late January. In May, the whole matter was deferred to await the first interim report of the ADRC, which was ready only on 25 June. The reason for this delay was in large measure the government's reluctance to provide the resources necessary to implement the proposals. Yet delay only further exacerbated the situation. With every passing week the news about German rearmament forced the government to make piecemeal decisions which increased the overall costs of the air defence system. For instance, the number of fighter squadrons in Brooke-Popham's proposal was set at twenty-eight, which corresponded with the number approved under air expansion Scheme A. In May, with the approval of Scheme C, this number grew to thirty-five.[36]

The ADRC's interim report advocated the acceptance of the Brooke-Popham Committee's recommendations despite progress in scientific research. Scientists and the services had generally agreed on the shape of future research programs and believed that Watson-Watt's work was of crucial importance. However, none of this would result in any significant decrease in the resources necessary for an effective defensive system. The ADRC cautioned against any further public disclosure of on-going research work, particularly in regards to long-range early warning. Secrecy had to be maintained 'as long as possible.'[37]

Cabinet only finally met to discuss Brooke-Popham's proposals in early July. It was a contentious meeting, approval of any part of the programme was in dispute. Some argued that given the state of air defence technology, giving the go-ahead to any part of the system was premature. The cost of the scheme greatly concerned others. Other Cabinet members warned against indecision because 'at any moment the question might be raised in Parliament or in the Press as to what provision was being made for air defence of this nature and that from a political point of view it was important to take a decision.' The discussion concluded with a compromise. Approval was only given for the first part of Brooke-Popham's proposal.[38]

Also the government tried to bring on board the two most vocal opponents of government policy. This would allow them to be fully informed on the recent developments in air defence research without fears that they would make the information public. On 11 June, Lindemann finally agreed to join the Tizard Committee. While Lindemann would end, for the time being, his public pronouncements on air defence research he was determined to continue to press to control the agenda of air defence research. Cunliffe-Lister met with Lindemann around 20 June to discuss his role on the committee. At first Lindemann attempted to rehash the past, but Cunliffe-Lister insisted that they wipe the slate clean. He assured Lindemann that once he was fully briefed on the work of the Tizard Committee he would agree that 'everyone concerned attached the utmost importance to the work; that there was the most cordial co-operation between everybody; [and] that a vast amount of very valuable work was being done.' He stressed to Lindemann 'the vital importance of absolute secrecy.'

Given the assurances of the Air Minister, a person with more sense and less ego than Lindemann would have been content to wait until he was fully briefed on the state of air defence research before taking any action. Lindemann, however, was determined to make his views on every aspect

of air defence known before he was informed of the secret research work
so that it would be known that he had 'not divulged anything in respect
of which the Committee could be entitled to demand secrecy from me.'
At the end of June he sent Tizard an eight page note outlining all of his
ideas about air defence. He almost solely concentrated his discussion
on the need to destroy night bombers. With a series of bizarre statistical
calculations, for which 'the Prof' was famous, he dismissed anti-aircraft
artillery as being not only useless but far more dangerous to people on
the ground than to attacking aircraft. Lindemann believed that it was 'the
[aerial] mine-field rather than the gun that should be employed.' These
aerial minefields would be dropped from aircraft operating just ahead and
above an enemy formation or shot in front of them by a gun or rocket.
The mines would be attached to small parachutes or balloons. If laid accurately for a few minutes 'they would render a certain region impassable.'
It might even be possible to build non-exploding mines, consisting simply
of a heavy cable attached to a weight on one end and a parachute on the
other or one which would spray a gaseous mist or explosive dust which
would render the bomber's engines unserviceable.

While Lindemann believed aerial minefields should receive top priority he did have some better grounded suggestions on aircraft detection.
Unaware of Watson-Watt's work, he once again suggested the use of radio
reflection. Harking back to his early First World War research, he also recommended that research be undertaken using infrared detectors to locate
aircraft.[39]

Rowe examined Lindemann's note and found that, with possibly one
exception, there was not one good original idea contained in the entire
document. Most of his proposals had already been found wanting or were
theoretically unsound; others, such as aerial minefields, were already being
investigated. Tizard decided to meet with Lindemann before replying. The
lunch did not go well since Lindemann indicated that he was continuing to
discuss air defence research with outsiders and reiterated his longstanding
mistrust of almost everyone in the Air Ministry. After digesting the gist of the
memorandum and his conversation Tizard wrote a stern, almost threatening
letter to Lindemann, which demanded that he reconsider his conduct if he
were to be allowed to attend any meeting of the committee. He reminded
him that he would come under the authority of the Official Secrets Act.
He concluded by telling Lindemann: 'So my advice is, either come on the
Committee wholeheartedly, without misgivings, or don't come at all!'

Tizard feared that he had overstepped his authority and sent a copy
of his letter to Dowding asking him what he thought of his bold move.
Dowding returned Tizard's covering letter after having written on the
bottom: 'Thank You. Magnificent!'[40]

Lindemann was forced to swallow his pride and attended his first meeting of the Tizard Committee on 25 July. Much of the meeting was devoted to ongoing experiments on aerial minefields, one suspects this was done deliberately to prove to Lindemann that his fears were unfounded. The revelation that much of his own scheme was already being investigated did, for the time being, appease Lindemann. He also asked Tizard to give approval for research to be undertaken in his own laboratory on infra-red detection. Tizard had his doubts about infrared, but the committee approved preliminary investigations at Oxford during the autumn.[41]

In early July, Cunliffe-Lister proposed that Churchill be appointed to the ADRC in order to try at least to quieten him on air defence research. Baldwin agreed with the Air Minister's suggestion and deputised Hankey to brief Churchill on work of the committee. On 6 July Churchill wrote to Baldwin to accept the offer. He had a few ideas which, 'if of any value, would be quite unfit for publication, but would be of use to the nation.' However, he demanded that he 'must be free to debate all of the general issues of Strength, Air Policy and Air Programmes, etc.'

Two days later Baldwin agreed with this demand. He wrote to Churchill: 'Of course, you are free as air (the correct expression in this case!) to debate the general policy, programmes, and all else connected with the Air Services. My invitation was not intended as a muzzle, but as a gesture of friendliness to an old colleague.'[42] Neither man really believed these kind words, but they both hoped that, at the very least, top secret research would be removed from the public eye.

With the assistance of Lindemann, Churchill took quick action to try to put his stamp onto the ADRC. Churchill and Lindemann prepared a paper on air tactics and defence for discussion at his first meeting. Churchill expressed doubts that distant early warning of a raid was sufficient to guarantee success. He suggested that a very high speed aircraft, equipped with a radio, be developed to meet bomber formation while they were still over the sea. The interceptor would then send back information on 'the direction and force of the raiding force.' The interceptor he dubbed 'the Lambs' since 'everywhere that Mary went the Lamb was sure to go.' He then went on at length to urge that every possible effort be made to improve anti-aircraft technology. He concluded by expressing his belief that research would proceed 'at exactly the same rate as it would have been by the Ministry of Munitions in 1917 and 1918 and that neither waste of money nor disappointment will slacken our search.' At the ADRC meeting on 25 July it was decided that Churchill's proposal for the Lamb aircraft had merits and should be pursued, but he was advised that the Tizard Committee was already vigorously pursuing anti-aircraft research.[43]

Like Lindemann, Churchill remained unconvinced that all possible efforts were being made to improve the defences. On 8 August he wrote an emotionally charged plea to Cunliffe-Lister asking that a 'Follow-up' branch be established to ensure that all ideas presented to the ADRC were acted upon. His anxiety that nothing be left undone was fuelled by his belief that Britain was 'moving into dangers greater than any I have seen in my lifetime; and that it may be that fearful experience lie before us.'

In his reply of 18 August, Cunliffe-Lister informed Churchill that the Air Ministry already had a system for following-up all legitimate suggestions. As well, he tried to reassure Churchill by stressing the great progress that had already been made, particularly in the key are of 'detection and location.' Churchill's own idea about Lamb aircraft was a major contribution since he had 'hit on exactly what [was] needed as the complement to this.' A similar exchange of correspondence between the two men later in the month on the growing air threat and the aircraft industry seemed to have for the moment placated Churchill. For the rest of the year both he and Lindemann appeared content to work on their respective committees.[44]

As a result of these decisions, the government had, for the time being, ended the public debate on air defence. The political wrangling had not delayed the research programme. It was hoped that full attention could now be given to developing the new technologies and building up the system that could defend Britain from the scourge of the bomber. Already at a small isolated peninsula known as Orfordness a small group of scientists where making this a very real possibility.

6

Orfordness and Bawdsey

Orfordness is located some 90 miles north-east of London in Suffolk along the North Sea coast between Aldeburgh and Felixstowe. Despite being frequently referred to by those stationed there as the 'Island', it is 'the largest vegetated shingle spit in Europe.' The long narrow and low peninsula is formed by the river Alde, which at Aldeburgh turns south just a half-mile from the sea, flowing another 15 miles before reaching its estuary. The Ness contains 'a variety of habitats including shingle, salt-marsh, mud-flat, brackish lagoons and grazing marsh.' It is an important site 'for breeding and passage birds as well as for the shingle flora, which includes a large number of nationally rare species.' About 5 miles from its end, the river, now renamed the Ore, passes the historic village of Orford, which is dominated by a castle built by Henry II. Access to Orfordness is made by boat from the village quay.[1]

Today the site is owned by the National Trust, but from 1913 until the mid-1980s the military used it for a variety of purposes. In the First World War, it was a busy RFC airfield and armament testing station. Wimperis began his armament research there and in 1915 Tizard's Armament Experimental Flight was transferred to Orfordness from Uphaven. Armament work had recommenced there in 1929. In 1935, the site appeared virtually abandoned. Some of the huts still contained notices posted by the First World War station commander.

To A. P. Rowe, who in the early 1930s spent several summers on the Ness engaged in armament research, the site, with its 'pink thrift and yellow shingle and cries of terns', was 'surely one of the loveliest places in the world.' To one of the young researchers sent there to begin work on radio location, however, 'Orfordness was a very forbidding place – a land of freezing winds, shingle, mud flats and dykes and a few comfortless huts.'

Regardless of one's views on the attractiveness of Orfordness, no one could deny that it met all the requirements for the first radar research centre. As Rowe relates: 'Technical considerations called for a flat area of land; a coastal site was needed to provide the realistic conditions of air-craft approaching our coasts from the sea; the need for speed demanded the availability of electrical power and laboratory space and the need for secrecy was paramount.'[2]

The Cult of the Imperfect

On 13 April 1935 the Treasury approved £12,300 to establish the radar research facility at Orfordness. Well before this date, however, Watson-Watt and a handful of the Radio Research Station's staff were already laying the groundwork for the new research programme.

Watson-Watt directed the programme so that a practical working pro-totype would be ready as soon as possible. He dubbed his philosophy for undertaking radar research 'The Cult of the Imperfect.' This concept developed from his dealing with the limited research budgets provided to government scientific organizations in the interwar years. His often quoted slogan for radar development was 'Give them the third best to go on with; the second best comes too late, the best never comes.' Watson-Watt justified maintaining this approach even after monetary and manpower resources became, by the standards of the time, almost unlimited because 'nothing was comparable in importance to the provision of some measure of distant early warning and location.' He used 'harsh objectivity' to main-tain the need to rush an imperfect system into operational use and to deny 'every refinement, every embroidery' that might cause even a minor delay. This approach would make great progress possible where tolerances were sufficient to allow for substantial leeway. It worked exceedingly well in the early years when highly skilled scientists and technicians could closely supervise the building and running of a few large and comparatively simple radar sets. It would be strained to the limit when production needed to be accelerated and would fail altogether when more sophisticated types of radar were required. Whatever its limitations, it cannot however be denied that 'The Cult of the Imperfect' succeeded in providing Britain with an effective daytime air defence system by the summer of 1940.[3]

There was a variety of ways that Watson-Watt's philosophy affected early radar research. Most importantly, the type of equipment he envis-aged was deliberately intended to be as simple as possible to fulfill the primary function of providing distant early warning. His plan was to use a direct adaptation of the pulse radio techniques the Radio Research Station had been using since the mid-1920s to study the atmosphere. A transmitter would send out a very broad radio pulse which would 'flood-

Above: Orfordness
Transmitting Hut.

Right: Orfordness
Transmitter Hut Interior.

light' a large area of sky in front of transmitting aerial array. A separate antenna would send any reflected signals to a receiving set which would in turn be connected to a CRT display. The display would show any signals visually on a linear scale; the further away a target, the further along the screen its reflected signal would appear. Watson-Watt appreciated that this system might not be the best theoretical solution either for getting the maximum range of detection, or for accurate location of a target. It also had grave technical limitations in detecting low flying aircraft. But the system required the least long-term technological research and offered the best possibility of detecting the greatest percentage of approaching aircraft.

Part of this philosophy of the 'third best' was the decision that radar should operate at a wavelength of 50 metres rather than substantially lower wavelengths which had several theoretical advantages. The advantages of shorter wavelengths were of some importance, including probably increasing the maximum range, making more accurate height determinations of targets and that transmitters would require less power. The primary benefit of 50 metres was that it would allow for existing radio techniques to be used with far fewer modifications. Watson-Watt also believed that, since the longer wavelength was already being used for radio broadcasts and ionospheric research, it would be easier to preserve the security of the project by providing a convenient cover story for radar transmissions.[4]

In the first months of the radar research programme Watson-Watt confined himself to providing overall guidance, management, and inspiration to the small group of researchers he assembled, while continuing to act as the superintendent of the Radio Research Station. Initially the team he assembled to carry out the experiments consisted of just four members. All but one were current employees of the Radio Research Station's staff. Wilkens ran the day-to-day research programme, particularly after the transfer of the main work from Slough to Orfordness in mid-May. Wilkins, according to one of the early radar scientists, had 'a shy and modest disposition,' with a wry sense of humour and a facility for inventing nicknames. He was well liked by everyone and fell naturally into the leadership role. His supervisory and research work was so crucial in the early months that Watson-Watt dubbed him the 'mother' of British radar.

Wilkins was initially joined by L. H. Bainbridge-Bell, 'the most talented circuit designer at Slough.' In 1933, Watson-Watt, Bainbridge-Bell, and James Herd, the deputy director of the Radio Research Station, co-authored *The Cathode Ray Tube in Radio Research Work*, a masterly summary of the Radio Research Station's research accomplishments. George Willis, the third man on the team, was Bainbridge-Bell's long-time technical assistant.[5]

The final member was Dr E. G. (Eddie) Bowen, the first scientist specifically recruited for radar research. Bowen was born on 14 January 1911 in the small Welsh village of Cockett, near Swansea. His father was a steelworker and an organist in the local Congregational Chapel. At eleven or twelve years of age Bowen developed an 'obsessive' interest in the new wonders of radio. He won scholarships to attend the Municipal Secondary School in Swansea and then to Swansea University College. In 1930 he was awarded a first-class honours degree in physics and continued at Swansea, completing a MSc the next year. His interest in radio led his professor to arrange for him to undertake a PhD in physics at King's College (London) under the supervision of Appleton. He spent much time during 1933 and 1934 undertaking research using a cathode-ray direction finder at the Radio Research Station, where he came to the attention of Watson-Watt. His dissertation, which was titled 'The Penetration of Radiation', was closely connected with the Radio Research Station's research into the ionosphere. He completed his doctorate in the early spring of 1935.

With jobs scarce in a depression-ravaged Britain, Bowen took an unusual step for someone with his academic qualifications and applied for a position advertised by the Radio Research Station. He made the short list of eight candidates and was then interviewed by Watson-Watt and Herd. Bowen relates that the main questions revolved around whether or not as a Welshman he could possibly work with two Scotsmen as his bosses. They then asked him to sing the Welsh national anthem. Bowen replied that he would do so but only if they would first sing the Scottish national anthem, reckoning that this could not be done without the aid of a bagpipe. They did not take up his challenge, but his answer seemed sufficiently impressive to win him the position.[6]

Bowen joined the staff at the end of April, as a junior scientific officer making just £256 per annum. Bowen was

Dr E. G. Bowen.

unaware that he was about to be plunged into a top secret military project until his first day on the job when he was required to take an oath to uphold the provisions of the *Official Secrets Act*. He noted the relish with which Herd informed him that violation of any part of the act could result in 'literally "being hanged by the neck until life was extinct".' He was then given a copy of Watson-Watt's memorandum on radar to read overnight and ordered not to let the document out of his sight. When Bowen desired to go to his hotel's bar that evening he decided that it would be safer to find a secure hiding place than to bring it out in a public place. He carefully hid the envelope under his bed sheets. When he returned to his room he found the envelope apparently undisturbed but covered by a hot water bottle. That night Bowen had a nightmare about 'swinging on the end of a gibbet.' The next morning he made sure to have a chat 'with the chambermaid and was relieved to find she bore no resemblance to a Mata Hari but was a buxom country girl with an unmistakable Buckinghamshire accent.'

A. P. Rowe argued after the war that as a group the early radar researchers 'consisted broadly of people possessing the average ability of those who in pre-war days were prepared to work in government establishments.' With the possible exception of Bowen, he doubted any measure of genius was present until widespread recruiting from the universities began in late 1938. This assessment might be more a reflection of class prejudices, which valued scientific credentials over technical skills or theoretical knowledge over experimental persistence. Yet for Watson-Watt this was exactly the type of individuals that fitted well into 'The Cult of the Imperfect.' None of the early radar researchers were destined to win a Nobel Prize and most would never even be elected to membership in the Royal Society. But, if they were to succeed under Watson-Watt's tutelage, genius was not necessarily crucial. Instead they would have to be willing to engage in hard improvisational applied research with little hope of the normal rewards given to a publishing scientist.[8]

The Island

Preliminary work on readying the Orfordness site began in March when Watson-Watt, Herd, Bainbridge-Bell and Wilkins visited the site, accompanied by representatives of the Air Ministry's Works and Buildings Department. Two old brick buildings were selected as laboratories and positions were marked out for six 75-feet-high self-supported towers. Two of the towers were to be used for supporting the transmitter aerial. The other four were set out in pairs for receiver aerials so that a 'spaced aerial system of height (or angle of elevation) measurement' developed by Wilkins at Slough could be tested. Work commenced soon after funding was approved and advanced sufficiently that it was ready to be occupied by mid-May.

Meanwhile work continued at Slough on designing and gathering parts for the transmitter and receiver. Bainbridge-Bell, later joined by Bowen, developed the former and Wilkins the latter. Their task was to use the basic pulse transmission and reception techniques developed for ionospheric research, but to improve overall performance greatly. Since the reflected signal would only be 10^{-19} the power of the transmitted radio pulse, it was necessary to increase the transmission power by a factor of a hundred. Existing ionospheric transmitters had a peak power output of around 1 kilowatt and it was initially felt that radar required approximately 100 kilowatts to be effective.

The transmitted pulse width or duration of radio pulses would have to be dramatically reduced to avoid the superimposition of signals reflected from the ground with those reflected from aircraft at short ranges. The best ionospheric transmitters operated with a pulse width of 200 micro-seconds, which meant that the minimum range of detection was 50 kilometres. Pulse width needed to be reduced to just twenty microseconds to allow for an acceptable minimum range.

Less radical improvements were required for the receiver; still, Wilkins set out to improve performance by a factor of two. The third area of improvement was in antenna gain, but here less could be expected, particularly with the 50-metre wavelength selected by Watson-Watt for the early experiments. Virtually the only immediate improvement possible was 'to increase the antenna height, which would concentrate the signal towards the horizon instead of throwing it an upwards direction.'[9]

The radar team used their ingenuity as scavengers to put together the new equipment. The transmitter used a variety of surplus electronic components 'purloined' from the Radio Research Station and the National Physical Laboratory. The only indication that the project was of national importance was that the Admiralty was persuaded to provide several of their highest power transmitter valves or tubes. Bowen later described the apparatus as 'a marvel crudity.'[10]

On 13 May 1935, the radar group left Slough for Orfordness. The radar team was augmented by two technicians of the Radio Research Station to assist with the setup of the station. It was a beautiful spring day and the radar team greatly enjoyed the drive to the coast. The next morning at Orford, in the midst of a howling sleet and hail storm, an RAF work party assisted with the loading of the equipment into a power boat hired by the Air Ministry. Upon reaching Orfordness, loads were transferred

Crown and Castle Hotel, Orford.

onto a vehicle assembled from the chassis of a Model T Ford and an old fire-engine. The strange contraption carried the equipment along a rough track to the two reconditioned huts. It took several days of back-breaking labour to transport all of the equipment to the 'island.'

As soon as unloading was finished, work began on assembling the radar prototype. They made rapid progress, despite having to work around workmen completing repairs and alterations to the buildings. By 27 May, all the components of the set were in working order; two days later the system was given preliminary testing using the ionosphere as a reflective source. They only awaited the completion of the masts, which were finally ready in the first week of June, and the placement of the aerial arrays, finished a few days later.

Work on Orfordness soon settled down into a rough routine. Bowen found a room in Orford, the other three in nearby Aldeburgh. They met at the Orford quay about 8:30 a.m. and worked usually until 7:00 p.m., six or even seven days a week. Late night sessions were not uncommon and those not ready to leave by the time the boatman retired for the evening either rowed themselves back across to Orford or spent the night on a camp bed placed next to the equipment. Bowen fondly remembered the times he roughed it on the island, having a supper of cake and beer, and, usually, an identical breakfast. The townspeople soon were used to the strange comings and goings and dubbed the workers 'the Islanders.'

Watson-Watt and his wife Margaret would motor to Orford on many of the weekends. They stayed at the Crown and Castle Hotel and usually entertained Bowen, Wilkins and any official guests to supper in the restaurant. The men would then retire to the near deserted lounge to discuss the week's progress and the future course of the research programme.

To Bowen his time at Orford and the Ness were idyllic. He recalled in his memoirs: 'In spite of the crude conditions under which we lived and the lack of amenities, for me it was one of the happiest periods of my life. It was a time of high achievement and very obvious progress and we did not care a jot about the absence of refinements in our private lives.'[11]

By mid-June substantial progress had been made. The transmitter was giving 50 to 60 kilowatts power and the pulse width was reduced to twenty-five microseconds. These improvements resulted in the uncovering of several unexpected problems. Most important was the discovery of atmospheric layers lower than the ionosphere which reflected some of the 50-metre signals and which might obscure the reflections from aircraft. Watson-Watt's prediction that using a frequency already being utilized might hide the testing proved to be incorrect. The existing commercial signals occasionally interfered with the testing and the testing interfered with the commercial signals. This resulted in drawing attention to the

Orfordness transmitter. A shorter wavelength, which would mitigate most of these problems, was looking more and more attractive even before the first aircraft was located.

The Tizard Committee decided to observe these developments in the afternoon of Saturday, 15 June. Watson-Watt, always a showman, hoped to overwhelm the committee members by treating them to the first ever tracking of an aircraft by British radar. However, this plan was thwarted during preliminary tuning of the equipment by Wilkins, Bowen and Watson-Watt some two hours before the guest were due to arrive. A Singapore III flying boat operating out of Felixstowe made an unscheduled flight over Orford and then off over the North Sea. The official report described what happened as the three scientists stared at the CRT screen:

> Strong echoes were observed as soon as the craft was sufficiently far off to allow the echoes to appear clear of the ground ray; these were used for measurements up to 25km, at which distance the boat turned back towards Orford, and was followed in, with reduction of receiver amplification as required.

The Tizard Committee arrived later that afternoon. They proceeded over to the station expecting to view the new equipment, but Watson-Watt sprung his surprise and announced that a Valencia bomber was about to fly over and be tracked by the radar. With members of the Tizard Committee crowding around the CRT display, the bomber flew overhead and headed out to sea. Unfortunately, atmospheric conditions had markedly deteriorated in the three hours since the flying boat had made its unexpected appearance. Interference was caused by the newly observed low atmosphere reflective layers, morse shortwave signals and approaching thunderstorms. Still, intermittent signals could be seen from the aircraft on its outward journey from 12 to 27 kilometres away.

At 7:00 a.m. Sunday morning the committee arrived back at the huts to repeat the tests of the previous afternoon. A combination of very bad atmospheric conditions and poor weather prevented any successful observations. Watson-Watt later recalled that while failures during laboratory demonstrations were not uncommon, the performance in front of the Tizard Committee was 'sub-sub-show business.'

Watson-Watt and the committee returned to the Crown and Castle for breakfast and a meeting. Although the committee members were impressed by the rapid progress at Orfordness, Watson-Watt felt that as a result of the poor results of the premature display he needed to defend the programme. Much of the discussion centred around the relative advantages of using 50 or 5-metre wavelength and on how to proceed in the coming months.

It was an important meeting since several major decisions were taken in large measure to bolster Watson-Watt's flagging spirit. Tests of the existing system were ordered to continue on a daily basis for at least a month using aircraft from nearby RAF air stations. Authorization was given to acquire the very best shortwave radio transmitters commercially available so that experimentation could begin on wavelengths smaller than 50 metres. In order to ensure the very best equipment was used, Watson-Watt was asked by the Committee to travel to Germany 'in order to examine the ultra short wave reception work proceeding in connection with Civil Aviation.' His cover was that he would be undertaking the trip in order to improve British civil aviation. Watson-Watt told the Committee that whatever the wavelength used, far better performance could be expected by building 200-foot masts to support the antennae. They approved the building of the new structures as soon as possible. Finally, the Committee concluded the meeting by advising Watson-Watt that 'nothing had occurred to justify any relaxation of the urgency and importance of the general investigation.'

Despite this endorsement of the radar programme, Watson-Watt remained unsatisfied. He stayed behind when the committee returned to London later on that afternoon. On Monday he personally supervised another test, this time with a Scapa flying boat. With better weather and atmospheric conditions the aircraft was followed to and from its route, which took it some 46 kilometres from Orfordness. 'The responses at the maximum range reached by the aircraft were of such magnitude that a much greater range could have been obtained.' This test, along with Rowe's discovery that a senior RAF wireless officer found that atmospheric interference over the weekend had been the worst he had ever known, finally succeeding in easing Watson-Watt's concerns.[12]

Testing continued throughout the next month. As a result of continuing interference at 50 metres, the Orfordness staff began trying to use a wavelength of 25 metres. This lower wavelength was soon shown to be much more reliable and would be used for an increasing number of the tests throughout the summer. In conjunction with testing, the transmitter was almost completely redesigned and the cruder aspects of the receiver set were replaced with more reliable components.

On 16 July Rowe arrived at Orfordness, ostensibly to discuss the erection of the 200-foot masts, but in reality to check on the progress since the committee's last visit. Rowe was treated to a test using the set at 25-metre wavelength to track a Bristol 120 aircraft. He reported to the members of the committee: 'Up to the distance of 40 Kms. the aircraft response was obvious at a mere glance ... the aircraft response was, to my untrained eye, quite definite until a range of 53 Kms. was reached.' Watson-Watt was able

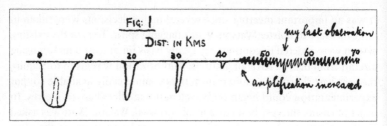

Rowe's sketch of radar display.

to follow the aircraft until it was 67 kilometres away, a fact that was confirmed by the radio direction finding system on the aircraft. Rowe carefully sketched what he had observed on the screen and included it in his report. Many other aircraft that flew through the test area were also tracked. Watson-Watt detected one unknown target at 53 kilometres and Rowe could clearly see the signal on the CRT screen at 45 kilometres range. Rowe was very impressed. He reported to the Tizard Committee: 'One month has elapsed between the first attempt at detection witnessed by the Committee, and the demonstration of 16 July. It would be difficult to exaggerate to the Committee the advance made in this short interval. There is no relation whatever between the results of the two demonstrations.'[13]

On 24 July Watson-Watt and Wilkins conducted another test using a single Wallace aircraft as the target. The Wallace was followed out to 34½ miles before disappearing from the screen. While the men were waiting for the Wallace to make its return run, an unexpected echo appeared on the screen more than 20 miles away from Orfordness. From its fluctuations and size Watson-Watt and Wilkins deduced, using the information displayed on the CRT display alone, that this new echo was not produced by one aircraft, but instead a small formation of three that had wandered into the test area. Watson-Watt later explained that he had used a simple rule square root law 'which was known in optics and radio about the interference from three overlapping responses.' Then as the two men watched the screen, the pattern changed. They looked at each other 'sagely' and said: 'Now one of them has broken away and there are two left.' The observations were confirmed by the Wallace pilot who reported that he had seen a formation of three Hart bombers in the area and that one of the bombers had indeed left the formation. This was the very first radar observation of numbers in a formation, change in a formation, and the recounting of a formation.[14]

While testing of distance early warning continued for the rest of the summer, the attention of the researchers turned to developing means to fix accurately the three-dimensional position of an aircraft. In late August,

Wilkins began experiments on means to determine the angle of elevation
of the reflected signals from the receiving antenna. This angle combined
with the distance and direction of the target plus a corrective figure based
on the curvature of the earth would provide the altitude of an aircraft.
By mid-September Wilkins was measuring the height of an aircraft flying
at 7,000 feet 15 miles distant, with an average error of 1,200 feet. He
had developed the method for determining height in 1932 when the
Radio Research Station was asked to develop a technique for determining the angles of elevation of shortwave signals received from the United
States. Wilkins' technique involved comparison of the ratios of the signals
received directly from the target to those that reflected off the ground from
'two horizontal antennas set up in the same vertical plane at two different
heights above the ground.' The intensity of the two signals were compared
in the field coils of a goniometer and when adjustments in the coil resulted
in the two signals cancelling each other out, the angle of the target could
be determined if the target was in the direct line of site of the antenna.[15]
If the target was not directly in the line of site, either to the left or to the
right of the best viewing angle, then knowledge of the target's direction or
azimuth was necessary if the height was to be accurately calculated. It was
a neat technique but would prove more difficult to perfect than anyone
anticipated.

Direction or azimuth was initially considered a more difficult problem than height finding. At first, Watson-Watt had envisaged determining
direction of a target by use of the range cutting method, where a single
receiver would compare the signals from three widely spaced transmitters. Bowen pointed out that while this might work in theory for a single
target, it would be close to impossible to track multiple targets using this
technique. Rowe confronted Watson-Watt with the crucial need to find a
simple solution or risk the failure of the entire research programme.

Watson-Watt proved that necessity is the mother of invention. Sometime
in the autumn, while on the train for one of his weekend visits to
Orfordness, the answer suddenly popped into his head. It was an elegant
and simple solution; one of those ideas that when explained led Wilkins to
wonder how he could not have thought of it himself. The receiving antennae were simple dipole or half-wavelength aerials. Watson-Watt realized
that if two dipole antennae were mounted at right angles to each other
or crossed on the same horizontal plane, the difference in signal strength
would give a reasonable sense of direction. If the target was directly in
the line of site of the array the signal strength would be equal, the further
away from the line of site the greater the difference between the signals
received by the two antennae. The direction could be determined by the
use of a goniometer and a CRT display.

Wilkins carried out the first experiments using Watson-Watt's direction finding method at the Radio Research Station at Slough in November. He used a cathode ray direction finding set, an aircraft radio navigation aide, which was attached to crossed dipoles. An oscillator attached to an aerial mounted on top of a 90-foot tower provided the signal. The results of the tests, wrote Wilkins, 'demonstrated the principle to be sound.' He returned to Orfordness and began testing the technique with the prototype radar set by the end of the year. Here too the solution was easier to implement in theory than in practice, but in late 1935 there was great confidence that the problem was well on its way to being solved.[16]

Bawdsey and the Radar Defence Chain

In just a few months the Orfordness team had proven that they could provide long-range early warning and held out the hope that they would soon be able to provide accurate detection of the position of any attacking bombers. It was by any standards an extraordinary technological accomplishment, but Watson-Watt realized that it was only the first steps in a very long journey.

Watson-Watt was planning the next stage of the radar programme before any progress had been made on height or direction finding. In his meeting with Rowe at Orfordness on 16 July he outlined the shape of a radar air defence system covering the coast from the Tees to Portsmouth. He envisaged thirty stations spread out at 100 kilometre intervals along the coast. The estimated cost of each station was £7,000. About £200,000 would be required to build all of the stations, plus the cost of purchasing land and providing power connections. Watson-Watt held out the possibility that as range was extended it might be possible actually to intercept aircraft as they crossed the coast, something that acoustical mirrors could never achieve.[17]

The results of the latest work at Orfordness and Watson-Watt's proposals for a radar chain were reported to the Air Defence Research Committee for discussion at its meeting on 25 July. They ordered the Air Ministry, with assistance from the Tizard Committee, 'to examine the position and formulate plans for the establishment of the radio detection and location method' in regards to personnel and equipment requirements.'[18]

On 14 August the Air Ministry formally embraced radar as being the key to an effective air defence system when it suspended work on the Thames Estuary acoustical mirror programme. Discussions between Rowe and Watson-Watt had already begun to trace out the rapid expansion of the still tiny research team. Orfordness, however, could not accommodate any large scale expansion and it was obvious that a new site was necessary to house what would soon become Britain's largest single military scientific research programme.

The search for the site of a new radar research centre began in earnest in the second half of August. Watson-Watt needed to find a facility which was large enough to house the greatly expanded research activities and was in close proximity to Orfordness and the sea so that it was possible to use the existing radar apparatus in conjunction with a second set at the new laboratory, and, ideally, some elevation above sea level in order to increase the range of detection. Wilkins told Watson-Watt about a large estate named Bawdsey Manor, located along the coast 10 miles south of Orford, which he believed might make an ideal laboratory site. Wilkins had discovered Bawdsey during a drive around the area on the first weekend after their arrival of Orfordness.

Intrigued by Wilkins' description, Watson-Watt immediately set off for Bawdsey. A quick examination of the estate showed that it would be more than ideal for an expanded research establishment. Discreet inquiries indicated that the estate might be for sale, at the right price.

Bawdsey is located near the mouth of the River Deben, about half-way between Orford and Felixstowe. The estate was the property of Sir Cuthbert Quilter, who had made his fortune at the end of the nineteenth century as a promoter of one of the first private telephone systems in Britain. Quilter began construction of the manor home in 1890 and work

Bawdsey Manor.

Bawdsey Entrance.

proceeded in fits and starts until 1904. No one architectural scheme or style was followed; construction simply began on each phase of the house as profits flowed in from the telephone business. When profits dropped construction ceased. There are at least four distinct styles of architecture present in the house, including the early Dutch in the older parts, the mock-Tudor Red Tower, the Oriental White Tower, and the neo-gothic front. The house is situated on a 70-foot promontory overlooking the sea, one of the few pieces of high ground along this otherwise flat section of the Suffolk coast. It was an extremely large building well able to accommodate any foreseeable growth in the radar programme. There were no fewer than eight large reception rooms on the ground floor, plus a large entrance hall, a conservatory and a suite of domestic offices. The reception rooms were mainly panelled in oak and all had floors of solid English oak. On the first floor there were three reception rooms, eight principal bedrooms, a dressing room, six other bedrooms and a nursery. The second and upper tower floors contained twenty-six more bedrooms. The whole of the property, wrote an Air Ministry inspector, was 'built regardless of cost in first-class materials' and was carefully maintained throughout.

The estate consisted of 168 acres of land. Near the manor were a squash court, cricket pavilion and an outdoor swimming pool. The formal gardens included an Italian style garden with a lily pond, pergola and grass paths; a circular sunken rose garden with fountain; and a large walled kitchen garden with a range of greenhouses and potting sheds. Outbuildings consisted of sixteen cottages for the estate workers, a model farm, a model dairy, a power station, laundry, garages, stables and a clock tower.[19]

Watson-Watt informed Wimperis of his desire to acquire Bawdsey in a visit to Orfordness in late August. Wimperis reported to Tizard:

I have inspected [Bawdsey] at a respectable distance by road and air. Some of the land there we might be able to buy and build up a receiving station which would also act as a radio research and training centre for the whole scheme. It is likely enough, however, that the owners who want to sell the entire property would insist on us taking the house as well. This Watson-Watt would rather like as it would save building. It is a colossal establishment with acres of gardens, outbuildings, cottages, farms, and so on, and if we do decide to take it over, the question of organization and administration would take some thinking out![20]

With such grandiose schemes now in the offing, the Tizard Committee asked Watson-Watt to prepare a report summarizing the progress to date, future research directions and requirements, and the shape of the proposed radar defence chain. On 9 September Watson-Watt informed the Tizard Committee that radar had followed aircraft up to a distance of 92 kilometres, detected the approach of an expected aircraft at over 60 kilometres, and seen an unplanned approach of a plane at 50 kilometres. Accuracy of detection was within 1 kilometre. Most of the tests were conducted on aircraft flying at over 10,000 feet, but some were undertaken with planes as low as 5,000 feet. Improvements in maximum ranges could be achieved by building the much taller 200-foot masts and placing them on high ground at or near the coast.

Watson-Watt then sketched out the shape of a national radar defence system. He recognized that 'at this early stage in a new technique' there were substantial uncertainties in any such proposal. He felt, however, that the 'emergency circumstances' then faced by Britain warranted moving forward quickly on a scheme based on just four months of research. The cost of the chain was identical to that which he had reported to Rowe in July, about £200,000. The vast expense was more than compensated by the huge benefits for national defence.

This scheme had its limitations. Watson-Watt explained that in order to incorporate height sensing into the chain there would be an additional cost of £2,000 per receiving station. Providing an accurate direction was more problematic since the radar used a 'flood-lighting scheme involving no "search" whatever, i.e., no manipulative action at the transmitter or receiver,

Bawdsey Receiver Tower and Research Block.

for plan-location.' Another limitation with the system was its inability to track aircraft after they crossed the coast. Further research was also required on much shorter wavelengths, as small as 50 centimetres, which might provide a means of more accurate positional detection, and on automating the control of the receiver's CRT display or other detection gear.

In order to perfect the existing system, and to expand the current research programme 'the provision of a suitably situated central research and development station, of large size and with ground space for a considerable number of masts and aerial systems' was an 'urgent necessity.' The rate of progress in the research was dependent on having a large staff of 'exceptional quality,' although numbers of qualified applicants could be very limited. Watson-Watt also advised that the research centre serve as a training centre for the staff of the new radar chain. Watson-Watt envisaged the new research centre being run by a civilian research director.[21]

Watson-Watt's position paper was enthusiastically received by the Air Defence Research Committee at its next meeting on 16 September. They approved of the building of the radar defence chain and the acquisition of Bawdsey Manor and urged the Air Council, under the guidance of the Tizard Committee, immediately to begin work on the new defensive system. Given that the Air Minister chaired the committee the Air Council's acceptance of the proposal was a forgone conclusion. Sir Warren Fisher, the Treasury representative, assured the ADRC that his department would give prompt approval of the necessary expenditures.[22] In hindsight, the ADRC's extraordinarily rapid decision was the single most important act of this committee and by itself can be seen as justifying the campaign of Lindemann, Chamberlain and Churchill's for its formation. Given the limitations of the technology in September 1935, the ADRC's decision was bold, verging on reckless. The prompt action must be seen as a reflection of the atmosphere of crisis that prevailed over the bomber threat, and the growing confidence that was being established between the decisions makers in the armed forces and government and scientists.

Few concerns were raised about proceeding with this huge commitment of resources to such an untried technology. Remarkably, no one even mentioned the inability of radar in the autumn of 1935 to fix a target's direction and height. On 25 September, at the first meeting of the Tizard Committee held after the ADRC's decision, Lindemann raised the only serious question about the limitations of the new technology. He asked if the signals could be jammed by enemy radio interference. Watson-Watt was convinced that jamming would prove very difficult to implement and could easily be overcome. After all, while the early test conducted at 50 metres had suffered from interference from other transmitters, it had still proven possible to track aircraft. Lindemann remained unconvinced.

By the end of 1935, Bawdsey was purchased and the Air Council had authorized the construction of five stations of the chain to defend the Thames Estuary. The first two new stations at Bawdsey and Clacton were to be finished in the spring of 1936, so that they could be tested with the existing equipment at Orfordness in June. The other three installations at Shoeburyness, Eastchurch and Dover were to be finished in time for the first large-scale test of an entire segment of the chain during the 1936 annual air exercises.

On 19 December the treasury approved the expenditure of £62,000 to purchase land and to construct the five radar stations. Of this figure, £17,000 were transferred from funds already authorized for the acoustical mirror programme, effectively delivering the coup de grâce to any further work on the technology. 'For reasons of secrecy' the Treasury suggested hiding the rest of the expenditures on the radar chain in a general heading of 'Miscellaneous Air Defence Works' in the published budget.[23]

This was only one of several moves to preserve the secrecy of the expanding programme. Watson-Watt filed a patent application for radar on 17 September 1935. For reasons of national security, the patent was declared a state secret and withheld from publication. Watson-Watt fully endorsed this decision, although after the war he would deeply resent the British government's refusal to compensate him and other radar scientists for the economic value of their discoveries.

Another issue frequently linked to security concerns was the naming of the new technology as RDF. It has become a legend that RDF was chosen to denote British radar in order to deceive enemy agents into believing that the towers being erected along the coast were part of a system of Radio Direction Finding. Several of the early radar pioneers claimed to have had some role in creating the term and historians have generally accepted their accounts.[24] In an interview given in July 1961, Watson-Watt stated that he originated it when 'a very charming retired colonel,' who was the represent-ative of AT & T in London, asked him: 'What are those odd things that you are setting up on the East Coast?' Watson-Watt, not having an opportunity to think up something more clever, replied: 'Oh, these are radio direction finding stations, to assist the fighter defences.' Since this seemed to convince the colonel, who had some knowledge of radio, Watson-Watt and the Tizard Committee realized that the letters RDF might also fool any 'informed spy as meaning radio direction finder.' Watson-Watt added: 'So RDF was delib-erately designed to make a suggestion of the false without a denial of the truth. It was partly for radio direction finding but also for much more.'[23]

While a wonderful story Watson-Watt's account is pure fiction and an instructive example of the risks of relying solely on his often heavily embel-lished recollections. The truth, as is often the case, is much more mundane. On 23 August Wing Commander J. O. Andrews, a staff officer at the Air Ministry, reported to Wimperis that the Deputy Chief of the Air Staff had requested 'a name for this system of detection which [did] not immediately indicate its method of operation.' Andrews suggested that the use the term 'R.D.F. (a compression of R.D. and D/F.) to serve as verb, noun or adjec-tive, as required.' If Wimperis agreed this term would be used in all further official documents. On 17 September Wimperis concurred with Andrews' recommendation and British radar became known as RDF. This is con-firmed by the earliest document which refers to the term. This document is a paper written by Wimperis on 24 September 1935 titled 'Notes on pro-posed methods of plotting information received from R.D.F. Stations.' RDF seemed suitable at a time when the Orfordness device could measure nei-ther direction nor height. As we have seen, Watson-Watt always intended that radar could accurately pinpoint the location and heading of aircraft, and therefore, it is highly unlikely he would have originated this term.

There were several other significant decisions made in late 1935 and early 1936 that were to influence the radar programme profoundly in the years ahead. One of the hallmarks of Bawdsey and its successor laboratories was that they were under the leadership of civilian scientists. Many have argued that by placing a civilian scientist in charge British radar research freed the programme from many of the constraints of close military supervision. As long as Watson-Watt headed the programme, scientists were left to get on with their work with a minimal amount of bureaucratic constraint. His successor, A. P. Rowe, while instituting a more hierarchical structure, still allowed a much more open atmosphere than existed in traditional military research centres. There was limited compartmentalization of research within Bawdsey. This allowed a remarkable flowering of technology and of new concepts on how science could be utilized in the military and, eventually, in the world at large.

Yet civilian control came about by a fortuitous example of bureaucratic intransigence rather than design. In his September 1935 memorandum Watson-Watt had recommended that a civilian scientist command at Bawdsey, the clear implication being that he would naturally assume the position. He looked forward to ruling over the estate from his private apartment in the manor house just like a Scottish Laird. Wimperis, however, had his doubts. Right from the beginning he saw the administration of Bawdsey, including the research programme, a future training facility and the maintenance of the estate itself, as a huge task. Wimperis believed that Bawdsey should be commanded by 'a capable and efficient' recently retired senior air force officer. He even had the names of two possible candidates. Watson-Watt was only going to be offered a half-time position as head scientist, splitting his time between Bawdsey and Slough.[26]

Wimperis' choice of administrative structure was approved by the Air Ministry but ran into opposition from the Treasury. Treasury officials did not see the need to have two highly paid individuals in very senior positions at Bawdsey. They told Wimperis that in every discussion up to this point they had assumed that Watson-Watt would be in overall charge of both research and administration. The Treasury officials believed that the scientist would eventually be needed at Bawdsey full-time. Faced with the unyielding position of those who controlled the purse strings, Wimperis was forced to acquiesce and agreed to appoint Watson-Watt as the Superintendent of Bawdsey in December. By this date Watson-Watt calculated that he was already devoting 80 per cent of his time to the radar programme.[27]

On New Year's day 1936 a discussion was held between Watson-Watt, Wimperis and Air Commodore Bowen on liaison between Bawdsey and the air force. Watson-Watt was adamant that he alone should have direct contact with Fighting Area Headquarters, the RAF signals personnel and, if necessary, the Deputy Chief of the Air Staff. Wimperis agreed with this arrangement, which meant that all important administrative and liaison functions at Bawdsey were under the control of Watson-Watt.[28]

Overall supervision of Bawdsey research remained with the Tizard Committee, although initially financial administration was to be undertaken by the Department of Industrial and Scientific Research and the National Physical Laboratory. Those employed at Bawdsey were to remain as DSIR or NPL employees, with the Air Ministry transferring funds to cover their salaries. This arrangement was found to be cumbersome, however, and in the spring of 1936 all Bawdsey staff were transferred to the Air Ministry and charged to the Directorate of Scientific Research.[29]

The winter of 1935-36 was devoted to hiring staff, setting up Bawdsey and building the new radar stations. Although much delay was experienced, particularly in building the new taller masts, there was great hope among almost all those with knowledge of the new technological miracle that tests that summer would prove that the bomber would not always get through. Certainly, progress had been remarkable. From a theory tested using the crude apparatus used at Daventry in February 1935, one year later a huge new radar research laboratory was in the process of being established. Only Frederick Lindemann and Winston Churchill remained unconvinced that dramatic progress had been made.

IN THE YEAR 1936 AT BAWDSEY MANOR
ROBERT WATSON-WATT
AND HIS TEAM OF SCIENTISTS DEVELOPED
THE FIRST AIR DEFENCE RADAR WARNING STATION
THE RESULTS ACHIEVED BY THESE PIONEERS PLAYED
A VITAL PART IN THE SUCCESSFUL OUTCOME OF
THE BATTLE OF BRITAIN IN 1940.

Bawdsey Marker.

7

The Tizard-Lindemann Dispute

During the winter of 1935-36 the debate on air defence research appeared once and for all to have abated. Churchill and Lindemann, while still pressing for increasing the pace and scope of the research effort, seemed relatively content to pursue their agendas peacefully within the Tizard Committee and the ADRC. There was no repeat of the parliamentary debates and questions on the top secret research programme.[1]

Churchill even remained relatively quiet over other events, muting to a certain degree his criticism of the government over its apparent mishandling of a series of international crises. In October 1935, Italy invaded Abyssinia, provoking a split in the Baldwin government and revealing the inability of the League of Nations to take any strong action against aggression by one member state against another. On 7 March 1936, Germany renounced the Locarno Treaty and sent troops into the Rhineland. There was much sympathy for Germany's actions in Britain and little desire to cause an international crisis. The crisis revealed the unpreparedness of all the British services for war. Without full mobilization the army could not send a single division to France. The RAF was little better off. It could immediately muster just seven bomber, six fighter, and three Army and Navy cooperation squadrons. Only one-third of the aircraft were available to implement the scheme for the defence of London. There were no aircraft to defend other cities and a very small proportion of the necessary searchlights and virtually no modern anti-aircraft guns.[2]

According to his official biographer, during this period Churchill toned down his attacks on government policy because he hoped to rejoin cabinet and to assume the new post of Minister of Defence. Churchill's hopes rested with growing criticism of the Committee for Imperial Defence, which coordinated defence policy through its various sub-committees. Only the Prime Minister had executive authority within the CID. Many, including Churchill and Austen Chamberlain, urged Baldwin to create a new Ministry of Defence, headed by a powerful cabinet minister, to control defence policy. In late February, Baldwin decided instead to establish a new, far less powerful post, the Minister for the Co-ordination of Defence. The new minister was not to have a department, but instead would act as the Prime Minister's deputy on defence issues by relieving him of much of the burden in supervising the day-to-day running of the CID and assist in making defence policy. Despite the lack of true executive authority Churchill coveted the position. Baldwin and other senior cabinet members, however, retained a deep seated animosity towards Churchill, which had only been made worse by his frequent public attacks concerning rearmament. Instead, Baldwin appointed to the new position the comparatively obscure figure of Sir Thomas Inskip, the Attorney-General.[3] After hearing of Inskip's appointment, Lindemann is reported to have restated a line originally uttered by Disraeli, that it was 'the most cynical thing that has been done since Caligula appointed his horse as consul.'[4]

Throughout this period, evidence from a variety of intelligence sources indicated that the rearmament of the Luftwaffe was taking place at such a pace that in numbers of aircraft it had already surpassed the RAF, despite the various expansion schemes. This in turn led the Defence Requirement Sub-Committee of the Committee for Imperial Defence to recommend in late 1935 yet another expansion programme, Scheme F. Approved in February 1936, the scheme had two basic goals which the Air Staff thought more important than simply achieving numerical parity. The first was to provide a much more powerful bomber strike force to act as a potent deterrent to German aggression by the replacement of all light bombers with more effective medium bombers. The second goal was to make the RAF war ready by providing adequate reserves of men and material. No additions were provided in the number of fighter squadrons, but it must be remembered that there was a general acceptance that major expenditures were going to occur if the radar programme and other research projects lived up to expectations. Moreover, it was realized that these scientific developments might lead to 'a considerable alteration in the scheme of [the] Air Defence of Great Britain.'[5] As well, a new generation of radically better fighters were being brought into service. In February 1936, Hawker Aircraft received a contract to build 600 of its eight-gunned, monoplane Hurricane fighter. Orders for the even better Supermarine Spitfire were soon to follow.[6]

The State of Air Defence Research

The growing reliance on science to solve Britain's defensive problems weighed heavily on the mind of A. P. Rowe in late 1935 and early 1936. He had initiated the study which led to the establishment of the Tizard Committee and he was not only the secretary of that committee but of the ADRC as well. Rowe was much more than a recorder of minutes; along with Tizard, Wimperis and Dowding, he was in effect a part of the senior management team that oversaw the widening research programme. During the 1935-36 Christmas and New Year's holidays Rowe took some time off from the Air Ministry for what should have been a well deserved rest. At home, however, Rowe used his time away from the increasingly difficult daily management chores to reflect back on a year of tremendous achievement. He realized that the work of the Tizard Committee offered the possibility that Britain could soon be as secure from air attack as the Royal Navy had made it from continental invasion since the Spanish Armada. Unable to simply relax, Rowe decided to prepare a detailed 'Appreciation of the Present Position' which was intended to serve as a template for another interim report.

Radar was the key technology for successful air defence, but Rowe realized that, even if the chain of stations worked as Watson-Watt had promised, it only solved part of the bomber threat. By guiding the new eight-gunned fighters to visual range of approaching bombers, daytime attacks in good weather could be defeated. This could only be accomplished if research was undertaken to ensure that the most modern methods were used for transmitting and recording the massive amount of information that the radar chain would provide to ADGB. At night the radar chain offered little advantage. There was still little data on the effectiveness of existing technologies, such as searchlights guided by short-range acoustic detectors, in detecting modern aircraft. If clouds were present searchlights were ineffective unless new ways of using them were found. Experiments were required to see if searchlights could be used to illuminate clouds. This might allow night fighters to see bombers as a dark silhouette against the clouds. Lamb aircraft, as proposed by Churchill, could track enemy bomber formations so illuminated in order to direct fighters to their prey.

Searchlights would be useless if bombers flew in the clouds. Detection of these aircraft could only occur if some form of small short-range air-craft-mounted detection scheme could be developed. There were several possible technologies being considered for this vital task. Foremost was a form of acoustic detection, experimental work for which was underway under Tucker's supervision at ADEE. Another was Lindemann's scheme for infrared detection. Finally, Rowe proposed that at some future date research could be undertaken to develop a small aircraft mounted radar apparatus. Watson-Watt had proposed such a device, but Rowe felt that the small number of radar researchers should focus their effort on the early warning system.

Without an aircraft mounted detection system, other techniques were necessary. Anti-aircraft gunnery, even the modern 3.7-inch gun still not

Short-Range Acoustic Detector.

in widespread service, would be limited by its inability to be accurately directed at aircraft in anything but the best daytime weather conditions. Since these were also the best conditions for fighter interception, Rowe questioned the value of gunnery. The only hope of greatly improving anti-aircraft fire was through a radar tracking system, research for which would be commenced by a War Office research team during the coming year.

The only alternative means for destroying enemy aircraft at night or in bad weather would be to develop some means to utilize the chain radar ability to provide an approximate interception point and then to make a large area of the sky deadly to a passing formation of aircraft. Four different technologies were under investigation. The first, using tethered balloons or helicopters to provide an impassable barrage of wire, Rowe dismissed as being tactically inflexible. The second was Rowe's favourite concept of death rays. Even a year after Watson-Watts' paper demonstrating their impracticality, Rowe listed recent claims by yet another group of boastful inventors. The others technologies held out more promise but only after extensive research.

During 1935 the Tizard Committee had implemented research into two different types of aerial minefields. Rowe believed this work so vital that he 'suggested that all forms of wire barrages should be investigated on high priority until one of them proves successful.' The first was the short wire minefield, a wire attached to a small parachute at one end and a small explosive charge on the other. Hundreds of these mines would be laid just ahead of a bomber formation. The second, the long wire barrage, consisted of long cables trailed by aircraft flying just above a bomber formation. The wires needed to be of sufficient strength to cripple any aircraft that ran into them. Rowe considered this idea 'to be one of the most important contributions to Air Defence.'

Lastly, the Tizard Committee was considering the possibility of using friendly bombers to fly above an enemy formation and drop small bombs into their path. This idea had been proposed many times going back to the Great War, but the difficulty of achieving a direct hit had always appeared to make the idea impracticable. Now, Rowe informed the committee, it was proposed to use acoustic proximity fuses to detonate bombs by detecting the sound of a nearby aircraft. If these fuses functioned properly, bombing could be a potent anti-aircraft weapon.

The paper ended with an analysis of some of the tactical problems that might be subject to further investigation; including how differences in speed and altitude of fighters and bombers influenced the rate of interception, and the average distance of sighting of aircraft at different altitudes and in various weather conditions. Rowe stressed the need for scientific analysis and proposed the creation of a special staff to undertake these studies.[7]

Rowe was careful not to attempt any 'prophecy' in his memorandum, but instead provided a 'headquarters' assessment of what had been accomplished and what still needed to be done. In a private letter to Tizard, however, he was willing to express his hopes for the future if they were 'given more than ordinary luck.' He told Tizard: 'Being neither young nor old I do not know whether this is a dream or a vision.' In Rowe's utopian scheme for a defence system radar would solve most of the problems. During the day the chain and fighters would ensure every raid was intercepted. For night fighting Rowe told Tizard:

> Let us be optimistic and say that short-range RDF [radar] will work in fighters (Watson-Watt is hopeful). Then the combination of RDF at the coast & RDF in fighters provides the element of surprise, simplicity, uncanniness & destruction; the fighters may have guns or bombs. This leaves only attack in clouds (to be met by mobile wire barrage aircraft fitted with RDF) & low attack (very difficult at night).

If airborne RDF worked 'few aircraft would get past the coast without having fighters already on their track.''This is a dream,' Rowe confided to Tizard, 'but if RDF location from aircraft is possible, it might very well be realized.'[8]

The reaction to Rowe's report was generally very positive. Air Marshal Joubert, Commander-in-Chief, Fighter Area Headquarters, was hopeful that the research programme would 'completely revolutionize' home defence. Joubert agreed that there was a need to establish scientific teams working on improving communication and information at Fighter Command Headquarters.

Rowe's report was the basis of discussion at the first two meetings of the Tizard Committee in 1936. Investigations had already begun in a variety of new anti-bomber weapons, including aerial mines. Research on the short aerial mines involved determining the minimum weight of explosive required to destroy an aircraft by detonating charges 'against wing structures.' For the long aerial mines preparations were commencing at the Royal Aircraft Establishment to test the weight of cable required by 'catapulting airframes from a cliff top against a cable suspended from moored balloons.' The highlight of these meetings was Watson-Watt's appearance on 25 February. He outlined the possible details of an aircraft-mounted radar, dubbed RDF 2, which would home fighters onto a target from 5 kilometres to 150 yards. Work on airborne interception radar would soon become the second major research project at Bawdsey.[9] These discussions led to the completion of the committee's second interim report in early March. The major difference with Rowe's report was that airborne radar was now listed as being the most likely solution to short-range location.[10]

It is remarkable that after only one year so much experimental work was already underway, and similarly striking is the generally optimistic tone of most of the members of the committee and of senior RAF officers familiar with the research programme. Those who have written about this period have often ignored the importance placed on technologies other than radar. While radar was seen as key to a future air defence scheme, these other inventions were also considered crucial. Only a small proportion of the committee's time was spent on radar, which would however ultimately prove to be its most important legacy. In hindsight things like aerial mine-fields and air-to-air bombing seem foolish, but at the time they seemed to offer the only practical solution. Little of the work on these bizarre schemes had any long-term applications, yet it should not be surprising that, in pushing the limits of knowledge, several well intended projects failed. Some of Rowe's other ideas, however, were of greater long-term importance. The establishment of scientific teams to assess information is the root of wartime operational research. Scientists would no longer simply develop equipment but would increasingly become involved in determining the way the armed forces used them.

Not everyone agreed with the prevailing upbeat mood. Neither Lindemann nor Churchill thought that all was well, nor did they find it easy to confine their concerns within their respective committees. On 7 February Lindemann wrote to Lord Swinton, as Cunliffe-Lister was now known, to provide an independent assessment of the current research situation. While not openly critical of work of the Tizard Committee, he did not fully endorse it. Lindemann, whose views on air defence had changed little since joining the committee, emphasized the importance of finding a solution to night bombing. Lindemann believed that only the development of a kite balloon barrage and aerial mines could prevent German aircraft from destroying London.[11]

Churchill expressed his frustrations with what he considered a lack of progress in a draft letter for Hankey, the Secretary of the CID, which he sent to Lindemann for comment on 26 February. Churchill feared that 'the crisis which is approaching may be upon us long before any practical results are achieved.' His views on experiments that needed to be pursued mirrored Lindemann's, with the addition of work on the development of improved anti-aircraft shells.

Lindemann applauded Churchill's letter but added a paragraph concerning Watson-Watt's successful work on detection. 'My paragraph,' he told Churchill, 'endeavours to draw the moral that this method of handing over experimental work to an enthusiastic believer is the only one by which results can be achieved.' Churchill agreed to include Lindemann's additions to the final draft of the letter, but it remains doubtful that it was ever sent to Hankey since a Cabinet post still remained a possibility.[12]

Encouraged by Churchill's letter, Lindemann decided to produce an exhaustive twenty-seven page summary of the events of the last eighteen months. It is not necessary to examine the details of this rambling monologue since his position on all aspects of air defence research have already been discussed elsewhere. Lindemann's attitude to the current state of research was more succinctly expressed in his covering letter to Churchill when he sent him the memorandum:

> So far as I can see they have been to all intent and purposes nil, save for Watson-Watt's very successful and commendable detection and location work. I do not know whether it would be worth your while to have a private talk with him and perhaps with A. P. Rowe, the Secretary of the Committee. I have reason to know that they are both most dissatisfied with the slowness and lack of drive which have been exhibited.[13]

Thus, the stage was now set for the famous confrontation between Lindemann and the rest of the members of the Tizard Committee. For the time being Lindemann followed Churchill's lead and remained silent, although it is hard to believe that committee members did not sense his growing frustration and anger. Evidence outside of the minutes and reports of the Tizard Committee and the ADRC is scarce, but there is enough to confirm Lindemann's belief that not all was as rosy as the official record suggests. The second year of work of the Tizard Committee was not marked by the same tremendous progress as the first, and certainly Rowe and Watson-Watt were dissatisfied with some aspects of the management of the research programme.

Rowe expressed his deep misgivings in a 'secret and personal' letter he wrote at home and sent to Tizard on 22 February. His concerns, however, were not with the lack of activity, but with the way that the huge increase in workload had swamped him. He was worried that Wimperis was not putting 'forward any scheme of re-organization since it might appear that he is boosting his own end.' Yet more staff was desperately needed to supervise the growing research programme.

Only part of the problem was in personnel, the larger part was in needless bureaucratic paper work. Rowe felt that he was chained to his desk. Moreover, he felt that 'the time & effort required to convert a recommendation of the Committee into honest toil at an Experimental Establishment [was] absurd ... The only exception (& that an illuminating one) is the Orford work which Watson-Watt runs more or less himself.' Rowe asked

Tizard to put these proposals for reorganization into the second interim report. He concluded by confiding:

> I hope that this letter does not sound like a grumble; my only fear in that connection is that, for the first time in my life, I am under a doctor whose only remedy seems to be a fortnight's leave.
>
> The difficulty of the job is not the work but the worry of a bad organization. I am afraid I haven't the placid mind that can accept these things as inevitable. I wish I had.[14]

Little else is known about these managerial problems. Rowe makes reference to a memorandum Tizard sent to Lord Swinton on 29 January in which the need to expand the personnel of the DSR was discussed. Unfortunately, this document does not survive. Wimperis in his diary makes passing references to the need for reform of his office, indicating that Rowe was incorrect in assuming that his superior had not attempted to resolve the situation. Wimperis' proposals for expanding the DSR's staff were, in fact, turned down by the Air Ministry the day before Rowe wrote his letter.[15] The lack of a proper senior level management organization for research at the Air Ministry would become an increasingly serious problem in the years ahead.

Lindemann was also correct that Watson-Watt was deeply troubled in the late winter and spring of 1936. At the best of times Watson-Watt could be an obstinate and demanding character when he did not get his way. Rather than acquiesce to Air Ministry's decisions concerning his terms of employment, Watson-Watt fought a prolonged battle with the bureaucrats. They argued over his title, his powers and responsibilities, and his pay. Watson-Watt demanded that he be appointed to a new position of 'Director of Investigations on Communications (DIC),' with status and pay equal to the DSR. A deputy director would run Bawdsey. When the Air Ministry balked at this, he enlisted the support of the most influential people he could muster, including Tizard and Sir Frank Smith, the head of the National Physical Laboratories, and made a personal appeal to Lord Swinton. The situation remained unresolved and it led Watson-Watt directly into the dispute between Lindemann and Churchill.[16]

There were also difficulties in the radar programme itself. By March 1936 it became apparent that it would be impossible to fulfil the wishful schedule outlined in late 1935. Research at Bawdsey was hampered by the slow pace of recruitment of new staff and in undertaking alterations to the site necessary before research could commence. Sometime in early 1936 the radar system went through a major redesign; Watson-Watt's original concept of having alternating transmitter/receiver and transmitter stations

Bawdsey Transmission Towers.

was abandoned. Instead all stations would both transmit and receive their own signals. As a result, stations would be twice as far apart as originally envisaged. There also were long delays in completing the transmitter and transmission towers, caused by a combination bureaucratic lethargy, delays by private contractors and an unrealistic construction schedule.

The Collapse of the Tizard Committee

While there was, therefore, a basis for Lindemann's concerns, it is certain that the accomplishments of the Tizard Committee far outweighed any of its shortcomings. Tizard knew about and was working to correct problems hampering air defence research. Rowe's complaints are what might be expected in any organization trying to contend with an unprecedented growth. Watson-Watt's were in large measure caused by his own inflated ego and the impossibility of achieving the hopelessly optimistic goals he had laid down in late 1935. The radar programme was still moving forward. Research into other air defence technologies was well underway by the spring of 1936. Moreover, Tizard generally agreed with Lindemann's assessment that preventing night bombing posed the biggest challenge. Yet that summer the Tizard Committee collapsed when Blackett and Hill resigned over what they believed was the intolerable conduct of Lindemann. In September, the old committee was dissolved and a new one reconstituted with all of its original members except Lindemann, who was replaced by Professor E. V. Appleton.

The clash of wills between Henry Tizard and Frederick Lindemann is the single most written about event in the development of the air defence system, even including the discovery of radar. The Tizard-Lindemann dispute has all the required ingredients of a great drama. Two former friends, both important men with strong personalities and their own power base, duelled for control of the nation's most vital military research programme. Once Lindemann was made a member of the Tizard Committee his clash with its chairman was inevitable given their respective personalities.

Tizard was the consummate intelligent, politically astute, bureaucrat, working within the system to build consensus and always willing to accept legitimate compromise. He could be very forceful when he felt the need and had a very refined sense of how those taken into his confidence should behave. In return he became the willing confidant of anyone who had an opinion on air defence research. Any violation of gentlemanly conduct could spark Tizard's anger.

Lindemann simply could not fit himself into Tizard's approach. Working with Churchill, Lindemann hoped to seize control of air defence research. It would be unfair, however, to cast any aspersions on his motives; Lindemann was not seeking personal power, but did what he truly felt was best for the nation. He and Churchill believed war was imminent and quite rightly viewed Britain as still being vulnerable to attack. Gentlemanly niceties had to be dispensed with for the greater good.

Just as it was inevitable that this falling out would occur, it was also certain what the outcome would be. Lindemann had few friends in the inner circles of power and if Tizard demanded satisfaction then only he could come out the victor.

The end of the uneasy truce between Lindeman and Tizard was the unintentional result of one of the first statements made in the House by Inskip in his new capacity as Minister for the Co-ordination for Defence. In a debate on funding the new ministry, on 22 May Inskip briefly mentioned the on-going work of scientists, which he believed was bound to play an increasing part in the nation's defence. He told the Commons: 'It is my business to try to quicken the tempo of and the application of the results of their research to actual instruments of defence, as well as to extend the range of their researches.' Churchill seized upon this statement as providing him with on open invitation to outline in a 'secret and personal' letter his and Lindemann's concerns about the 'slow, timid and insignificant' pace of progress. He informed Inskip that he was having 'a great deal of difficulty' in preventing Lindemann from resigning his membership on the scientific committee.[17]

When Inskip replied on 29 May promising he would investigate Churchill's complaints but mentioning no specific actions, Churchill decided further prompting was necessary. He was motivated to make a quick response by Lindemann's reporting of the tempestuous meeting of the Tizard Committee held that morning. In the deliberately clinical minutes of the meeting it is difficult to judge the tone of the discussion. There is much evidence, however, that Lindemann was unable or unwilling to hide his opinions from the other members of the committee. The major piece of business concerned experiments undertaken by the Royal Aircraft Establishment on balloon barrages and on both types of aerial mines. Dr Harold Roxbee Cox presented the RAE's recent experiments and its plans for future work. According to Tizard, Lindemann without cause accused RAE scientists of 'slackness and deliberate obstruction.'[18] In July, just before the final showdown between the other committee members and Lindemann, A. V. Hill composed an ode, in the style of the Earl of Derby's translation of the Iliad, which with vicious sarcasm laid the complete blame on Lindemann. Contained within the verse is our only detailed account of what occurred on 29 May. Hill wrote:

Von Alpa-plus [Lindemann] arose and thus began,
'Oh ancient Sigma [Tizard] eminent in war
And in the council wise: thy present words
No Trojan can gainsay, and yet the end
Thou hast not reached, the object of the debate.
This city cannot be immune from war
Until a hail of parachuting mines
Descend unceasingly at its eastern gate,
So shall the long-haired Greeks remain at home
Nor lay their infernal eggs upon our streets.'

Thus angrily, and round his body flung
His cloak, and on his head a billycock,
Then passing cocked a snook at Lambda-Mu [Roxbee Cox],

Last called his shiny Rolls of eighty steeds
And soon without the tent of Odin [Churchill] stood.
Him, from his godlike sleep, he sought to rouse
Loud shouting: soon his voice his senses reached:
Forth in his slumber-suit bearlike he came
And spoke to deep designing Alpa-plus,
'What cause so urgent leads you through the camp,
In the dark night to wander thus alone?'

To whom von Alpa-plus of deep design replied,
'Oh, Odin, godlike son of destiny awake:
For ancient Sigma's professorial crew,
With Hermes of the glancing wings [Joubert] and Rho [Rowe]
Who keeps the minutes but who wastes the hours,
Will not be happy till the long-haired Greeks
Upon this city lay their infernal eggs,
They have no mind to fill the sky with mines
Attached to parachutes: and precious days they waste
In vain experiments with RDF.

The next day Churchill sent to Hankey, the secretary of the CID, a detailed official ADRC memorandum in which he outlined Lindemann's charges of gross negligence in the handling of aerial minefield research. By writing an official report Churchill forced the issue onto the agenda of the next ADRC meeting. Hankey and Swinton decided that Tizard deserved to see an advance copy of Churchill's correspondence before it was circulated to the other members of the CID Committee.[19]

Tizard's first reaction was a fit of anger, provoked not by the substance of the charges but because Lindemann had gone outside of the committee and the Air Ministry to voice them. He told Hankey to inform Swinton 'that in these circumstances he did not consider that it would be possible for him and Lindemann to serve on the same Committee, and it would, therefore, be necessary to decide who should go.'[20]

On 12 June Tizard sent Swinton a letter in which he reiterated his demand that either Lindemann be removed or that he accept his resignation. Also enclosed was a memorandum for the ADRC rebutting the allegations that aerial minefield research was not proceeding in a timely fashion. Tizard informed Swinton that he had considered 'whether the nation would lose or gain more by the severance of Lindemann's connection from the work' and that he felt that for the good of the nation he must be dismissed. Lindemann's 'querulousness,' his refusal to accept the views of the majority of the committee and his insistence that discussion be focused on 'relatively unimportant' issues obstructed the air defence research programme. Although he had offered some useful suggestions to Watson-Watt, Tizard believed, without any direct evidence, that outside of the committee Lindemann had decried radar. Tizard claimed that he done his best to accommodate him. Finally, Tizard warned Swinton that this action 'may be part of a political move in which Lindemann is the willing tool of others.'[21]

Swinton found the controversy 'all very tiresome,' but his sympathies were entirely with Tizard. He was well aware that if Tizard resigned so would Blackett and Hill. He arranged to have dinner with Tizard on 14 June, the night before the ADRC meeting, in order to try to get him 'to hold his hand for the present.'[22]

Swinton's diplomatic efforts, however, were not going to succeed. After reading Tizard's response to his paper on aerial minefield research, Churchill requested that Lindemann arrange a clandestine meeting between himself and Watson-Watt. Watson-Watt, in the midst of his dispute with the Air Ministry, was willing, after some cajoling by Lindemann, to go behind everyone's back and personally brief Churchill. The three men met on the afternoon of Friday 13 June in Churchill's Westminster apartment. Churchill drank brandy, Lindemann had milk and Watson-Watt sipped tea. Watson-Watt informed Churchill that his efforts were being stymied 'by the scale by which the normal machinery of the Air Ministry could operate at abnormal speed, as opposed to the higher speed by which emergency machinery might have been made to work.' Watson-Watt outlined his struggle with the Air Ministry to be placed in charge of all radio research, which 'indicated again the Air Ministry's unwillingness to take emergency measures.' The 'immutable attitude' of the ministry made Watson-Watt reluctant to make 'large-scale recommendations of great urgency' so long as they might consider any such proposals as manoeuvres to magnify the importance of his post. The evening after the meeting Watson-Watt returned to Churchill's flat and delivered to him a copy of a letter he had written to Sir Frank Smith at NPL, which summarized the afternoon discussion.[23]

Now fully convinced that Lindemann's concerns were legitimate, Churchill asked him to prepare a briefing paper on other problems in the research programme which he could use in Monday's ADRC meeting. Lindemann swiftly replied by outlining what he saw as widespread delays and failures. According to Lindemann what was needed was a permanent director of air defence research reporting directly to the CID, who would be given top priority and control of research laboratories. In Lindemann's opinion, only Watson-Watt, 'despite the handicaps and delays' had acted with the type of drive and zeal necessary.[24]

The scene was now set for a fiery confrontation between Tizard and Churchill at the ADRC meeting on 15 June. No detailed accounts of what transpired have survived, but it appears that Churchill attempted to ambush Tizard by extending the discussion from aerial minefields to all aspects of the air defence research, including radar. Churchill read passages from Watson-Watt's summary of their meeting. According to Wimperis, Tizard, supported by Swinton, did exceedingly well in deflecting the criticism and defending his committee and the Air Ministry. The majority of the committee supported Tizard. After it become obvious that he was losing the debate, Churchill reluctantly agreed with the majority opinion that the ADRC should send the whole matter back to the Tizard Committee. The scientists were ordered to prepare another interim statement on air defence research, which would hopefully allow them to reach some sort of compromise between Tizard's and Lindemann's positions.[25]

Inskip and Swinton orchestrated the ADRC's decision. They realized that it was highly unlikely that peace would come to the Tizard Committee, but they were reluctant to act as arbitrators on matters 'which only the scientists could properly understand.' They were also justifiably fearful that any decision that they might impose on the scientific committee, which would inevitably go against Lindemann, might lead to public disclosures similar to what had occurred in 1935. On 16 June Churchill informed Swinton that he was having second thoughts about the wisdom of the ADRC's decision, since Tizard would simply rally the rest of the committee against Lindemann. This would likely force Lindemann to resign. If this happened Churchill warned 'the matter is bound to become one of public controversy.' He asked Swinton to appoint himself or another 'impartial person' to hear both sides present their case.[26]

When Swinton rejected Churchill's request, another more threatening letter followed on 22 June. Now Churchill vehemently denied ever having agreed to allow the matter to be decided by the scientific committee. He told him:

It is always difficult to have a public controversy about unmentionable topics. I am however quite sure that if instead of serving on your Committees Lindemann and I had pressed our points by all the various methods and channels open to us, these ideas would have received better treatment then they have received.[27]

The threat greatly concerned Swinton, who had other worries about Churchill. Churchill was also demanding that he be able to use his position on the ADRC as a platform for a debate on further air force expansion. Swinton did not disagree with Churchill's message on expansion, but he strongly disapproved of using the ADRC for this purpose.[28] Swinton desired to restrict Churchill's access to secret information, but this was not his decision to take. Instead, he took steps to curtail any further leaking of information by those working for the Air Ministry. On 17 June he summoned Watson-Watt to meet with him. Swinton indicated to the radar scientist that as a result of his disclosures to Churchill, any support he had from him in his dispute with the Air Ministry was now lost. Swinton asked Watson-Watt to waive his feelings and to accept the ministry's position on the conditions of his employment. Swinton closed the meeting by telling Watson-Watt that 'when a man had been a bloody fool he preferred to tell him directly, and then let him get on with his job.'[29] Watson-Watt was forced to subordinate his personal feelings to that of 'the national interest' and accept the Air Ministry's appointment as Superintendent of Bawdsey. He was to receive the maximum possible salary and benefits at his rank, £1,250 per annum plus a £100 cash allowance, an unfurnished flat in Bawdsey manor rent free, and forty-eight vacation days a year.[30]

Churchill, however, was quite right that sending the matter back to the Tizard Committee could have but one result. Tizard, when he was roused, could be savage. He knew he had the support of the other members of the committee, which meant that he felt that there were no constraints on his actions. He did not follow Inskip and Swinton's advice to seek compromise; instead he deliberately inflamed the situation. On 17 June he wrote to Lindemann, ostensibly to advise him that he had asked Rowe to prepare a new interim statement as requested by the ADRC. Tizard condemned Lindemann for his 'ill-founded criticism' which could only have the effect of 'retarding progress.' He told his former friend: 'I wish we could have settled such differences of opinion that exist in a friendly manner in our own Committee, but you have made this difficult, if not impossible.'[31] In anticipation of Lindemann's resignation, two days later Tizard asked Wimperis to nominate Appleton as his successor.[32]

Lindemann, after consulting with Churchill, decided not to let Tizard goad him into resigning. He told Churchill he wanted the government to demand his resignation since this would justify him 'in raising Cain' to force an acceleration of the research work. Instead, with Churchill's concurrence, he replied to Tizard with an equally forthright letter. He bluntly told Tizard that his methods would only work if there was 'ten or fifteen years time' in which to prepare and, with the exception of Watson-Watt's work, that up to this point the committee had not made 'any appreciable advance.' Lindemann warned that he was determined to continue to use 'every means at my disposal to accelerate progress.' He concluded: 'I am very sorry if this offends you, but the matter is too vital to justify one in refraining from action in order to salve anybody's amour propre.'[33]

Failing to force Lindemann out and with no sign that Swinton would fire him, Tizard now decided to practice some diplomacy. On 5 July Tizard replied to Lindemann, telling him he was not personally offended by his criticism and that they both needed to try 'giving co-operation a further trial.' Tizard finished the letter by reminding Lindemann that they had lost sight of the primary goal. He wrote: 'I am much more interested in defeating the enemy than in defeating you.' The same day Tizard confided to Swinton that he had little hope that his efforts to heel the rift would succeed and reiterated his belief that either he or Lindemann would have to go.[34]

Any hope of appeasing Lindemann ended on 10 July when he received from Rowe the preliminary draft of the new interim report which was to be discussed at the next committee meeting. Lindemann not only disagreed with virtually everything in the report, but was particularly offended by the method by which it was prepared. He informed Rowe: 'It is quite intolerable that a draft purporting to represent the views of the Committee should be produced without these views ever having been ascertained, that it should be circulated on Friday and that we should be asked to send amendments by Monday, so that the report can go straight on to the main [ADRC] Committee.' He demanded an opportunity to discuss the report and suggest amendments or, in lieu of this, to circulate his own rebuttal paper to the ADRC.[35]

When Lindemann produced his report, Tizard now had no choice but to call the committee together for a showdown with Lindemann on 15 July; its first meeting since 29 May. In the days leading up to the meeting the tension grew. Lindemann announced that he was seeking the Conservative nomination for a by-election for one of the Oxford University seats in the House of Commons. There were fears that he was doing so in order to have a parliamentary platform to press his views on air defence research. It was during the week leading up to the meeting that Hill wrote his satirical ode. The poem closes with the announcement of the meeting.

At last with downcast visage Sigma [Tizard] spoke:
'The Game is up without von Alpha-plus [Lindemann],
Of wily counsel and deep design,
Who speaks with politicians and the Press,
And soon may be M.P. for Oxenbridge,
All hope is gone and many-murdering Death
Will hunt his victims in our streets.' To which
Theta [Blackett] of bright ideas, Phi [Hill] of none,
Rho [Rowe] of the Minutes, weary Omega [Wimperis],
Had nothing printable to add. But set

A Day to meet Geheimrat Alpa-plus
And pray for mercy from his mighty friends,
From Odin [Churchill]. Godlike son of destiny,
And from himself, the man of deep design,
Then ancient Sigma and his stag-eyed crew
Will make submission to von Alpha-plus,

(Except for Lambda-Mu [Roxbee Cox] who hanged himself).
Your presence is requested at 11:
The number of the room is 008.

We cannot know exactly what was said in Room 008, since the confrontation was so tempestuous that the secretaries that assisted Rowe in keeping the minutes were for a time sent out of the room. The length of the minutes of an average meeting ranged from four to eight typed single spaced pages; the minutes for this meeting consist of just two brief paragraphs. The first states that the committee deferred the consideration of the minutes of the previous meeting, the second laconically recorded what occurred.

> The Committee considered a draft report on the rate of progress of its investigations, copies of which had been circulated to members of the Committee. With the exception of Professor Lindemann, all members of the Committee agreed on an amended form of the report which the Secretary was instructed to forward to the Air Defence Research Sub-Committee of the Committee of Imperial Defence. The Chairman suggested to Professor Lindemann that he should prepare a minority report if he so desired.[36]

Lindemann reported that his refusal to accept the majority's viewpoint was based on a fundamental difference of opinion over the priority given to short aerial mine research. He wanted to push research ahead as fast as possible, the others desired to delay until there was proof that minelaying aircraft could intercept bombers with sufficient accuracy.[37] According to Tizard, the other committee members tried to accommodate Lindemann's concerns, agreeing to alter the draft report considerably. In the end all effort to forge a consensus collapsed when Lindemann 'repudiated' the amended report. If Lindemann hoped to force Swinton to demand his resignation he was mistaken. Instead, Hill and Blackett, disgusted with Lindemann's behaviour and determined to support Tizard, submitted their resignations to Swinton that afternoon. The Tizard Committee had ceased to exist.[38]

The Aftermath of the Dispute

Swinton's sympathies lay entirely with Tizard. He knew that the committee had to continue and that Tizard was indispensable. He tried once more to find some sort of compromise and invited Hill and Blackett to meet with him on 22 July. On 17 July Hill again wrote Swinton to reiterate his position. He told Swinton that one of the determining factors in his decision to resign was his concern that this 'unpleasantness' had taken a toll on Tizard greater than all his work on air defence, for he was 'a sensitive and not too robust person, in spite of his very robust character and mind.' The meeting with Swinton lasted some 1½ hours. Determined to protect Tizard and to be rid of Lindemann, neither Blackett nor Hill could be persuaded to withdraw their resignations.[39] However, Swinton told them that he absolutely refused to accept the resignations. 'Poor man,' Hill told Tizard, 'he is in rather a fix, but he had better get out of it now, rather than later.'[40]

In the meantime, Lindemann decided to continue the fight. He arranged to have an edited copy of his rebuttal report circulated to members of the ADRC. Churchill hoped to use the report to defend Lindemann at the next ADRC meeting on 24 July. Swinton and Inskip had already decided on what course to take. Instead of discussing Lindemann's paper, Swinton, as Chairman, ensured that the committee focussed on the need to reconstitute the scientific committee with members 'who could work effectively together.' Wimperis described the gathering as 'a useful peaceful meeting. Winston in a chastened mood.' There was no doubt that Churchill and Lindemann had been thoroughly routed, if not humiliated.[41]

There was a lengthy delay before the scientific committee was recast, but this was the result of the need to bring the matter before Cabinet. Swinton fully endorsed Tizard's views and, when the new committee for the Scientific Survey was approved on 2 September, there was only one change in its membership; Appleton replaced Lindemann. 'When this was done,' Swinton told his Cabinet colleagues, 'it might be expected that Professor Lindemann would raise the matter publicly, but the other scientists could be relied upon to answer him effectively.' The official notice went out to the members of the new committee on 9 September.[42]

Despite the fears, Lindemann did not go public with his concerns. At first it was because he was not informed that the Tizard Committee had been reconstituted without him. As late as 23 October Lindemann still had not received official confirmation, even though the new committee had already met twice.[43] Even after he was finally notified in early November he remained silent. It may be that he was distracted by his disastrous campaign to get elected to the House of Commons. It is also possible that his sense of duty precluded risking national security. Another factor may have been that a clever political manoeuvre by Baldwin had deprived Lindemann and Churchill of much of their support in Parliament for further public debate. In late July, the Prime Minister arranged a meeting with a delegation of MPs and Lords for a private question and answer session on air defence.[44]

Despite his removal from the committee, Lindemann continued to have influence mainly through Churchill. Churchill was disgusted by the treatment of his good friend because, in his judgement, the nation was 'losing one of the best scientific brains in the country.' He blamed himself for rousing passions against Lindemann and on 19 November he asked Baldwin to accept his resignation from the ADRC.[45] Baldwin did not accept Churchill's resignation, probably because it was made just four days before he was scheduled to have his second private meeting on Britain's defences with the parliamentary delegation. Churchill would be among those attending and Baldwin feared providing him with further ammunition with which to attack government defence policy. Churchill had roundly denounced the rearmament programme in parliamentary debates on 12 November. Baldwin was quite right in predicting that Churchill would play a leading role in the meeting on 23 November. By not allowing Churchill to leave the ADRC Baldwin precluded any public discussion on air defence research and the reconstitution of the scientific committee.[46]

With Churchill remaining on the ADRC it was not possible to ignore Lindemann completely. He continued to press for an acceleration of research into aerial minefields and successfully convinced the new Tizard Committee to provide funding so that R. V. Jones could continue his work on infrared detection. Lindemann would remain looming in the background as a self-proclaimed watchdog over air defence research.[47] He would emerge from the shadows again some two years later.

The Tizard-Lindemann dispute was much more than just a clash of personalities, for it determined who would control the research agenda. If he had wrestled control, Lindemann would have brought chaos to the research programme in the same way he had destroyed the Anti-Aircraft Research Sub-Committee in the 1920s. Undoubtedly this would have adversely affected the radar programme. Churchill played a dangerous game, relying on his own judgement to determine what classified information could be revealed to the public in the furtherance of his quest to speed rearmament. It is remarkable just how willing he was willing to threaten to reveal air defence research secrets, including radar, to pursue his laudable fight. It is doubtful in this case the ends justified the means. If the Germans had learned about British radar at this early stage it was possible that they could have developed effective countermeasures to disrupt the early warning system. Certainly this was a major concern of Bawdsey scientists in 1936. Churchill and Lindemann would press their case once more some two years later but, before we examine this second showdown, it is necessary that we see how the air defence system was being transformed by the scientists at Bawdsey and a group of RAF officers at the Biggin Hill aerodrome.

8

Transforming the Air Defence System

Since Churchill and Lindemann failed to seize control of the research agenda, in large measure the main consequence of the Tizard-Lindemann dispute was that it distracted attention from the establishment of an effective air defence system. If Lindemann and Churchill had succeeded there can be little doubt that research on integrating radar into the air defence system would have been greatly disrupted. Tizard's departure from active participation in the research programme would have occurred just before he made his most important scientific contribution to the research programme – the development of new radar guided interception techniques. The importance of technique, rather than hardware, is often lost in histories of the development of radar. Radar is, however, simply a tool which provides information which is only useful if procedures are developed to allow for a successful interception of an approaching aircraft. Much of the research programme that was sparked by the successful experiments at Orfordness in 1935 involved the marrying of the radar with other new technologies and the existing air defence system.

Close collaboration between scientists and airmen was key to the success of the research effort. Scientists, particularly Tizard, were brought into the inner sanctums of policy making. It was realized that the scientific-military partnership was leading to a complete transformation of the air defence system. The full extent that scientists were embraced and the system transformed can be seen in an extraordinary letter which Air Marshal Joubert sent to Tizard on 31 March 1936. Joubert had learned that Tizard was being called to 'explain and amplify' to the Home Defence Sub-Committee of the CID the conclusions of the CSSAD's second interim report. Joubert asked Tizard to use this opportunity 'to attack the whole of the present [air defence] system.' Joubert believed that radar meant that the three zones of the air defence system could be done away with and 'the whole of England beyond the coast-line' should be opened for aircraft operations. Now was the time 'to start with a clean sheet and decide what we are going to do.' Unfortunately we do not know the outcome of Tizard's testimony before the committee, but that this letter was written reflects the growing influence of Tizard and the conviction of senior RAF officers that a revolution in air defence was taking place.[1]

Yet while many of the scientists and airmen who knew of the work supervised by the Tizard Committee shared Joubert's views, they were also aware that huge hurdles needed to be overcome if a new air defence system was to be made effective. The main problem facing them was how to replace the existing methods of intercepting bombers. The tactics of the First World War, as modified in peacetime exercises, was to 'concentrate defending Fighters over certain Sectors adjacent to the area to be defended, utilizing them to form a *barrage* into which it was hoped the raiders would fly.' Since it was not possible to provide accurate interception vectors the fighters were concentrated at the rear of the fighting zone. Yet the air force officers understood that owing to the 'high performance of aircraft, this procedure would give the Fighters insufficient time to deal with a raiding formation before it was actually over its objective.'[2]

New tactics would require sufficient information from radar to allow for the vectoring of defending aircraft close enough to the approaching enemy to effect a successful interception. Given that bombers would soon be flying at close to 300 mph and fighters would be travelling even faster, closing rates between aircraft could approach 600 mph. It was not even certain that it would be possible given these great speeds to make an interception even if radar could provide relatively accurate information on the height, bearing and speed of bombers at up to 60 miles from the coast. While the Bawdsey research team would have to be trusted to make radar work as promised, the RAF had to develop the equipment and procedures needed to use this information effectively. If a successful interception was to be achieved, ground controllers had to know at all times the location of defending aircraft and means had to be developed to quickly and accurately determine the correct course to meet approaching bombers. While in theory radar could provide this information, there was no provision in the design of the chain radar system for tracking aircraft over British airspace. The first and foremost concern was with detecting aircraft at long range approaching the coast. Enemy aircraft needed to be intercepted close to the coast or once over land tracked by the observer corps using binoculars and ears.

Interception Experiments   *helped to make radar a success*

In one of those happy coincidences that marks this story, the RAF was just
finishing the development of new equipment that could eventually keep
track of defending aircraft. In the 1920s and 30s, considerable research
went into developing radio navigation aides to assist pilots lost in bad
weather or at night. Medium frequency radio direction finding (DF) was
introduced into the RAF in 1924. An aircraft's radio transmissions could
be used by a single DF station to home an aircraft back to base. If two
different ground based DF stations were calibrated to work together using
a clock, then the location of an aircraft could be calculated using simple
triangulation. In 1934, experiments showed that direction could also be
determined from high frequency transmissions, the same radio band that
radio-telephone sets on fighter aircraft operated on. Tests of the new technique were conducted in 1935. They proved that fighters could be located
'with surprising accuracy.' Two Chandler-Adcock high frequency DF sets
were installed at Biggin Hill and Hornchurch air stations by the summer
of 1935. This relatively crude apparatus consisted of a standard radio
receiver attached to a directional aerial. The antenna consisted of two vertical aluminum plates connected by a horizontal bar, the centre of which

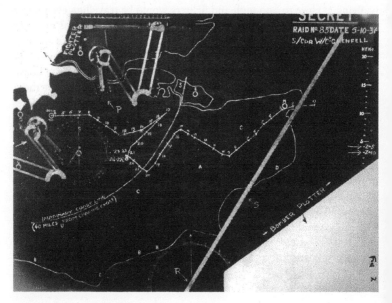

Biggin Hill Experiment Map.

this (DF) helped to develop the radar system.

was mounted on a vertical spindle. The operator rotated the antenna by using a wheel attached to the spindle. He listened to the signal through headphones, with each ear piece receiving signals from the different plates. When the signal was equal in each ear the direction of the aircraft could be determined. A more accurate and less difficult high frequency DF set using a cathode ray tube to provide a visual display was developed by the Radio Research Board. It was installed at RAF base Northolt in early 1936.[3]

Information on early experiments using DF to control fighters is rather sketchy. It appears that the first interception experiments were initiated by Tizard during the autumn of 1935. Tizard suggested that the tests consist of a flight of aircraft, posing as enemy bombers, ordered to 'fly on a definite course, speed and height.' As soon as the first information of the location of the approaching force reached headquarters defending aircraft would be sent up to attempt an interception. They would take their orders from ground controllers using radio-telephones and be tracked using DF. When ground controllers thought that the fighters were in the proper position to intercept the bombers, they would be ordered to release a smoke trail which could be observed from the ground in order to gauge the relative position of the fighters from the attacking formation. This was vital information since the effectiveness of aerial minelaying required that controllers be able to position aircraft with sufficient accuracy.[4] Experiments using DF to track fighters in 1935 had 'negative results.' The chief problem was that early DF sets were found to be very difficult to use to get an accurate plot.[5]

A more extensive series of tests was carried out in the late winter and early spring of 1936 using the Biggin Hill and Hornchurch DF sets and the new cathode ray tube set at Northolt. Again very few of the details of these tests are known, but from 6 February to 27 April fifteen test flights were made with an average error at interception of just over 2 miles, which ruled out 'any successful dropping of wires.' The test revealed that there were significant advantages to the cathode ray tube DF since, like radar, information was displayed visually. This allowed fixes to be made in just fifteen seconds, as opposed to one minute with the aural sets. However, because all government scientists familiar with CRT displays were involved in radar research, it would be some considerable time before additional DF sets of this type could be procured from private industry.[6]

The first full scale test of the new DF equipment took place from 27 to 30 July. The existing Biggin Hill, Hornchurch and Northolt DF stations, the new DF station at North Weald, and eight fighter squadrons based at the three airfields participated in the exercise. The tests were intended to show how DF could be used in both homing and positioning. In order to allow for the latter, special telephone lines were laid between the control room at Biggin Hill aerodrome and the DF stations at North Weald and Northholt. These exercises proved that DF was invaluable for homing an aircraft back to its base. Positioning of aircraft proved more problematic. Using information provided by the three linked DF sets, controllers at Biggin Hill were able to give to their own two squadrons fixes on their position in 45 to 75 seconds. This proved to be very useful in providing the sector controller with information about the location of the aircraft under his command. Yet, DF sets operated on distinct frequencies and changes of frequencies required a minimum of 2½ minutes, even with the superior CRT set. Since there was only one DF set per sector, if this set was being used to work with another sector's DF set to provide positioning, it was unavailable to provide homing bearings for aircraft in its own area. As a result, the chief signal officer involved in the test recommended that each sector be provided with two DF stations of the CRT type.[7]

Even before this round of DF tests Tizard had proposed to the Air Ministry that they undertake a new, far more extensive series of experiments on the practicality of interception. Tizard informed the Deputy Chief of the Air Staff on 13 July that 'the experiments should be undertaken both to determine what results can be expected if RDF proves successful and to assist in methods of backing up RDF when enemy aircraft crossed the coast.' Tizard suggested that two mutually related experiments be undertaken. The first was 'to determine the percentage of occasions which interceptions could be expected in normal daylight weather when warning and approximate position is given 15 minutes from the coast, and to obtain data as to the time that [would] elapse between receipt of the warning and interception taking place.' The second was a repeat of the earlier experiments in which it was determined how close fighter aircraft could be directed by the ground to an approaching bomber. Like Tizard's earlier proposal this test was necessary to determine the practicality of aerial minefields, but primarily 'the information was required to determine what range [would] be required of a RDF set for installation in fighter aircraft for use at night or in poor visibility.' This is particularly significant for two reasons. It shows that at this early date Tizard was beginning to place reliance on airborne radar to solve the night interception problem, even though it did not yet exist. Moreover, given that this meeting took place just two days before the final meeting of the original Tizard Committee, it also indicates that Lindemann may have been right in his suspicions that Tizard was unwilling to press ahead with the development of aerial minefields.

Tizard envisaged these experiments as being undertaken in such a way as to simulate radar, since no sets were yet available for operational tests.

In order to do this, bombers would have to fly a predetermined course and report their positions to ground controllers. It was also agreed that the trials should be undertaken on as realistic a scale as possible using entire squadrons or at least flights of both bombers and fighters. Since the latest generation of aircraft had not yet reached service, bombers would have to allow fighters a 30 to 40 mph speed advantage, which approximated the anticipated advantage the Hurricane would enjoy over German bombers. Tizard believed that these tests had to commence as soon as possible, and he offered to work on 'advising on the exact nature of the experiments' through his entire two month summer vacation if they could be started in early August. The experiments would require the full time attention of a sector commander, who would be provided with a fully staffed Operations Room.[8]

As a result of the DF tests, Biggin Hill was the only sector that fulfilled these requirements. On 27 July the Air Ministry approved Tizard's proposal and on 4 August the famous Biggin Hill Experiments, which would develop the techniques which would be crucial during the Battle of Britain, began. Placed in charge of the tests was Wing Commander E. O. Grenfell, commander of Biggin Hill. Assisting him in working out interception techniques was Squadron Leader R. L. Ragg, a navigation specialist and former experimental test pilot. The interceptors were supplied by No. 32 squadron flying Gloster Gauntlets, the latest biplane fighter, under the command of Squadron Leader Arthur McDonald.[9]

Testing began on 4 August, the first round apparently a repetition of earlier interception experiments, which involved having the fighters intercept the bombers over Biggin Hill. The fighters were already in the air, instructed to fly anywhere within a range of 25 miles from Biggin Hill, east of line between Hornchurch and Horsham. This line was selected in order that the fighters could be fixed by DF. The bombers were instructed to fly 40 miles south-east of Biggin Hill and ordered by radio-telephone to begin their approach to the airbase along a predetermined course at a continuous speed and height. The fighters were then fixed by DF, a procedure made far easier by having one of the aircraft's radios set to continuously transmit a signal which could be received by the DF sets, but at a frequency different than the other fighters' radios. After issuing the signal to proceed to the bombers, the control room staff calculated the course and speed required to bring the fighters over the airfield at the same time as the fighters. The fighters were tracked by DF and their course and speed adjusted in order to bring them to the airbase as close as possible to the same time as the bombers. At the moment the control staff calculated was the estimated time of interception over the base, the word 'Fire' was transmitted over the radio-telephone. Upon reception of this signal the fighter leader fired a flare. Observers on the ground then calculated the distance between the bombers and fighters. While these crude initial tests were a far cry from the realistic simulation desired by Tizard, they did begin the process of developing the tools required by the ground controllers to keep track of the fighters during an interception. For instance, Ragg developed special protractors and rulers that allowed the control room staff easily to record the track of the fighters on the control room map.[10]

Tizard was anxious to move on to the next stage of the experiments. On 7 August Tizard, Wimperis, Rowe and Grenfell met with the senior officers of the new Fighter Command, including the Commander-in-Chief, Hugh Dowding, to discuss a more realistic interception test. Fighter Command was established in July 1936 when the unwieldy ADGB command was broken up into four smaller mission-specific divisions: fighter, bomber, coastal and training commands. Dowding, only three weeks into his new post, was determined to command not only the operational units of Fighter Command, but also to continue to control the development of innovative technology and techniques that would make the air defences of Great Britain effective.

A near verbatim transcript of the crucial meeting of August survives and it provides a rare window into how Tizard and Dowding worked together to outline some of the most crucial experiments ever undertaken by the British armed forces. Tizard began by 'emphatically' disclaiming any desire to have executive control of the experiments. Grenfell was to remain in command but he was constantly to maintain 'direct touch' with Tizard.

There then followed a discussion on the nature of the experiments, with Tizard explaining that he desired to develop procedures that would allow fighters to intercept bombers at the coast, when radar lost track of them. Dowding expressed safety concerns about having single engine light bombers, like the Hinds, flying far out over the sea and proposed that the bombers only pretend to by flying over water, but in fact make their approach over land. Tizard agreed and told Dowding:

> The actual 'coast-line' is immaterial. Our first experiment is this. Fifteen minutes warning is given in which you get the first important warning of approach. If you want a rough position there I can give you an approximate distance up to 20 miles but no estimate of height. Five minutes later you get a position giving 5 miles error and height within 3,000 ft. error. Five minutes after that you get a 2 mile error and 2,000 ft. height error and after that one minute checks on this.

Using this information fighters on alert status would be ordered to take off and climb on a specific compass heading. As more information on the bombers' position was received, similar to that was hoped for from radar, the fighters' course and altitude would be altered. At the interception point the fighters would discharge a trail of smoke which the bombers would be able to use to record the closeness of the interception. Dowding agreed with Tizard's proposal and ordered the experiments to proceed immediately. The easy give and take between 'Stuffy' Dowding and Tizard is obvious here and would be one of the hallmarks of their cooperation in the years ahead.[11]

One other important result of this meeting was that it was agreed that a scientist on the staff of the Directorate of Scientific Research would be assigned to work in the control room with Grenfell and Ragg. Rowe selected Dr B. G. Dickins, then working on mechanical engineering problems at Farnborough, to join the Biggin Hill team.[12] Dickens would meticulously record the results of the experiments in a series of reports which make it possible to trace the evolution of the experiments in some detail.

The Biggin Hill experiments went through a rapid evolution over the seven month period in which they took place. Each test had certain common features. Fighters began the tests on the ground ready to take off within five minutes of the first warning that bombers were approaching. The object was to guide fighters from the ground so that they could intercept the bombers at or near an imaginary coastline. The goal was to improve interception techniques, which as they developed allowed for an increase in complexity of the next round of tests. Each experiment required a simulation of various aspects of the air defence system. As Tizard had described earlier, since radar would be unavailable, the bombers, positions were provided to the control room in a way it was hoped the new technology would supply information in the near future. Many of the tests were dummy runs, in which the controllers directed a completely imaginary interception. During tests using actual aircraft, the imaginary coastline would be brought closer to Biggin Hill to provide a similar time available for interception as would be the case when the new generation of faster service aircraft came into use.

By the end of August it became clear that DF was not ready to use for plotting the position of the fighters. The lack of sets, coupled with a need to use the few available to track bombers so that accurate simulated radar plots could be provided to the ground controllers, meant that an alternative system was developed. In this case, it was discovered that during the brief period required for interception, not more than half an hour, a fighter's position could be relatively accurately plotted on the control room map by a system of dead reckoning (DR). As long as the fighter leader and the control room carefully synchronised their watches, the fighters immediately obeyed instructions from ground controllers, and there was no significant difference between the winds affecting the bombers and fighters, DR worked relatively well.

At first bombers flew at a constant altitude, course and speed. As the controllers mastered this situation, the bombers were allowed to make a few alterations in course. In the next phase the bombers were permitted to make any number of course changes; then they were allowed to alter their altitude. In the final tests the controllers were required to control interceptions of multiple raids. Later dummy runs involved simulating the new generation of aircraft, with fighters moving at 300 mph and bombers at 240 mph. They also began incorporating inconsistencies and errors in the track of the bombers in order to reflect the anticipated limitations of radar.

A minimum of four staff were required, more if multiple interceptions were to be attempted. They included the fighter plotter and the bomber plotter, who recorded the tracks of aircraft on the map; the recording clerk who logged all relevant information and commands; and the sector commander, who controlled the operations of the fighters. He was provided 'with suitable facilities for estimating the necessary courses for the Fighters to steer in order to intercept the Bombers, and [was] in direct R/T communication with the Fighters.'

Those involved in running the experiments soon learned that the keys to a successful interception were precision and speed in recording information on the positions of bombers and fighters, and presenting it in a readily understood fashion. To allow for this the control room was dominated by the plotting board, a large 6 x 4-foot table 'upon which was painted in white a skeleton map of the area over which the operation was to take place.' The map was 'as free from detail as possible – a thin outline of the coast – the position of the aerodrome from which the aircraft are operating – the position of half-a-dozen prominent towns or landmarks (denoted by capital letters) – and a compass rose near the aerodrome and one round point of reference from which the position of Bombers may be plotted in terms of distance and bearing.' On one side of the map a chart was drawn on which the altitude of all aircraft could be marked.[13]

Improvement in the interception techniques was a collaborative effort of all involved, including Tizard who was at Biggin Hill almost daily in the first two months and then on a regular basis as the test progressed. The keys to placing fighters successfully in the path of attacking bombers lay in providing them with the right course to set, and the ability to alter it as the bombers changed their heading. At first they tried to use instruments or complex graphs to find the correct course. These methods proved too cumbersome. However, sometime in the last half of September, during only the second experiment in which the bombers were allowed to alter course, they discovered a much more simple, reliable and faster method – the principle of equal angles. *Radar + defence ang system in developing*

This principle of equal angles was suggested by Tizard as an easy way to calculate an approximate interception course. This involved the creation of an imaginary isosceles triangle. The base of the triangle was a line drawn between the position of the bombers and fighters; the course of the bombers formed the second side. The angle between the base and the bombers' flight path was equal to the angle between the base and the third side of the triangle, which provided the course for the fighters. This angle became affectionately known to some RAF squadrons as the Tizzy Angle. Since the fighters were travelling faster than the bombers, they always arrived at the interception point first, which gave the fighters a great tactical advantage. After only five tests using the Tizzy Angle, Grenfell discovered that in many cases the controller could estimate the approximate course, without actually drawing the triangle on the plotting board. The controller simply recreated the triangle to calculate a new interception course each time the bombers changed course. Almost immediately upon the introduction of Grenfell's method interception times decreased by several crucial minutes, *Key* even though the bombers were allowed to make more course changes during their approach then previously.[14] *Good stat*

By the end of 1936, the fighters were intercepting 90 per cent of all raids which maintained a constant altitude, no matter how many course changes were made. However, when the bombers altered their altitude the rate of interceptions declined to just 60 per cent. The main factor for this was that the more the bombers and fighters flew at different altitudes during the interception, the greater the difference in the relative wind speeds affecting the two formations. This threw off the dead reckoning position of the fighter formation sufficiently to lead to a large number of failures. It indicated that in the future sufficient DF sets had to be provided so that the fighter's position could be accurately plotted.[15]

In his concluding report on the first round of interception experiments at Biggin Hill, written in February 1937, Dickens judged that if RDF worked as promised most daytime standing patrols could be eliminated and most raids intercepted. Night-time interception remained a more elusive problem, since a successful interception during the experiments was said to have occurred only if the fighters spotted the bombers. During the day this could be more than 10 miles, and only rarely could the controllers get the fighters to less than 2 miles of the bombers.[16] Even at this distance fighters would be unable to spot night raiders. Yet all involved in the experiments considered that a major breakthrough had taken place. Those who came to Biggin Hill to observe their progress were aware of the significance of what was being accomplished. On 5 October 1936, Hankey witnessed two highly successful interceptions. He wrote to Tizard:

> I paid a visit yesterday to Biggin Hill and I saw a demonstration of the interception problem. I was very much impressed by the way the thing has been developed and with the degree of accuracy and apparent certainty with which the fighter seems to be able to intercept the bomber. Actually I watched the plot of two raids; each one intercepted within sixteen minutes, under conditions which were, I think, more exacting for the fighters than would be the case in practice.

Hankey was even more enthusiastic when he wrote to Swinton about his experience. He told the Air Minister: 'On Monday I went to Biggin Hill, where I witnessed a most impressive demonstration, more absorbing than any cinema or theatre; both runs 100% successful. If Tizard and Co. can get RDF right we shall have gone a prodigious way towards an effective defence. But it is going to take time. The enthusiasm at Biggin Hill was infectious.'[17]

While work on improving interception techniques would continue right up to the Battle of Britain, the basic techniques of the air defence system were worked out during these Biggin Hill experiments. If Bawdsey could get radar to work, it was now possible to develop a system to defeat daytime air attacks on Britain.

Bawdsey – A Year of Failure

At just about the same time the team at Biggin Hill were developing the techniques necessary for successful interceptions, the radar researchers were scheduled to begin the first operational trials of the new radar station at Bawdsey. Initially, the tests had been scheduled for June and were to involve three of the new radar stations. Repeated delays in construction, however, meant putting off the experiments. By the end of the summer it was apparent that only the Bawdsey radar would be anywhere near ready for operations in 1936. Watson-Watt was anxious to finally prove that his invention would deliver as promised and decided that no further delays were possible. Beginning on 17 September, Watson-Watt scheduled a series of seven trials involving radar tracking multiple formations of aircraft approaching Bawdsey from the sea. Always a showman, Watson-Watt threw caution to the wind and invited to Bawdsey as many VIPs as possible to witness the very first test. Included in the list were Swinton, Dowding, Tizard and Wimperis.

In the days leading up to the exercise Bawdsey was the scene of frantic activity. Three new receivers were completed and installed in the stables. A new transmitter was put in place in a hut on the grounds of the estate. Scientists and technicians rushed up and down the two 240-foot towers putting in place the antenna apparatus and connecting cables. At the same time work crews were still building the transmitting tower. There was not time to test or calibrate any of the new equipment, antennae or connections before the arrival of the dignitaries on the morning of the 17th.

Things began to go badly right from the start. Upon the arrival of Swinton, Watson-Watt took him and the other guests to the stables to view the receivers, neglecting to leave behind anyone to guide any latecomers to the demonstration. The last dignitary to arrive was Dowding, who finding no instructions at the front gate, wandered into the near empty manor house looking for Watson-Watt and the rest of the party. After visiting the lift shaft and the ladies washroom, he wandered into the office of the accountant who demanded to know who he was and what he wanted. Finally ushered into the stables, Dowding joined other guests standing behind and looking over the shoulders of Bawdsey's three best radar operators, Wilkins, Dewhurst, and a recent recruit to the program, R. Hanbury Brown. The three were sitting and peering intently into the cathode ray tube displays awaiting the first indication of approaching aircraft. Watson-Watt gave a running commentary about what the visitors could expect to see.

It soon became apparent that something was very wrong. None of the operators could detect any sign of even a single aircraft. Watson-Watt knew that a flight of nine large lumbering flying boats, followed at a ten minute interval by a single Anson maritime patrol aircraft, should have already been detected. He quickly switched from promoting radar to trying to explain the mitigating circumstances for their failure to track the approaching aircraft. However, Watson-Watt could not prevent the dismay and disappointment of his guests when they began to hear the approaching aircraft before a single useful radar track had been sent to fighter command headquarters. Hanbury Brown recalled: 'As the first major demonstration of the use of radar in air defence it could not have been worse; a good sound locator would have done better.'[18]

After this disastrous display most of the dignitaries departed. Dowding, however, stayed on for lunch. The food was awful and, because they were late, served cold. Also joining the luncheon was E. G. Bowen, who that morning had been working on the top of the manor's Red Tower oblivious to events unfolding below in the stables. He discovered the rest of the staff in a suicidal mood and Watson-Watt seething with anger.

Bowen had been making adjustments on another radar set intended to be used for his new project of designing equipment small enough to fit inside aircraft. He had built a large transmitter working along the lines of the Orfordness device, but which sent out signals at the then remarkably short wavelength of 6.8 metres. He coupled this with a new type of miniature receiver and display unit suitable for mounting on aircraft. As Bowen tuned his equipment that morning he suddenly saw 'a glorious display of echoes' filling his screen right out to the limit of the range of the equipment of 50 miles. His test equipment had detected the approaching flying boats. In an effort to cheer up Watson-Watt, Bowen told him of his success. Watson-Watt immediately insisted that Dowding accompany him and Bowen up to the top of the Red Tower to view the airborne equipment. Fortunately, everything was still working and at that very moment the flying boats were beginning their afternoon flights. Dowding was able to see clearly the echoes of the outward bound aircraft.

Still, this minor victory did little to salve the wounds of the Bawdsey staff, particularly when on the next day the main radar again failed to provide any operationally useful information during two more major tests. Various technical explanations were given for the failures. Bowen believed that the transmitter was not giving off any useful signals at the proper wavelength of 26 metres. Wilkins thought the trouble was caused by a new coaxial cable connection between the receivers and the antenna which he had recently laid down. Watson-Watt blamed the failure on the Works Department which did not finish the towers in time to allow for the antennae to be properly calibrated.[19] In fact, all of these were likely contributory factors, but the main problem was that Watson-Watt had displayed a complete 'lack of good judgement' in proposing the September exercises. No VIPs should have been invited until several successful tests had been completed. This was the gist of bluntly worded letter Tizard sent to Watson-Watt on 20 September. He warned that unless the situation changed he would 'have to dissuade the Air Ministry from putting up other stations.'[20]

Watson-Watt realized that he had to rectify the situation quickly. All the staff at Bawdsey worked night and day to get the system working. Further flight testing was delayed and only recommenced once a full day had been devoted to calibrating the new antennae using formations of flying boats. On 23 and 24 September the Bawdsey staff was able to redeem themselves partially in two simplified tests using single formations of flying boats. In both tests the Bawdsey radar was able to track the flying boats up to forty sea miles away. Range and bearings were provided at one minute intervals, and, while not always accurate, they demonstrated that with further refinement radar could be an invaluable tool. It was not possible to provide any estimate on the height of the targets.

Despite the better results Tizard remained dissatisfied and worried about the reaction of the air force. Certainly there was cause for concern.

Shoto Douglas, the Director of Signals, wrote in an internal RAF memo that he believed that the situation was caused by Watson-Watt and his staff trying to run before they could walk. If they did not learn caution, Douglas warned 'the indifferent results achieved may prejudice the Service against what promises to be a most valuable aid to interception.'[21]

In early October, Tizard asked Appleton, who had made five visits to Bawdsey since late August, to report on the situation. He explained that the September exercise results were affected by the attention of the staff being divided between large scale planning for the home radar chain and the erection of the permanent equipment at Bawdsey. As a result the equipment had not yet been thoroughly tested and provided inferior performance to the lashed-up equipment used the year before at Orfordness. He thought that the Bawdsey staff should avoid planning for the radar chain since 'at least a year's intensive research at Bawdsey would be necessary before general plans could be laid.'[22]

Tizard did not immediately push for any further explanation, but when the problems at Bawdsey persisted into early 1937 and when another series of air exercises scheduled for January was cancelled, he began to demand answers. Appleton, who was now acting as the Tizard Committee's liaison with Bawdsey, had no doubt that the problem lay squarely with Watson-Watt. He reported to Tizard that Watson-Watt's good qualities were well known, but that it was only now becoming apparent that no one had realized his weakness for always 'wanting to change course.' Watson-Watt was not satisfied that his staff was being doubled, nor was he happy not constantly to be doing new experiments. But, Appleton told Tizard, 'I don't think that anything is amiss except that WW is not building up an enormous organization as rapidly as he desires.' Watson-Watt also had some real limitations in what we would today call people skills. He felt the committee to be dictatorial and had problems with his relations with other higher officials. Rowe made similar observations after a two day stay at Bawdsey in late January 1937. He reported that Watson-Watt could not be made in 'either heart or body' to remain at the research laboratory and was spending altogether too much time at the Air Ministry involved in determining radar policy because he had no interest in the day-to-day management of the laboratory.[23]

There can be no doubt that Watson-Watt was responsible for many of the problems that bedevilled Bawdsey in its first year of operation, but there also were factors beyond his control. The most persistent headache was in finding suitable staff for the laboratory. Throughout 1936 Watson-Watt endeavoured to find qualified personnel without any systematic hiring programme or an advertising campaign. Instead hiring both within and without the civil service was done informally, often by personal contact.

Hanbury Brown was recruited by Tizard directly from Imperial College where he was a doctoral student. Brown had first encountered the Rector of Imperial College in 1935 when he was confronted about the amount of time he was spending flying as part of the new University of London Air Squadron. Tizard wanted to know how Brown, a young man of nineteen, could fly at times he should have been attending lectures. Brown explained that he did not have to attend lectures since despite his age he already had a degree in engineering and was working for his PhD in radio engineering. '"Ah," said the Rector, "in that case I think I have a job for you. Come and see me in a year's time", and off he went like a rocket.'

Brown thought Tizard would soon forget all about him but some three months later at the Royal College of Science their paths crossed again. Tizard told Brown that the Air Ministry was starting some interesting work and that people with his sort of qualifications were badly needed. He would not provide any details of the work but assured him that 'it was worth doing and an opportunity not to be missed.' He was told to report for an interview at the ministry in ten days time. No formal application form was required. To Brown it all seemed rather mysterious.

At the Air Ministry Brown was interviewed by Watson-Watt, who was already known by Brown because of his research in cathode ray tubes. 'As usual he talked a lot,' but he did not provide any details about the secret research. Two weeks later Brown received a formal offer for a probationary appointment as an Assistant Grade III 'at the princely salary of £214 per year!' If he accepted he was ordered to report to Slough. No indication was given as to the nature of the job.

Brown hesitated to accept mainly because the salary was low for someone who already held a first-class degree in engineering and it meant abandoning his PhD. Still Brown decided to put his faith in Tizard and reported to Slough. Brown was immediately told to report forthwith to Bawdsey. He arrived there on 15 August 1936 still not knowing what possibly could be going on in this 'fairy castle on a distant shore.'[24]

Other recruits to Bawdsey tell similar tales. In 1936, Donald Priest was a recent graduate of the University of London with an honours degree in engineering working for the Air Ministry on improving aircraft-to-aircraft communications systems. One evening he was instructed to meet with a Major Lamb at the officer's mess. While Priest feared that he might be fired, Lamb instead began by outlining Priests' qualifications, which included being an avid ham radio operator. Lamb then told him: '"Now Don listen carefully. We have just the right job for you. Show up here on Monday morning. Be prepared to drive 150 miles. You'll get sealed orders telling you where to go. Be prepared for a long stay. I think you'll love it."' The order contained directions to Bawdsey Manor. Within a week or two of having learned the details of the radar secret, Priest realized that Lamb had been right in saying that it was the job for him.[25]

Yet enthusiastic and well trained young scientists like Brown and Priest were few and far between. In 1936, war still seemed unlikely and many top notch university graduates were not inclined to accept the comparatively low wages of civil service scientists or to forego an academic or industrial career to serve the nation. Moreover, the demand for staff grew exponentially throughout 1936. At the end of 1935 it was estimated that Bawdsey laboratory would require in addition to Watson-Watt, six scientific or junior scientific officers, fourteen assistants, plus ancillary staff. By the end of March 1936, these figures had grown to ten scientific or junior officers and twenty-six laboratory assistants. In July, Watson-Watt asked that approval be given to increase the staff complement yet again, so that by mid-1937 Bawdsey would employ three principal scientific officers, nine senior scientific officers, seventeen scientific officers, ten junior scientific officers, and seventy-two laboratory assistants.[26]

The Treasury demanded that as much as possible recruitment to Bawdsey be made among existing government employees. The only exception was at the junior most levels where some recruitment from outside was permissible. This greatly limited the pool of available candidates. In the midst of the crisis caused by the failure of the September air exercises, Watson-Watt pleaded for permission to hire two scientists who would bring much needed industrial design experience to the Bawdsey team. The senior of the two was a Mr G. E. C. Bailey, former chief design engineer at the Plessy Company, then working as a consultant for Phillips and one of the leading experts on cathode ray tube design. Phillips had just offered him a long-term contract at £900 per year, but Watson-Watt believed he might accept a £750 per year offer to work at Bawdsey as a senior scientific officer. The second candidate was Harold Lardner, Bailey's assistant at Plessy and who had remained behind when his boss had left the firm. Watson-Watt was convinced that Lardner could be persuaded to accept a salary of £600, some £80 less than he was making at Plessy, to work at Bawdsey.

Bailey was never hired and his invaluable expertise was lost to the radar programme. It is unknown if the Treasury refused Watson-Watt's request, or if the scientist declined to take a 20 per cent pay cut to work for the government. Lardner, however, joined Bawdsey on 16 October. He soon became one of the most important members of the team.[27]

By January 1937, Bawdsey's scientific staff numbered some thirty-seven men. Of these fourteen had been transferred from Slough, six were old Orfordness hands and six more were recruited from other government departments including the War Office, the Admiralty and the Post Office. In total only eleven were hired from outside the civil service, and only one of these, Lardner, was employed at anything other than the most junior grade. There was a particularly acute shortage of senior-level staff. Tabor Paris, Tucker's former deputy at ADEE, was brought in as the sole Principal Scientific Officer even though he was also in charge of establishing the War Office's own radar research programme at Bawdsey. Paris was soon doing triple duty as the deputy director, picking up some of the administrative slack when Watson-Watt made his all too frequent trips back to London. Watson-Watt divided the staff into nine different research groups. These included two groups responsible for transmitter and receiver designs under the supervision of Lardner. Wilkins was head of the Aerials and Arrays group, Bowen the Airborne Radar team , and Dewhurst was made responsible for chain home radar investigations. The only other new senior recruit to the staff, Edmund Dixon, an experienced Post Office engineer, was placed in charge of Development and Production.[28]

A Year of Triumph

During the first few months of 1937, the Bawdsey staff began to put the failures of 1936 behind them. The work environment created by Watson-Watt's management approach, or perhaps more correctly, lack thereof, proved to be ideal for the mainly young and enthusiastic group of scientists and technicians isolated at the remote laboratory. Most of the men were single and lived in the manor house, which also housed the main laboratories until a purpose built structure was constructed in 1938. The married staff either lived in Felixstowe and took the ferry across the River Deben to work, or a lucky few were assigned cottages on the estate. There was none of the routine or rigid discipline of the civil service to hamper their research. They were not required to maintain a regular schedule, but their dedication to the programme led the scientists and technicians voluntarily to work far longer hours than the typical government employee. When the weather was good, time out was taken for a midday swim. Sometimes their activities were somewhat more sophomoric. The grounds were, and still are, infested with a huge number of rabbits. Bowen recalls that he and some of the other staff members took to keeping a rifle in their laboratory. When a bunny revealed itself on the lawn below they would open the window and open fire. In the manor's dining room, which was set up along the lines of an RAF officers' mess, no dress codes were enforced. After dinner some of the men worked in the extensive gardens trying to maintain them, played tennis, learned to rock climb on the cliff walk or had a cricket match on the magnificent pitch located just to the side of the main house. When the weather was not conducive to outside exercise they

Bawdsey Manor
Dining Room.

frequently engaged in a game of billiards on the exquisite billiard table
purchased by the scientists from the Quilter family for £25. After these
workouts many would then go back to work. Frequently, half the staff
would be working in their laboratories until well past midnight. Working
on the weekend was a normal part of the routine.

All of the memoirs of the scientists who served at Bawdsey during
Watson-Watt's tenure as superintendent speak of the comradery and sense
of purpose that infused their lives. Bowen wrote that the atmosphere 'has
been compared with that of an Oxford or Cambridge college, where some
of the best technical discussions took place late at night in front of a roar-
ing fire.' Looking back after more than fifty years, Brown recalled that
his first year at Bawdsey was one of the happiest he had ever spent. In
large measure the fond memories were more than just a product of the
unusual freedom they enjoyed. By the early spring of 1937, major progress

was finally being made in a large number of radar and related research projects.[29]

Bowen's group continued to lay the groundwork for airborne radar. Dewhurst's team was sent back to Orfordness in late 1936 to design and build a new chain home (CH) radar station designed to work at the much shorter wavelength of 13 metres. It was hoped that this wavelength would give superior performance and be less prone to the radio interference which continued to plague the set operating at 26 metres. Edmund Dixon was given the special task of developing techniques for quickly and reliably transferring data from radar stations to headquarters. A number of investigations were underway into ways to prevent deliberate jamming of radar signals. Still another project involved building a portable inland radar station. Other staff worked on building and installing equipment for the chain radar stations at Dover and Canewdon and fine tuning the one at Bawdsey.[30]

This dizzying array of projects was the direct result of Watson-Watt's continuing optimism for the long term success of his invention. Despite remaining unproven, Watson-Watt pressed for approval of further expansions of the radar chain. He won tentative authorization from the Air Ministry for completing two more chain stations by the end of the year using equipment assembled at Bawdsey. Approval of a further twenty stations (later reduced to fifteen), however, was withheld until the Bawdsey station could prove itself.[31] The next round of tests was scheduled to begin in mid-April. As the date approached the primary focus of the Bawdsey staff was in making sure there would be no repeat of the September fiasco. Everyone understood that these tests would be under intense scrutiny. Success would lead to the building of the entire chain home radar system; failure might lead to the outright cancellation of the program.[32]

Several important improvements in the design of the radar station were incorporated before the upcoming tests. It had become apparent that as the transmitting antenna was placed higher up in the towers, in order to increase maximum range, gaps appeared in the vertical coverage. These gaps were caused by the way radio waves were reflected by the ground, which resulted in a signal that was shaped like two giant tear drops or lobes. One lobe was from the direct transmission signal, the other the signal which reflected off the ground. In between the lobes a gap formed which was filled by installing a second complete set of transmitting aerials located part of the way down the mast. Radar operators were provided with a switch to change between the main array and the gap filling antenna. Another major problem that hindered aircraft detection was that transmissions from the antennae were sent out equally to the seaward or landward side of the station. This phenomenon had actually been useful

Work on the radar system was immense in the existing station. Developing the Chain techniques [handwritten annotation]

in some of the early tests, since many RAF aircraft were not equipped for trips over the sea, but it became a matter of some urgency that operators not be confused by aircraft flying behind the station. In late 1936, Wilkins developed a reflecting dipole antenna which focussed transmissions in one direction. The reflecting dipoles were installed by the end of March.[33]

One other important design change was also incorporated before the exercise. The work at Orfordness had proven that operating at a new shorter wavelength of 13 metres greatly reduced interference from other radio sources. This new frequency was adopted as the norm for all future chain stations.[34]

Proven-works [handwritten annotation]

The month-long air exercises commenced on 19 April. Two squadrons of aircraft were allocated to fly out over the North Sea in three or more formations. Although the results were far from perfect, the radar performed well, detecting most flights at long range, and in some cases as far as 100 miles away. While height finding was not always reliable, particularly at less than 8,000 feet, the radar was frequently able to give the altitude within 2,000 feet. Once detected, formations were tracked for long periods, usually enough to ensure interception using the Biggin Hill methods.

There were, however, serious problems. The radar operators could not provide accurate information on the number of aircraft. They often had difficulty in distinguishing between two or more closely spaced formations. Most of theses shortcomings were the direct result of the exercises simply being too complicated for the inexperienced operators and plotters. This became apparent on 23 April when a special demonstration was given to Swinton. Unwilling to risk repeating the earlier catastrophe, Watson-Watt saw to it that the demonstration took place only after several days of successful tests and that the Minister would see a simplified version of the programme. With only three widely spaced formations in the air, the Bawdsey operators had no difficulty in detecting and accurately tracking each one.[35] Rowe, in his official report, stated: 'There is presumably no doubt that the performance of Bawdsey on this occasion would have aided interception to an extent which three years ago would have seemed almost impossible.' Rowe ascribed problems in the more complex tests as being the result of the operators having limited experience with interpreting the CRT display when formations were close together and concluded that further training and testing would alleviate these shortcomings. The problem of bad plots was also the direct result of the lack of filtering of incorrect information before it was sent to the control room. Rowe recommended that a full-sized control room be built at Bawdsey to develop methods to combine radar information with the techniques developed at Biggin Hill and to train staff in filtering and other techniques. He concluded that 'nothing occurred during the Exercises to justify delaying work on a chain of RDF stations.'[36]

After the April exercise senior staff officers required little persuasion to approve the full chain. On 11 June the Deputy Chief of the Air Staff urged Air Marshal Sir Edward Ellington, the Chief of the Air Staff, to give immediate permission to begin work on the entire chain since it would be ready only in September 1939. If they waited 'until we have had trials with the five stations before deciding to go ahead with the chain of 20 stations [they would] not get them at the earliest before the spring of 1940.' On 21 June the Air Ministry approved the full scheme and sent the matter forward to the Treasury. Construction of the fifteen additional stations would cost £1 million, with a further £160,000 per year required to operate the entire chain. There was a pressing need to move as swiftly as humanly possible and no quibble over money; even these huge amounts that would have staggered the government just a few years before, could not stand in the way. On 12 August the Treasury concurred and approved the required funding.[37]

Up to this point the Bawdsey staff had assembled or were in the process of building all of the radar equipment for the first five chain stations. It was realized by late 1936 that as soon as the radar chain was approved private manufacturers would be needed to build the necessary equipment. In the procurement of other advanced technology military equipment, like aircraft or warships, private industry was contracted to manufacture entire weapon systems. The private companies often developed their own designs and technology to fit loose specifications set by one of the military services. When results from other government research programmes were utilized by a private manufacturer, there was generally full disclosure of information. Radar manufacturing followed a very different model. No one company was made privy to the entire radar secret. Instead a select handful of electronics manufacturers were given specifications only for individual major sub-components such as transmitters or receivers.

This compartmentalization of knowledge about radar stemmed directly from continuing concerns about secrecy. In late 1936, the ADRC recommended to the Air Ministry that it make Metropolitan Vickers the principal contractor for radar production. The company was one of Britain's most important electronics firm. Air Marshal Sir Wilfrid Freeman, the Air Member for Research and Development, however, 'expressed some anxiety of taking a firm which has such a widely distributed international relations' as Vickers into their confidence. Freeman continued to have reservations even after he was personally assured by senior Vicker's officials that no foreigners, particularly Americans, worked at the firm's research centre and that the company was quite capable of keeping secrets from its business partners in the United States. Freeman was finally persuaded to approve issuing contracts to the company when it was decided that it would only be given the responsibility for the manufacture of transmitters. Only the Vicker's research director would be briefed on the work at Bawdsey. He would be sworn not to reveal, to members of his own research and development team, any information not directly relevant to the transmitter project. Another firm would manufacture the receivers and still others would produce ancillary electrical components. The Air Ministry works department was responsible for supervising private construction companies to build the towers and buildings required.[38]

This arrangement left the Bawdsey staff with responsibilities that far exceeded that of ordinary scientists. They would be heavily involved in site selection, writing up specifications for all equipment, overseeing manufacturers, setting up the electronics at the stations and, finally, in testing and calibrating the radar. Moreover, by preventing open contact with industrial researchers and manufacturing experts, Bawdsey was deprived of much of the expertise that would be required for the smooth transition from research to development and from there to production. Eventually serious problems would result from this organizational deficiency.

None of these problems was apparent in the summer of 1937. Instead, the success of the April exercises was matched by other equally important advances during the rest of the year. A major redesign of future radar stations was under way that summer. The first five stations, now dubbed as being of an interim design, were being equipped with three 240-foot-high wooden towers: one for supporting the transmitter antennae and two for the receivers. Studies about ways to minimize jamming showed that each station would have to be able to operate at four distinct frequencies. The receivers and transmitters were redesigned so that switching between frequencies could be accomplished relatively swiftly, but there was no way to adjust the antenna arrays. Thus four separate sets of receiver and transmitting antennae were required, one for each wavelength. Given that this had to include the gap filling system for transmissions and the need to place antennae at different heights for determining elevation on the receiver array, the number of towers was now inadequate. It was decided to equip each station with eight, divided evenly between transmitting and reception. In order to improve range performance further, the Bawdsey staff designed a new transmitting tower built of lattice steel construction, some 350 feet high. Wooden towers continued to be used to support the receiver antennae. The additional towers had the additional advantage of making the stations far less vulnerable to critical damage from bombing.[39]

In late June, Bawdsey received permission to construct a new permanent building for full-sized filter and control rooms. In the meantime a temporary filter room was organized next to the existing small control room in the stables. According to Edmund Dixon, the filter room was intended to investigate

> the problem of ensuring that the most probable data on position, track and speed, and composition of each individual track and speed, and composition of each individual raid are extracted from the incompletely accurate and incompletely concordant reports of several RDF stations, and that the identification of defensive and friendly formations is made secure as may be.

Only this secure information was to be transmitted to the control room so that the controller was presented with an unambiguous view of the tactical situation. The development of filtering techniques was 'clearly one of considerable complexity, involving preliminary experiment and subsequent operational exercise,' in which filtering had 'to be done by personnel intimate with the relative dependability of each individual in RDF intelligence.' Experiments on filtering techniques would be matched by research on the best layout, equipment and techniques for the control room.[40]

In July, the second interim station at Dunkirk became operational. Unlike at Bawdsey, RAF personnel manned the new station from its inception. When Bawdsey was first established it was envisaged that the huge estate also be used to provide training for the large number of personnel to man the chain radar stations and operation rooms. In July 1936, Squadron-Leader Raymund Hart was assigned to Bawdsey to set up a radar training centre. Hart was among the best trained and most highly experienced signals officers in the RAF. He was one of two British officers sent to the elite *Ecole Supérieur d'Electricité* in Paris just after the end of the First World War. He graduated at the top of his class. Before his appointment to Bawdsey, Hart was assigned to the headquarters of No. 11 Group, where he became familiar with various aspects of the air defence system. Hart's training centre was running by February 1937. That summer Hart settled on a standard curriculum which required six weeks training at Bawdsey. Hart believed that a further three months of operational duties was required 'for a crew to attain a reasonable degree of efficiency in taking readings.'[41]

The Dunkirk station and the filter room were first tested in an air exercise held between 9 and 11 August. Also in use for the first time was the third station at Canewdon, which was located between the other two. The new station was only equipped with a transmitter, but Dunkirk and Bawdsey made readings using this signal, as well as their own. Hart ran the filter room, which received results from both Bawdsey and Dunkirk.

The resulting filtered tracks were transmitted to Fighter Command Headquarters and the new No. 11 Group, responsible for the defence of southern England. The Headquarters compared the tracks to those picked up at the coast by the Observer Corps. Both stations and the filter room performed well, despite the inexperience of the Dunkirk operators and technical problems at Bawdsey. It appeared that the faith in radar was justified.[42]

Progress with the chain stations was being more than matched by developments in other projects. Bowen's team working on airborne radar made remarkable progress in the first eighteen months of work. The technical complexity of developing airborne radar dwarfed that of the large chain radar sets. Miniaturisation was required in order that the sets took up no more than 8 cubic feet and weighed no more than 200 pounds. Only a limited amount of electrical power was available to power the radar. In order to accommodate an antenna that could be mounted on an aircraft and not produce a large amount of drag, the sets had to operate at wavelengths far shorter than was possible using existing technology.

Given these daunting requirements, Bowen's group made very rapid progress. They were greatly helped in this by the unwitting assistance of EMI, the recording and electronics giant. In the summer of 1936, a highly advanced receiver developed by EMI for experiments with television was made available to Bowen. The receiver initially operated on a wavelength of 6.7 metres, had excellent sensitivity and consisted of a chassis just 3 inches wide and 15 or 18 inches long on which was mounted seven or eight valves. Bowen never discovered how the receiver made its way to Bawdsey, but he suspected 'it came through the backdoor of the EMI Company.' Since Bowen was barred by security restrictions from directly soliciting EMI's assistance, he was not only deprived of their expertise but also of any further television receivers. Until the end of 1938 this one chassis was the only receiver available for airborne tests. When it was combined with a small CRT tube the whole apparatus weighed less than 20 pounds. The EMI receiver helped push forward the research by as much as two years, the failure to utilize the company's talents further would delay the successful development of airborne radar by almost as long.

By the late summer of 1936 Bowen had built a transmitter which could transmit pulses at 6.7 metre wavelength. At first it was not possible to miniaturize the transmitter sufficiently to fit inside an aircraft. Bowen set up this experimental radar in the top of the White Tower at Bawdsey Manor. It was this apparatus that Dowding was shown after the disastrous demonstration of the chain radar set in September. From the ground the radar could detect targets at forty to 50 miles range, but there was no proof it would work in the air. Bowen came up with a brilliant improvisa-

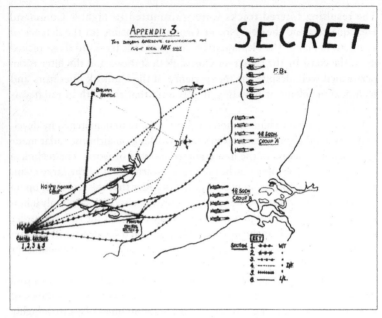

Diagram of radar exercise in autumn 1936.

tion. He decided that there was no reason the transmitter had actually to be in the aircraft; instead the airborne receiver would utilize radio pulses sent out from a transmitter on the ground. Bowen dubbed this radar RDF 1½, because chain radar was known by the code name RDF 1 and complete aircraft mounted set was dubbed RDF 2.

Testing of this hybrid radar commenced in the autumn of 1936 using a Hereford bomber which operated from the nearby airbase at Martelsham Heath. On its first flight the radar detected aircraft at between 8 to 10 miles distance. Bowen argued that RDF 1½ worked so well that it should be adopted rather than waiting for the development of a complete airborne radar system. Watson-Watt overruled him, citing serious difficulties in getting accurate range measurements unless the fighter was directly between the transmitter and an enemy bomber.

Bowen's group was, therefore, forced to continue development of the transmitter to make it small enough to fit inside an aircraft. In March 1937, Perc Hibbard, a member of the group, was able to build a small transmitter operating at 6.7 metres. That month it was tested in the Hereford and clear signals were detected from prominent objects on the grounds, such as wharves and the cranes at Harwich, at 3 or 4 miles range.

While technically successful, better resolution was required. This involved a variety of technical issues; one of the most difficult was to push the wavelength even lower in order that adequate transmitting and reception aerials could be fitted. Using new valves imported from the United States, Bowen's team was able to build a small transmitter which could send out pulses with the then remarkably short wavelength of 1.5 metres. The EMI receiver was modified to work with the transmitter. In the meantime the airborne radar group was given control of two Avro Anson maritime patrol aircraft.

On 17 August Dr A. G. (Gerald) Touch and Keith Woods, two members of Bowen's group, boarded one of the Ansons for the first flight of the new radar. They received clear echoes of ships off Felixstowe at ranges between 2 and 3 miles. On learning of these results Watson-Watt sought a way to bring the breakthroughs in airborne radar dramatically to the attention of the Air Ministry. An opportunity soon presented itself when he learned of an exercise scheduled for early September in which Coastal Command aircraft would attempt to locate units of the British fleet which were sailing through the North Sea from the Straits of Dover to Scotland. Without asking for permission from the Air Ministry, Watson-Watt ordered Bowen to see if his radar-equipped Anson could find and locate the ships.

In the early evening of 3 September, Bowen, Woods and Sergeant Naish, the pilot and experienced marine navigator, took off flying south-east along the coast from Martlesham to see if they could catch the fleet before it passed through the Dover Straits. At 6:50 p.m. 10 miles south-west of Beachy Head, Bowen looking at the CRT screen, saw very large echoes some 6 kilometres ahead. The radar had detected the battleship HMS *Rodney*, the aircraft carrier HMS *Courageous*, and four destroyers.

The next morning at the crack of dawn the three men again took off in the Anson to try to re-spot the warships. After being in the air for an hour and a half they detected a large echo 5 or 6 miles away. It was the *Courageous* and an accompanying destroyer. Upon approaching the ships 'all hell broke loose below.' Signal lights flashed in all directions, guns were fired – no doubt firing blanks – and aircraft started to take off from the *Courageous*. For the first time they detected echoes from aircraft which were between themselves and the sea at ranges of a little over half a mile.

Once they had observed the ships for some time they flew off to look for the missing battleship. After several hours of a futile search and low on fuel they began to head for home. It was only then that the radar scientists realized that the weather had soured and Naish informed them he had been unable to see the water for more than an hour. Dropping down to as low as 20 feet above the sea they were able to see a navigational buoy. Now orientated Naish could plot a course for home. Naish then climbed

*The anson
when a plane was
lost*

into the haze, levelling off at 12,000 feet. The radar let them know when they crossed the coast and using dead reckoning they estimated when they were near Martlesham. The plane was brought down to less then 300 feet before they could see the ground and find their way safely home. They had less than fifteen minutes of fuel left when they landed.

Upon landing, Bowen called the Air Ministry using a number given to him by Watson-Watt. He was connected to the Wing Commander in charge of the exercise. Bowen reported their sighting of the carrier. The Wing Commander confirmed the accuracy of Bowen's information, but demanded to know what squadron the Anson was assigned, since he had never been informed that the Bawdsey flight would be participating. The Wing Commander was astonished to learn that the Anson was operating independently. He was even more amazed when he asked Bowen why they had not responded to Coastal Command's radio signal sent out at about the same time they had sighted the *Courageous*. The scientists replied that their aircraft was not fitted with a radio. 'You will be interested to know,' the Wing Commander said, 'that due to the deteriorating weather the exercise was cancelled two hours ago.'

News of the success of airborne radar quickly spread. Bowen spent much of the next month giving demonstration flights to VIPs including Tizard, who was overjoyed with the tremendous progress that had been made. More important were the immediate results from these early tests. Bowen realized that airborne radar had two different but closely related applications. The first was a radar for detecting ships, dubbed ASV for Air-to-Surface Vessel. The second was for air to air interception, which was suitably named AI. ASV would be easier to perfect because it was discovered that the sea provided few interfering reflections. AI would be more complex because at first it appeared that the maximum range would directly depend on the altitude of the aircraft. The ground gave off a reflective signal which obscured the far weaker reflections from target aircraft at distances greater than the altitude of the radar.[43]

At virtually the same time as these historic flights were taking place yet another group of scientists at Bawdsey was beginning to show promising results. In this case it was the War Office researchers under Paris sent to work side-by-side with the Air Ministry staff at Bawdsey. The War Office's was interested in different applications of the new technology, particularly in directing anti-aircraft guns. In order to achieve the high precision needed for this task Paris' team turned to an idea Watson-Watt had first proposed in his very early memoranda. In order to maintain constant watch over a large area the chain system utilized an approach that Watson-Watt described as being akin to floodlighting. He also had suggested a system that would be similar to a single searchlight or the signal from a lighthouse using a rotating antenna. This is a type of radar we now are more familiar with seeing near airports. The prototype of this new radar was tested at Bawdsey in the autumn. The antenna was mounted on a platform located 100 feet up one of the wooden chain radar towers. In lieu of other targets, ships off the coast were tracked with ease. This left Paris to propose that, since the War Office was also responsible for coast defence, his team also be allowed to develop a radar for this purpose. By the end of the year the War Office had authorized Paris to proceed. Given the designation CD, this new project would have important ramifications for air defences in the years ahead. For once perfected, CD would form the basis of two vital components of the air defence system.[44]

By the end of 1937 Bawdsey had triumphantly recovered from the failures of the previous year. It had grown into a multifaceted research, development and educational centre on whose success the fate of the nation might very well rest. The scientific staff continued to grow until it reached forty members of Assistant III and above before the new year. In addition there was a large number of support staff and the nine War Office researchers. Joining them late in the autumn was Rowe, transferred to Bawdsey from the Air Ministry as Deputy Superintendent to bring order to the administrative chaos left by Watson-Watt's frequent absences.[45]

The work at Bawdsey and Biggin Hill meant that there was an excellent likelihood that an effective air defence against daytime raids could be developed to counter the growing might of the Luftwaffe. The only question was if there was sufficient time to turn this potential into reality before war commenced. Night raiders remained a major problem, however, since despite impressive progress, Bowen's AI radar was still only in the preliminary experimental stage. Senior RAF officers and their political masters would have to take this into account when making another round of important policy decisions about the size and scope of British air power as Germany continued to accelerate the growth of the Luftwaffe.

9

Appeasement, Air Policy and Air Defence Research

In late October 1937, the Air Staff prepared a report for Neville Chamberlain, who had replaced Baldwin as Prime Minister in the spring, on the RAF's situation if Britain were to be attacked on 1 January 1938. It painted a bleak picture of the war readiness of the nation's air force. Most squadrons' operational efficiency was reduced by the removal of experienced personnel to train new recruits and to staff new units. Out of Fighter Command's thirty squadrons only two were flying Hurricanes and there were only two Spitfires. The rest were equipped with a variety of already obsolete biplanes. The 'ideal' air defence scheme called for by the Home Defence Committee in 1937 required forty-five squadrons. The organization of the command was also deficient. By late 1937 the command consisted of three groups: No. 11 responsible for the south; No. 12 which looked after the midlands; and No. 13 for the north. Each group was to be divided into sectors which controlled two fighter squadrons. Only No. 11 Group was 'considered to be in a reasonable state of readiness.' The Group Headquarters for No. 12 Group lacked a control room and just one sector was fully equipped for war. No. 13 Group existed on paper only; even the site for its headquarters had not been selected.

Ground defences were equally unprepared. A balloon barrage designed to protect London from low flying aircraft was completely unready and the earliest even an improvised barrage could be ready was July 1938. Anti-aircraft defences were undergoing a major reorganization. Gone were the artillery zones of earlier schemes; instead, AA weaponry would be concentrated at vital points such as factories, ports or aerodromes. There were still very few modern guns. As an alternative to anti-aircraft defences for its airfields, the Air Ministry was actively pursuing a programme of acquiring satellite air stations in order to make it possible to disperse aircraft.

As for radar, only three stations, Bawdsey, Canewdon and Dover, were fully operational. One other at Great Bromley might be ready by the end of the year. Without further experimentation the 'information provided by these three stations would have only a very limited value.'[1]

The air striking force of bombers was also barely ready for war. There were a growing number of modern bombers such as the Blenheim and Whitley, but by far the most numerous was the obsolete Hind light bomber.

For the Air Ministry the short term prospects were gloomy, but it appeared that in the long term the situation might grow even worse because, despite increased production of modern aircraft, it was apparent that Germany was well ahead in the air armaments race. In 1937 and 1938, the main strategic concepts of the Air Ministry and its goals for rearmament would be challenged by Chamberlain and the Cabinet. The debate on air policy would be significantly influenced by the technological developments in air defence technology. Intimately connected to the air policy debate was Chamberlain's determination to pursue his infamous policy of appeasement with Germany and his desire not to financially ruin the country by excessive defence spending.

Air Defence and Air Policy

In the same month as the RAF reported on its war preparedness it proposed a significant increase in the rearmament programme. The Air Staff warned that unless major increases in aircraft procurement were immediately approved they would be unable 'to achieve that equality in Air striking power with Germany which represents the policy of His Majesty's Government.'[2] The Air Ministry proposed to rectify the situation with Scheme J. The new scheme called for an increase over previous approved plans of twenty-two squadrons (462 aircraft) of bombers and eight squadrons (112 aircraft) of fighters. Included was a whole new class of heavy bombers which would be capable of attacking well into the heart of Germany.

Chamberlain and Sir John Simon, the Chancellor of the Exchequer, balked at the huge cost involved. Sir Thomas Inskip, in his role of the Minister for the Co-ordination of Defence, had been trying to reconcile the demands of all three services for some time. He feared that the Air Ministry's requests would upset his efforts and could not be accommodated without significant disruption of the economy. For the next year the debate over the shape of aircraft rearmament would continue between the Cabinet and the Air Ministry. The champion of the air force's position until his abrupt resignation in May was Lord Swinton. His successor, Kingsley Wood, would prove an equally able advocate of the RAF's position. There is no need here to recount in detail the tumultuous debate which raged on as a series of air force expansion programmes were proposed beginning with Scheme J and continuing until Scheme M was approved by Cabinet on 7 November 1938. What is important to note, however, is that Scheme M marked a dramatic departure in policy by favoring the building of defensive fighters over strategic bombers and the abandonment of any notion of numerical parity with the Luftwaffe. The number of fighter squadrons would increase from thirty-eight to fifty by the spring of 1940. The bomber force would grow more slowly from seventy-three to eighty-five squadrons, with a completion date of this part of the programme being the spring of 1942.[3]

The debate took place against the backdrop of increasing German rearmament, the annexation of Austria in March 1938 and the notorious Munich accord in September which dismembered Czechoslovakia. Nor was the debate confined to the Cabinet room. Churchill and his supporters on the back benches of the Conservative Party would continue to demand an acceleration and increase of rearmament. They would be joined by the opposition Labour Party, which turned away from its support of collective security through the League of Nations to attacking the government for its failure to keep pace with the German arms buildup. In March, in the House of Commons Clement Atlee, the Labour leader, demanded that an official inquiry be undertaken on the slow pace of air force expansion. Chamberlain refused to hold an inquiry but keenly felt the pressure, particularly in Parliament.

In May, Chamberlain requested Swinton's resignation ostensibly because he needed an Air Minister able to represent his department in the Commons. Swinton was replaced by the able and enthusiastic Kingsley Wood. According to Swinton's biographer, the resignation was caused by a variety of issues of which air expansion was only one. Yet Swinton was placed in this position in large measure because he was unable to present to the nation the extraordinary developments in air defence research which were on the verge of providing unprecedented protection from bombers.[4]

The official historian of grand strategy, N. H. Gibbs, believed that the debate on the nature of air force expansion 'was largely unaffected by technical developments, radar among them.' He cited the lack of mention in the papers of the Cabinet, the CID and the Chiefs of Staff, which were the principal sources used for the study. In an unusual footnote the official historian, however, recorded that Swinton after having read the draft of the manuscript criticized the view that radar played no role in the decision to switch priority from bomber to fighter production. Swinton claimed that 'as soon as radar was discovered and proved, the theory that the only defence was counter-attack was dead.'

Swinton's view is unquestionably correct, for it is hard to believe that radar did not have a profound influence on those involved in the decision to give priority to fighter production. The existence of radar deprived the Air Ministry of one of its most effective arguments favoring the offensive strike force over defensive fighters. Senior RAF officers understood that once the radar chain was effective it would be possible to fight a defensive air campaign, yet they argued that you could not provide an effective deterrent or bring the war home to the enemy by defensive means alone. Nor is it surprising that radar is hardly mentioned in the high level planning documents. The building of the chain radar system had already been approved and there was little need to revisit the question in regards to air expansion. Moreover, the lower level internal memoranda reveal near unanimity of opinion in the Air Ministry and in Cabinet about the influence of the new technology.

In large measure the debate between the Air Ministry and the Cabinet reflected the public discourse in the Commons and in the press which also did not involve radar. This can be seen in the Cabinet meeting of 7 November 1938 which led to the approval of Scheme M. The lengthy discussion mainly involved the final outlining the Air Ministry's position and the government's reasons for imposing its own plan to concentrate on building fighters. There is only one passing reference to air defence research during the meeting and it came during the discussion concerning public relations. Sir Samuel Hoare, the Home Secretary, referred to scientific aspects of air warfare, and argued that since 'Air Defence was now a matter of the utmost importance, and he thought that perhaps even more attention might be paid to this aspect of the matter.' Wood replied that he would refer to 'this matter in is statement in the House of Commons' and that 'he had already taken steps to strengthen the staff concerned with scientific research.'[6]

Wood, however, must of had second thoughts about using the successful but secret research for political gain. There is no record in Wood's subsequent address to the House or in the ensuing debate of any mention of this classified work. Wood was well aware that one of the great successes of this period was the removal of serious discussion on air defence research from the public stage. In 1937-38, there was no mention in the media or the Commons of radar, Bawdsey or the work of the Tizard Committee. When air defence was mentioned it invariably involved mundane matters such as the number of barrage balloons, the amount of available anti-aircraft guns or various aspects of civil defence, such as the provision of gas masks and bomb shelters.

This veil of secrecy was accomplished despite the fact that all along the south and east coasts land was being acquired and huge towers were being erected. On 2 February 1938, in order to forestall any press disclosures, the secretary of the Admiralty, War Office, Air Ministry and Press Committee wrote to all news organizations to inform them of these 'conspicuous landmarks' that were being 'erected as a part of the air defence of Great Britain' but that would 'not be included in the Air Force List or other public documents.' The journalists were advised that 'the Air Ministry [was] anxious, in the national interests, that no reference to the construction, existence, function or purpose of these stations shall appear in the press, and that no photograph or films of them shall be taken or published.' Indicative of a time when journalists understood that national interest could take precedence over profit, the media complied with this request and posed no threat to the radar secret.[7]

With the Labour party joining the growing chorus attacking the pace of rearmament, the government took steps in February to ensure that Atlee was privately informed about the existence of radar. With the press and the official opposition muzzled, Britain's most important secret was removed from the debate. This of course precluded anyone in the government defending their record by mentioning the remarkable work being undertaken at Bawdsey. The secret would have been totally secure except that twice in 1938 Churchill would use the threat of going public with his privileged knowledge in order to try to coerce the government to increase the pace and scope of air defence research and to bring Lindemann back in from the scientific wilderness. Such a threat if carried out had the potential of crippling the new air defence system even before it went into operation.

While the development of the radar air defence system unquestionably played an important role in determining the shape of the Royal Air Force, there remains the larger question about whether or not it also helped shape government diplomatic policy during the crucial lead up to war. Watson-Watt, predictably egocentric, argued in his memoirs that Chamberlain 'was in large measure governed' by radar development in his decision in September 1938 to continue his appeasement policy and capitulate to German demands over the control of the Sudetenland region of Czechoslovakia. Watson-Watt cites as his sources for this CID reports in the summer of 1938, which briefed the Prime Minister on the military situation, and on the memoirs of Hoare about the shaping of policy in the inner-circle of Chamberlain's Cabinet.[8]

Hoare, however, makes no such claim. Instead he states that while he was 'conscious of the gaps in our air defence system and the need for time to repair them' the 'over-riding consideration' of the government was 'that the very complicated problem of Czechoslovakia ought not lead to a world war, and must at any price be settled by peaceful means.' The only way in which defence consideration determined Chamberlain's policy was in the geographic impossibility of preventing Germany from occupying any part of Czechoslovakia it desired.⁹

Chamberlain, however, must have been somewhat influenced by the realization that by delaying a military confrontation with Hitler for a year Britain would have an effective air defence system. Yet, it would be unwise to place too much weight on the influence of radar on diplomatic policy. If the Prime Minister was thinking about radar at all it was in the larger context of overall defence preparedness. There were many other more important factors that led Chamberlain to continue his ill-fated policy and historians continue to debate his motivations to this very day. What is certain was that the Chamberlain government was unwilling to defend itself from the criticism that followed the Munich agreement by revealing the nation's most closely guarded secret. Instead, the government would have to contend with trying to keep its fiercest critic from divulging the work of the Tizard Committee and, inadvertently, betraying the radar secret to the Germans.

Churchill and Lindemann Return

Of Churchill's many attacks on the government, none were as unwarranted as those involving air defence research. Using information from a variety of confidential sources from within the government and outside, he was well briefed on most issues. He continued to sound the alarm about Hitler's intentions and the need to take a firm stand on any further German territorial ambitions, particularly in the Sudetenland. His attacks on the Air Ministry in the House were a major factor in Chamberlain's decision to replace Swinton. While his personal attacks on the Air Minister were unwarranted, Churchill was quite right in his concerns that Britain was losing the air armaments race. Although his strategic vision was sound, Churchill was, however, misguided in his appreciation of the work being supervised by the Tizard Committee. In large measure this was the result of his almost total reliance on Lindemann's counsel in all scientific affairs.

On 16 May, just after Swinton's resignation, Lindemann sent to Churchill a copy of a possible letter to *The Times* or *The Daily Telegraph* about the shortcomings of radar. Lindemann's attack on radar was based on ideas proposed by his student R. V. Jones. Jones proposed a possible way 'to blind the RDF operator by strewing numbers of oscillators in the appropriate region.' Such oscillators of thin wires fifty to a hundred feet long' would provide a huge numbers of false echoes. Developed during the war under the code name Window, it was, at least at first, an effective radio countermeasure. Lindemann believed that the Germans also possessed radar and that it seemed absurd to think 'that the enormous volume of signals of peculiar type emitted by our station should not have been observed and interpreted by the German scientists.' Undoubtedly they would adopt some 'simple counter method' which would be disastrous if this is not taken into account. Bawdsey had in fact been working on anti-jamming for some time, but Lindemann did not consider that any such research might already be underway. Lindemann urged Churchill to approach Kingsley Wood about the slow pace of air defence research before 'Tizard and Co [can] explain to him that all is for the best in this particular world and it will be impossible to get him to stir things up.'[10]

On 9 June Churchill wrote an inflammatory letter to Wood in which he condemned virtually every aspect of air defence research. He started by expressing deep regrets that he and Lindemann had accepted Baldwin's invitation which had 'imposed silence' and prevented them from putting pressure on the government to speed research. Churchill informed Wood that, 'In all my experience of public offices I have never seen anything like the slow-motion picture which the work of this Committee [ADRC] has presented; and I fear it is typical of the whole group of committees which have been in existence during these vital years.' He attributed Lindemann's dismissal to his demands for 'more vigorous action.' Churchill warned Wood that if he asked ministry officials to respond to Churchill's charges 'you will no doubt be given excellent answers to these complaints and quite enough to baffle protest in Parliament, but this will not remove the glaring reproach that we have nothing to shield ourselves with at the present moment.'[11]

Wood forwarded Churchill's letter to the secretariat of the Committee for Imperial Defence for comment. Hankey took the lead in gathering evidence to refute Churchill's 'animadversions.' Unlike Swinton in 1936, Hankey was determined to expose not only the untruthfulness of the allegations, but also to use his vast personal knowledge of the last quarter century of the workings of the CID to show that Churchill's record of running military research programmes as a minister before and during the First World War compared unfavorably to those under the control of the Tizard Committee. The information was compiled into an exhaustive twenty-four-page report to be used as a basis for Wood's reply and a briefing document in case Churchill decided to make his charges public. The report covered virtually every aspect of air defence providing a detailed history of each major project.[12]

In order to refine his memorandum further, Hankey solicited comments from various individuals in the Air Ministry and War Office. He also felt compelled to advise Tizard of the contents of Churchill's letter. Tizard lost his temper violently. On 16 June he wrote to Hankey:

I have received Churchill's outrageous memorandum. He is hindering the work in every conceivable way & doing nothing to help it. Unless a very firm line is taken by Ministers I must resign as a protest, and all my colleagues will do the same. I feel that I should like to see the Prime Minister to explain this. Can you take me to see him, or is it unpolitic?

Four days later Tizard called on Hankey, still 'boiling with indignation at Mr. Churchill's letter.' However, in the course of their conversation, Hankey managed to get him to quiet down and promise not to inform the other members of his committee of the allegations and Tizard agreed to write a reasoned reply for Wood and Inskip.[13]

Tizard's reply was both indignant and sarcastic. Churchill was accused of 'deliberately closing his eyes to a development which has really effected a revolution and added to an immeasurable degree to our powers of defence.' Tizard strongly resented Churchill's 'continual pin pricking,' especially as he was in a position which enabled 'him to use large and poisonous pins quite irresponsibly.' If they did not force those working on air defence research to waste time answering his charges, Tizard claimed that he would find Churchill's attacks 'highly entertaining.' After all, one had only to contrast the developments since 1935 with 'the state of affairs when Mr Churchill was First Lord of the Admiralty' before the First World War. 'As a result of his total lack of real scientific imagination and foresight,' Britain had entered the war without any defence against submarines and without any method of locating them.[14]

Hankey used Tizard's letter and other responses to compose a second draft of his memorandum, now whittled down to a mere twelve pages. This draft formed the basis of the discussion at a meeting on 24 June held to write a response to Churchill. Present were Wood, Inskip, Hankey, Freeman, Air Vice-Marshal Peirse and Tizard. They decided that it was 'not yet in the public interest to divulge the epoch-making inventions that have been adopted.' As a result it was decided that 'the obvious tactic' was to play for time and to reply with a brief letter short on specifics, but which made a thinly veiled comparisons with earlier research programmes.[15] By withholding the details of his memorandum from Churchill, Hankey informed Chamberlain, 'the Government will be in a position to deliver a withering counter-blast if need be.' Chamberlain agreed with this approach since it was right 'to keep cards up our sleeve until we are ready to play them with effect.' A showdown with Churchill over air defence research now appeared inevitable.[16]

The expected confrontation did not immediately materialize. His bluff called, Churchill hesitated to go that extra step which would force him to reveal to the public the secrets he had sworn to keep. It cannot be ascertained if he was deterred by the veiled threats to compare the current research efforts with those of his previous ministries. Churchill knew that Hankey had more than enough information to embarrass him about earlier research programmes such as the tank. There was another controversy, however, that must have been instrumental in Churchill's decision not to bring his concerns to Parliament. At virtually the same time as he received Wood's letter, Churchill's son-in-law, Duncan Sandys, found himself embroiled in a controversy about revealing information protected by the Official Secrets Act. Like Churchill, Sandys was an MP, but he was also a junior officer in the Territorial Army. Using information he gathered in his military role, Sandys threatened to reveal information on the state of anti-aircraft artillery in a written question he submitted to Leslie Hore-Belisha, the Secretary of State for War. When on 28 June Hore-Belisha ordered that a court of inquiry be held to determine the source of the information, Sandys asked the Speaker of the Commons if he could be compelled to give evidence. This resulted in a Special Commons Select Committee being formed to investigate 'the whole question of the application of the Official Secrets Acts to Members of the House in the discharge of their Parliamentary duties.' The special committee did not issue its report until 18 October and until then the question of whether parliamentary privilege overrode the Official Secrets Act remained unclear.[17]

Churchill was implicated in the 'Sandys Storm' and he likely felt it unwise to cause an even more serious controversy. Instead he took more than a month to reply to Wood, writing on 26 July not with anger but with resignation: 'When I think what might have been done in this period to make us all safe, I find it difficult to express my grief.'[18] With this Churchill focused all of his considerable energy on the developing Czechoslovakian crisis, determined to persuade Chamberlain to stand firm against Hitler's demands.

It was only after the disgraceful Munich accord that Churchill once again turned his attention back to air defence research. Churchill's hand was greatly strengthened on 18 October when the report on the 'Sandys Storm' by the Commons Select Committee on the Official Secrets Act was published. The committee concluded that Hore-Belisha was responsible for a breach of parliamentary privilege when the army council ordered Sandys to testify about the sources of his information.[19] This appeared to free Churchill from the constraints of the Official Secrets Act.

On 30 October, Churchill wrote again to Wood. In his opinion the ADRC was failing and was 'in danger of becoming an actual barrier upon swift progress.' He offered no specific examples of the committee's shortcomings, but instead he ventured to offer a solution: the appointment of Lindemann to the ADRC. He told Wood: 'This would enable me, with his assistance, to put before you a continuous stream of valuable ideas and also to criticize with technical knowledge, which I do not possess myself, the progress or non-progress on existing lines of enquiry.' Churchill left no doubt about the consequence of failing to acquiesce to his demand:

> Unless I can feel that the work of this Committee is going to receive a new impulse from your own personal direction, I should consider it my duty to free myself from the silence I have observed for three years upon this branch of our defence and, with due regard to the necessity for secrecy, to endeavor to get something done by Parliamentary action.[20]

Lindemann, Tizard and the ADRC

Wood consulted several Cabinet colleagues about Churchill's demand. Inskip, upon learning of the threat, decided that if possible it would be in the government's best interest if Churchill could be accommodated. He realized the main stumbling block was Tizard, who would have to agree to again serve on the same committee as Lindemann. If Tizard resigned from the ADRC, he would also likely quit the scientific committee and take the other members with him. The scientists' work was too important to national security to risk this, and their departure might prove just as politically embarrassing as Churchill's parliamentary questions. On 15 November Inskip informed Tizard of the contents of Churchill's letter. Inskip indicated that he was inclined to appointment Lindemann, but only if Tizard agreed. Inskip tried to reassure Tizard that Churchill had provided an assurance on behalf of Professor Lindemann of his 'good behavior.' Inskip asked Tizard to meet with Wood and himself to discuss the situation. He concluded by telling Tizard: 'Mr. Churchill has his own methods but he is sincerely anxious to help and he feels very strongly that Professor Lindemann has a contribution to make which we ought not to lose.'[21]

The meeting was held on 21 November at Inskip's parliamentary office. It was a frank exchange in which Inskip and Wood admitted that 'the reason they wanted to give in was that if they didn't Winston Churchill would go off the Committee and "make a stink", and they [didn't] want a stink right now.' Tizard replied that if Lindemann was coming on, then he was going off. 'Whereupon they both completely defeated [Tizard] by saying that in that case they would not have him on and that they would let Churchill go off and stand the consequences.'[22]

Tizard asked for a few days to think it over, but he realized that Inskip and Wood had created a situation wherein his keen sense of duty left

Radar

him with no choice but 'to surrender.' Writing to Inskip three days later to confirm his reluctant consent, Tizard made one demand of his own: if Lindemann was to join the ADRC, then another scientist more knowledge-able on air defence issues must also be added. Inskip immediately agreed and accepted Tizard's recommendation that A. V. Hill be made the addi-tional scientific member.[23]

The government had avoided an open break with Churchill on air defence research. Churchill may have felt he had won a victory, but it proved to be a pyrrhic one. The main consequence of appointing Lindemann was the very rapid diminishing of the power of the ADRC. It ceased to be a policy making body and became one whose main focus was to provide an outlet for Lindemann and Churchill to vent their criticisms of the research programme. Thus Churchill, by using political blackmail, only succeeded in emasculating his own committee and enhancing the power of the Tizard Committee.

To Tizard the second showdown with Churchill and Lindemann was even more bewildering than the first and had a far less happy outcome. Tizard was deeply bitter about these events and believed that he was receiving very mixed messages about his own performance. Just two weeks before he was forced to accept Lindemann's appointment to the ADRC he had been asked by Air Marshal Sir Cyril Newall, the Chief of Air Staff, if the Air Ministry could have first claim to Tizard's service if and when war broke out. After brooding about the situation for several months, Tizard came to the conclusion that Lindemann's appointment was the direct result of his own inadequacies. He saw the problem being caused by the requirement for him to divide his time between Imperial College and air defence research. Tizard decided that he had better resign from the latter rather then fail at both.[24]

Freeman responded for Wood, attempting to soothe Tizard's wounded pride. He told him: 'It seems clear to me that we cannot afford to let you go.' Rather than stop working for the Air Ministry, Freeman suggested he leave the college and accept an appointment with the CID. Tizard knew that the Treasury would never allow an offer to be made that would come close to his academic salary, but he was suitably flattered and agreed to stay the course.[25] Tizard realized that he would soon be subject to another round of criticism by Churchill and Lindemann. He was now certain that with the support of the Air Ministry he could deal with any complaints.

The attack on the Tizard Committee was delayed by Inskip and Wood's decision to no longer hold regular meetings of the ADRC. After unsuccess-fully lobbying Wood to call a meeting where Tizard could be confronted directly, on 27 March 1939 Churchill wrote the Air Minister outlining his and Lindemann's evaluation of the situation. Churchill pleaded with

Wood to intervene and speed research. 'What we need,' he told Wood, 'is an exceptional priority. We go buzzing along with a host of ideas and experiments which may produce results in '41 '42 '43. Where will we find anything that can operate in June, July and August of 1939?' While not denying the importance of radar, he argued that immediate action was required on aerial mines or parachute bombs which would augment the existing defences. Churchill concluded:

> After nearly four years we have not been given any definite results. The story is lamentable, and only the public interest prevents it being told in Parliament.
>   Now I do beg you to use your authority. It costs so little; it is such a minute fraction of your sphere of responsibility. But unless you insist that the small primary steps shall be taken at once, day after day, nothing will happen except this vast prolonged grimacing.

On the surface there was a certain validity to Churchill and Lindemann's charges, with the exception of radar not a single significant technological development appeared anywhere near introduction into service. Even when the chain radar system was completed Britain would remain almost completely defenceless to night attack. Nor was their advocacy of aerial minefields preposterous, since many other experts considered they would have an important role in a well rounded defensive system. Yet Tizard and his supporters were confident that in any public showdown they could prove that the contention that research was not being undertaken with alacrity was false. It is only by considering the full scope of air defence research that the veracity of the two positions can be evaluated.

Invention and Contraptions

Tizard knew that his committee could withstand any outside scrutiny into its affairs because of the time and resources being devoted to non-radar research. While radar was always seen as the key to any air defence system, it was believed that it alone could not provide a means to destroy bombers at night. A Tizard Committee report on the state of air defence research produced in June 1938 lists fifty-eight different technical questions considered by the ADRC and Tizard Committee, only one of which was radar.[26]

Every possible avenue of research was explored in an effort to make another technological breakthrough. Remarkably, with few exceptions, every other major initiative taken by the Tizard Committee either failed or was not ready in time to be of use during the Battle of Britain. The non-radar research supervised by the committee has never been studied before.

When the success of radar research is contrasted with these other programmes, it plainly illustrates the extraordinary nature of the achievement the radar scientists. It also shows the key reason why the radar programme was successful; it was based on good hard scientific knowledge and well established technology. The developments at Orfordness and Bawdsey represent all that it is best about British experimental science, in the tradition of Rutherford and J. J. Thomson. The other air defence research reveals all the worst elements of British empirical science, experimentation based not on sound theory but on inspiration, endless tinkering and trial and error. Lindemann was quite wrong to charge that the Tizard Commitee was not vigorously pushing forward research. The problem was that few of the non-radar projects, including aerial mines, were based on good fundamental science.

The committee gave its support to any proposals that held out at least a small chance of improving the air defences. Yet it did weed out many hundreds of impractical, if not outright mad, submissions from inventors. Many proposed a form of heat rays to stop bombers. Most of these were simply tossed out, but at least one was given further scrutiny. In October 1936, the committee reviewed a proposal by the famous, but then aged, inventor Nikola Tesla to 'bombard aircraft with large numbers of minute particles travelling at very high speed.' The committee would have dismissed the proposal out of hand, but as a courtesy to the 'Hungarian Edison' a more complete evaluation was assigned to Blackett. He examined the details of the invention and reported back in December 'that there was nothing of value in them.'[27]

No record has survived about most of the weeded out proposals. However, Tizard kept one of them in his own papers. On 4 February 1937 Basil Liddell-Hart, the well known military writer and defence correspondent for *The Times*, wrote directly to Tizard at Imperial College enclosing a proposal for an anti-aircraft projectile he had received from the distinguished novelist and ingenious marionette developer C. S. Forester. Forester wanted an expert's opinion whether there was any value in his idea.

Forester's idea was 'to place across the path of an approaching aeroplane a cross or star of piano wire.' Spools of wire were to be placed inside a shell 'segmented so that it will break up with a minimum bursting charge at the top of its flight.' Rockets were to be attached to the end of the wires, the firing of which would 'proceed to pull out the wires to their full extent.' Forester concluded:

> I fancy that it might be possible by this method to stretch wires at least a
> quarter of a mile long in the path of the aeroplane. It is also just possible

that by the use of parachutes such as are employed in star shells the wires might be kept up in the air for a period long enough to increase their usefulness.

Tizard quickly wrote to Forester to tell him that his idea was 'no good.' It was enough to convince Forester that he had little future as a defence scientist. It was a good thing, for Forester returned to writing and later that year published the first of his famous Hornblower novels.[28]

A number of the inventors' proposals were sent forward to the Tizard Committee for discussion. Not all inventions could be so easily dismissed. In the summer of 1938, a Professor Goldschmidt came to see Tizard to describe his dynamic wire barrage. The device 'consisted of a continuous loop of wire travelling at high speed [which] was projected into the air, it being maintained in motion by passing over driven pulleys at the ground. By this method, the loop of wire would extend to heights of several thousand feet, any thickness of wire could be used, and the high speed would make it likely to cut into the wing of an aircraft which collided with it.' If practicable, the committee felt that 'the method had obvious applications for defence against low flying attack and would of course not be vulnerable to enemy action.' Goldschmidt was given a contract to construct a small scale model of the device at University College, London.

Goldschmidt met with the committee on 3 January 1939 and outlined his scheme for a full-sized prototype, capable of firing a ½-inch cable into the air and mounted on three railway trucks. Approval for the construction of the prototype was swiftly given, but five months later the decision was revisited when costs began to escalate. Goldschmidt had introduced a series of modifications to the device while it was under construction. The inventor was warned that while he could proceed, no more than £3,000 could be spent on the project.[30] After this date nothing more is known about the dynamic wire barrage.

Lindemann has often been blamed for initiating some of the more bizarre projects approved by the committee, but in truth he was more advocate of these technologies than originator. The one which he initiated was R. V. Jones' experiments with a short-range airborne infrared detector, an idea with a more secure scientific footing than most. This project commenced in late 1935. By the spring of 1937, Jones had built and tested an airborne infrared aircraft detector. The device, however, had limited range of around 500 yards, and would not work through clouds; nor could it provide any indication of the distance of the target.[31]

By 7 October further tests of Jones' infrared device had revealed that it could not be used to locate modern aircraft which had screened exhausts. The Tizard Committee decided that as a result of Bowen's successes with

AI radar, Jones' work at Oxford should be terminated. Jones agreed that his current detector was not practical but he asked for the opportunity to develop a new type of infrared detector that measured the difference in temperature between the aircraft and its surroundings. He believed that he could create a system in which pilots would be presented with the information in such a way that it would be 'akin to direct vision.'

Despite Jones' visionary proposal, the Tizard Committee ordered the project terminated in early 1938. Writing bitterly in his memoirs, Jones believed that the decision was in part caused by his association with Lindemann, the basing of the work at Oxford, and Watson-Watt's jealousy about a rival research programme challenging the supremacy of radar. There may be some element of truth these charges, but a more important a factor was that any practicable application of infrared was far in the future.[32]

Lindemann was not alone in promoting specific research projects. In late 1936, Blackett wrote one of his first papers for the committee on the defensive use of RAF bombers. He wrote to Tizard that 'England with her dense population close to the coasts and her democratic institutions, is far more vulnerable to [attacks on civil population] than, for instance, Germany, with a less dense and more highly disciplined and protected population.' He argued that the aggressor would always have the advantage. He proposed, therefore, a variety of ways bombers could attack other bombers by dropping bombs on them. 'It would', he wrote, 'avoid the social disadvantage of the present system, that 60% of the new air armaments have no great defensive value, except by the distasteful procedure of resorting to [attacking civilian populations], which may also turn out very ineffective.'[33]

Blackett's idea was not original, but an old concept revived in early 1936. ADEE was already hard at work developing an acoustic fuse that would detonate in close proximity to a bomber, greatly enhancing the chances of damaging the target. Blackett had little faith in any acoustic technology and was determined to find an alternative. It would lead to one of several very large non-radar projects supervised by the Tizard Committee. For while most of these bizarre ideas considered by the committee were dismissed with little or no discussion or minimal expenditure of money and resources, a few became major research endeavours.

Major Non-Radar Research Projects

On 25 August 1937 Blackett wrote a paper in which he calculated that if a fuse could be developed using optical means or sound to explode in close proximity to an enemy bomber, lethal damage during air-to-air bombing would be achieved 25 per cent of the time, as opposed to a mere 8 per cent using timed fuses. Blackett proposed that development of a new type of detonator which detected variations in the intensity of light should commence at once.

The development of proximity fuses for air-to-air bombing received support from senior air force officers. In early May, Dowding told Tizard that he agreed that RAF bombers might 'be able to take an effective and possibly a decisive part in the attack' of enemy aircraft, but that 'the effective participation of a bomber in the attack must depend on the production of the automatic fuse.'[34]

Research on developing a photoelectric fuse began at the Royal Aircraft Establishment by the beginning of 1938. Blackett took the lead in designing a prototype photoelectric (PE) fuse. It was hoped that the fuse would work using a photoelectric cell which would respond to changes in the intensity of light of as little as 1 per cent. Tests of the PE fuse begin at RAE in the late spring of 1938. A large bomber was flown at 200 mph 'over at different distances from the "eye" propped up at the angle' for a 2,000 feet drop.[35] By February 1939 the PE bomb was tested in a series of dropping trials. Two out of five bombs detonated as expected and it was felt that the causes of the three failures were known and could be rectified. As a result the Tizard Committee recommended that 400 to 500 of the bombs be produced.[36]

Despite this early promise, however, the photoelectric fuse failed to perform as anticipated. The PE bomb suffered from some fatal flaws. It was soon realized that even if the fuse could be made to function as intended, a new stabilized bombsight would have to be developed in order to allow the attacking aircraft to change course if the target aircraft took defensive action. Existing bombsights required that the bomber fly on a level and straight course and any change of direction or altitude of the target aircraft would make any such attacks futile. It was the lack of a stabilized bombsight that led Tizard to conclude in November 1938 that air-to-air bombing had little future.

Still, testing of the PE fuse continued because it was hoped to incorporate it into other anti-aircraft weapons. Problems, however, continued to plague the programme. Relying on light as a source of activation, no amount of experimentation could make the fuse useful for the night defences. The fuse also proved to be quite delicate, which made it impossible to install into conventional anti-aircraft shells. Problems also plagued Tucker's efforts to develop an acoustical fuse, which had the worrying habit of exploding prematurely. Both fuse projects were cancelled in 1940, giving way to a far superior third type of proximity fuse.[37]

This new fuse used a form of miniature radar. Work on the radio fuse began at Bawdsey in late 1938 under the direction of Australian-born scientist W. A. S. Butement. By the time the fuse was tested, the idea of air-to-air bombing had been abandoned, instead the fuse would be incorporated into anti-aircraft shells and rockets. The radio fuse was not to be tested until the summer of 1940, too late to have to play a role in the Battle of Britain. It would only be perfected much later on in the war.[38]

While the proximity fuse story ultimately had a successful ending, the development of aerial mines was a completely futile exercise. Research into both the short aerial mines and the long wire barrage was actively pursued after Lindemann's departure from the Tizard Committee in 1936. Between March and September 1936 a series of sixty-four flight experiments was made of the short aerial mine in which airplanes were flown into the mines consisting of dummy bombs suspended from parachutes. The analysis of these experiments led to the development of the ideal aerial mine consisting of a parachute 3 feet in diameter carrying 100 feet of high tensile steel cable and a bomb with a four ounce explosive charge. The whole mine weighed 3 pounds.

With the design of the mine determined, calculations were made on the quantities of mines which would be required in order that 'a reasonably large proportion of invading aircraft' would be struck by mines. The Air Staff were asked to examine the results and they were aghast at the large numbers of mines and minelaying aircraft that would be required to lay even a single field. They promptly cancelled the programme.[39]

Full scale testing of the long wire barrage, which involved aircraft towing behind them long lengths of wire, about to 2,000 feet in length, also went on during the summer of 1936. On 16 October Roxbee Cox reported to the committee 'that lethal damage was unlikely to be caused to modern aircraft by the means of a plain wire, except at very low altitude.' As a result this project was also cancelled by the end of 1936.[40]

The research on the short aerial mine, however, was not completely wasted, for it provided valuable data on the arming of balloon barrage cables. By February 1938 research on improving the lethality of the balloon barrage cables, combined with the improved interception techniques developed at Biggin Hill, revived interest in aerial mines. The balloon experiments indicated that it might be possible to use much longer wires of from 1,000 to 2,000 feet long. In the spring and summer of 1938, a series of experiments were held in which a plane was flown at high speed into dummy mines and the motion of the bomb was photographed by another aircraft. The experiments was designed to establish the lightest and longest cable which [could] be used, and also the conditions of impact under which the bomb [would] be likely to strike the colliding aeroplane.'[41]

By the end of 1938 the mine consisted of a 3-foot parachute and small one pound bomb at opposite ends of 2,000 to 1,000 feet of three cwt.

cable. Each mine weighed just 15 pounds. The mine worked by having the cable strike an aircraft's wing. The parachute would then drag behind the aircraft and pull the bomb up towards the wing where it would detonate. It was estimated that given recent improvements in the accuracy of interception, the minefield would have to be only 4 miles long, and the long wire dramatically reduced errors in altitude measurements. The barrage was described as consisting of '100 mines, spaced at intervals of about 210 feet.' It was estimated that the probability that an aircraft passing through the field of striking a mine and suffering lethal damage was only about seven per cent. 'The most serious factor of uncertainty' was the accuracy of interception that could be achieved, for if not correctly positioned the aerial minefield would be useless.[42]

By the spring of 1939 the design of the mine had gradually evolved and now consisted of three parachutes and a bomb connected by 2,000 feet of piano wire. Preliminary tests of interception techniques using 'minelaying' aircraft laying smoke to simulate the mines proved that it was more difficult than the normal interception technique.'[43] On 21 June 1939 the Tizard Committee assessed the situation. Tizard thought that the interception experiments showed that the mines would likely prove useless in defending large areas, but might prove valuable in defending isolated targets such as Malta. Tizard also related Dowding's view that the laying of a small number of minefields might have a 'deterrent morale effect.' The Committee recommended that a preliminary order for mines should be placed sufficient in quantity to give the manufacturer some incentive.[44]

In the next year progress on the aerial mine project was very slow. By February 1940 only 1,500 dummy mines had been constructed and 10,000 live mines ordered. Tests of interception techniques using obsolete Harrow aircraft continued to show that it was very difficult to position the minelayers into the correct dropping position. Since minelaying would only be practised at night, it was found that even when aircraft were in the correct position the minefield was usually not dropped at the correct time. The ground controllers could not get a clear three dimensional picture of the situation from radar and frequently mistimed the drop.[45]

It is likely that as the Battle of Britain began the whole project would have finally been cancelled, but by this time Churchill was Prime Minister and Lindemann his scientific advisor. Lindemann persuaded Churchill to order one million mines costing between £7 million and £10 million. A flight of minelayers were formed on 25 September 1940. Despite an almost complete lack of success in laying mines, the only results being several premature explosions which damaged minelaying aircraft and one dubious claim of a German bomber brought down on the night of 26 October, the flight was increased to a full squadron of twenty-four aircraft by December. Lindemann forced further tests in 1941 with the latest American-built Havoc night fighter. These proved that the weapon was useless. It was almost impossible to position minelayers correctly, and even when this occurred, the mines rarely worked as intended. Attempts to develop other delivery systems were equally futile. On the night of 27-28 December 1940, 658 mines attached to balloons were released over London, but the enemy ceased operations before they were fully deployed. In total the known casualties attributed to aerial minelaying was one doubtful claim of a single aircraft brought down, two horses injured by tripping over wires from the mines, and a few civilians hurt when the bombs exploded upon contact with the ground after the balloon experiment. Finally on 19 November 1941 the aerial minefield programme was cancelled, a total and ignoble failure.[46]

Little more successful was the third major attempt to develop new anti-aircraft weapons – the unguided anti-aircraft rocket. The rocket project was code named UP, for unrotated projectile, 'a descriptive name adopted at an early stage in the development' of the weapon. UP research began on 17 April 1935. The rocket was intended to be simpler to produce, more mobile and less expensive than the new 3.7-inch gun. Dr A. D. Crow, the Director of Ballistic Research at Woolwich, was placed in charge of a small team of three scientists to build a prototype.[47]

By 1 April 1936, Crow had developed a theoretical model and 'a mass of experimental work had been carried out' which proved that it was possible to build a simple solid fuel rocket propelled by a new special type of cordite, SC (Solventless Cordite). Approval was given to construct new laboratory space and to begin hiring additional staff. In November, the first of the new 2-inch UP was fired. The rocket weighed just 9¾ pounds, of which 2 pounds formed the warhead. The 2-inch UP was designed to have equivalent performance of the 3-inch AA gun.

By September 1937, a working design of the 2-inch UP was ready, but at this point the performance of the smaller rocket was judged insufficient to bring down modern bombers. This was not surprising, given that the 3-inch gun, whose performance the rocket was trying to emulate, had long since been considered obsolete. When this obvious error was pointed out by the ADRC, the Master General of Ordnance informed them that Crow had proceeded with the small rocket because 'it fitted in with the safety restrictions imposed by the then existing Woolwich facilities and, at the same time, was calculated to give a reasonable range and to provide a stepping off point for work on larger Rockets.'

Crow was instructed to begin designing the much larger 3-inch UP. The new rocket was 3 inches in diameter, 4 inches long and weighed 38 pounds, including 13 pounds of SC cordite. It was intended to be effective against aircraft from 6,000 to 27,000 feet. By June 1938, sixty scientific and technical officers were working on the project at Woolwich and a new laboratory at Fort Halstead. Testing commenced on the 3-inch rocket in September and continued into the summer of 1939. By May 1939, some 3,300 3-inch UP rounds had been test fired. Technically the rocket performed very well, but it was proved inferior to conventional AA gunnery. Yet the rocket certainly had some advantages which delayed the cancellation of the programme before the war. Its warhead contained some 4½ pounds of TNT, double that of the 3.7-inch AA shell, giving the weapon a substantially greater killing power. Since the rocket had a comparatively slow acceleration, the PE fuse, which was too sensitive to be fitted into a shell, might be utilized. Moreover, an aerial mine could be fitted into the rocket's warhead. Even more bizarre schemes were actually tested where 2-inch UPs were used to launch a wire curtain of cables supported by parachutes ahead of low flying aircraft. Of course few of these developments had any chance of success since they were based on useless technologies. Still the rocket programme continued and a number of them were pressed into service to defend airfields against low flying attack during the summer of 1940 and British cities from night bombers during the Blitz. A few German aircraft were brought down by the rockets and even the wire curtain, but on the whole the AA rocket's performance was markedly inferior to the new rapid fire Bofors gun against low altitude aircraft and radar controlled 3.7-inch guns for higher flying targets.

Rocket research, like the proximity fuses, would at least useful in the long run. In 1940, Crow proposed that the 3-inch UP be developed into an air-to-ground weapon. This proved to be highly successful when introduced into service in 1942. Aircraft equipped with the rockets proved excellent tank killers in the Middle East and subsequent campaigns. The weapon also proved very effective against submarines and in other anti-shipping roles.[48]

Of all the major non-radar air defence research projects, the largest and most costly was the Silhouette Detection scheme. Silhouette detection originated from inquiries made by the Tizard Committee into the effectiveness of searchlights. In September 1935, a Flight Lieutenant Acherely reported to the committee that a bomber could be seen as 'a silhouette against a layer of mist or dust particles illuminated by searchlights which had themselves been unable to detect the bombers.'

By the spring of 1936, RAE had demonstrated that silhouette detection appeared to work in laboratory experiments. The Tizard Committee recommended that a full scale test of an illumination scheme covering 10 square miles of sky should proceed. These tests were undertaken in the autumn, although sufficient searchlights were available to illuminate just 4 square miles using 100 kilowatts of light per square mile. The experiments showed that it was possible to see aircraft flying above an illuminated 800-foot-thick layer of clouds. Fighters flying 5,000 above the test aircraft could clearly see its silhouette against the bright clouds. There were serious limitations with the test, particularly the small area illuminated, which precluded any attempt to see if the fighters could intercept the silhouetted aircraft. A. V. Hill proposed, however, that a much larger test area of about 200 square miles using a new type of more powerful searchlight be constructed. This test area would be constructed south-east of London so that if the scheme proved effective the test area could be incorporated into a full scale illumination zone defending the capital.[49]

Hill's proposal set in motion serious discussion in the Air Ministry during the winter of 1936-37 about whether or not to proceed. At issue was the huge cost associated with setting up the test area, estimated at some £400,000, the subsequent expense of fully defending London if the project proved practicable, of between £4 million and £7½ million, or, if desired, the staggering price of at least £12½ million to cover much of the industrial heartland. This compares to the mere £1 million approved in the summer of 1937 to complete the chain radar system. Silhouette detection was so expensive because it was a huge electrical and civil engineering project. The test area would require fifty specially built 400 kilowatt lighting units, each capable of giving illumination over 160 degrees covering a cloud area of 4 square miles. Every searchlight would need a one acre site and accommodation in a glass roofed building. To provide the electricity, overhead high tension lines would have to be set up from existing substations.[50]

In order to try to persuade the Air Ministry, the Tizard Committee found that on average there were 111 nights per year where cloud cover was of the right height for illumination to work. When objections were raised about the huge amount of power required, the committee pointed out 'that the availability of power might have applications in peace time, e.g. for the electrification of farms, etc.'[51] This was sufficient to convince the Air Ministry. In April 1937, Swinton requested that the Defence Policy Sub-Committee of the Treasury support the request for the £400,000 for the test area.

The Treasury baulked at the huge costs involved in making the system operational if the experiments proved successful. The Treasury only finally agreed to provide funding in October 1937 after receiving assurance from the Tizard Committee that the project was 'highly desirable.'[52] An area on the north bank of the Thames in the eastern part of the North Weald sector was selected. In January 1938, a contract was tendered to General Electric to build a prototype searchlight. By November 1939, one month after the war began, the test area was finally ready for trials.

Yet even before the test commenced doubts, about the scheme were being expressed. Even if successful, the test area was tactically of limited value and the reduced London defence scheme would take at least another year to complete. However, setting up the searchlight stations was only one issue. As the Air Ministry reported in November 1939: 'Special provision for the large electricity supply necessary would have to be made, and this would be likely to be the governing factor in the time required to complete an installation.' Referring to the Air Ministry report, Major General H. L. Ismay reviewed the programme for the CID. He concluded: 'I am inclined to be very doubtful as to whether this scheme will be worth while [sic], not only because of the time lag, but also because of developments in other directions.' These developments were, of course, airborne radar which by autumn 1939, it was believed, rendered silhouette detection an unnecessary and costly failure.[53]

Of all the non-radar major research projects overseen by the Tizard Committee only one had any significant influence on air defences during the first two years of the war. From its inception, the committee took the lead in redesigning balloon defences. The balloon apron barrage of the First World War was useless against the far sturdier aircraft of the 1930s. A new style of balloon was designed by 1935, but there was much technical uncertainty about how it would perform in practice. The Tizard Committee oversaw an ambitious research programme which including assessing the strength of cable required to damage aircraft, the development of special wing cutting devices and methods to place bombs on balloon cables to ensure the destruction of any aircraft that ran into it. Overall the research programme significantly enhanced the effectiveness of the balloon barrages which protected London and other cities, and smaller vital sites from low altitude attack during the Battle of Britain. As a strictly passive means of defence, however, balloons were more useful for boosting morale as a visible symbol Britain's resolve rather than an effective anti-aircraft weapon.[54]

Lindemann and Churchill – The Final Round

Given the lack of results of non-radar research, it superficially appears that Lindemann's and Churchill's charges that the story of air defence research was 'lamentable' was accurate. However, it is one thing to point out the flaws, it is another to have any constructive solutions. Almost all of Churchill's scientific and technological concepts stemmed from Lindemann. Lindemann's ideas were not only unhelpful, they were counter-productive.

In March 1939, Lindemann launched his final offensive against the Tizard Committee by attacking the aerial mine programme. In his 27 March 1939 letter to Wood, Churchill enclosed a memorandum from Lindemann on the mines. The Prof envisaged using thirty minelaying aircraft flying at 20,000 feet over the sea to lay a curtain of mines 19 miles long. Radar would be used to guide the aircraft to the best location for laying the minefield. He explained: 'To keep such a nineteen mile curtain alive would require about ten tons of bombs per hour, containing five tons of explosive. Reckoning a six-hour night, this would imply 250,000 mines and 30 tons of explosive.' Such a curtain, Lindemann believed, 'would bring down one in five of the aeroplanes making the double journey.'[55] Lindemann wanted to drag the whole programme back two years and restart the development of short aerial mines. He believed that the thin wire which made up the long aerial mine could be easily cut by special devices attached to an aircraft's wing. 'The 2,000 ft. mine,' he wrote, 'has a 20 times greater chance of collision than a 100 ft, mine, but weighs some 20 times more.' According to Lindemann to equal the chances of hitting with the shorter mine, it was simply necessary to drop an equal weight of mines. Lindemann had designed a new type of miniature mine, which he dubbed the 'minelet', which would be used with the short-wire device 'for attacking propellers.'

Lindemann's proposal was never challenged. Instead, with Lindemann now on the ADRC, Wood decided that a new tactic would be used to control him and his political friend. Rather than instigate another debate which had the potential of either spilling out into the public domain or see Tizard and his followers resign, Wood decided that it was time to brief Lindemann fully on aerial mine development, invite him to upcoming tests and to allow his ideas to be put to the test. He ordered that the minelets be given a full scale trial.[56]

Right from the start there were problems with the minelet scheme. Lindemann delivered the minelets to Woolwich for tests in March. There were two types: 'one with a specially sensitive filling requiring neither fuse nor detonator, intended to go off on impact with an airscrew blade; the other with detonators but also with no fuse or safety arrangement of any kind.' On 17 April Pye reported that the devices were very dangerous because 'their use is based on a principle fundamentally different from that accepted for all other forms of bomb' and that any testing of them would require special permission from the Ordnance Board.[57]

When Lindemann learned that the superintendent of the test centre was insisting on following 'the most rigorous precautions,' he responded: 'In peace time, of course, this is a good thing, but it seems doubtful today whether a million to one chance of injury should be allowed to hold up progress in some appliance which might save large numbers of lives in war.'

At last in mid-May the minelets were tested by exploding them against fixed propellers. Lindemann had to admit to Churchill: 'The effects on metal air screws was disappointing. They were dented slightly, but not damaged appreciably.' Still Lindemann persisted in holding out hope that the minelets would be useful. He concluded: 'The effect was much more satisfactory on a composite wooden-screw such as will probably be used in Germany after the first few months when they cannot replace duralumin screws.'

With Lindemann and Churchill thwarted for the time being, Wood suggested to Tizard that he invite Churchill to Bawdsey to see the latest radar developments. Churchill arrived at Bawdsey on 20 June for a full tour. Tizard recorded in his diary that 'the demonstration went very well and he got a clear grasp of the method. He is still very keen on minelaying and attaches an exaggerated importance to it. We are going to have a talk about this later.'

Tizard was quite right; Churchill was truly impressed by what he saw. On 27 June he wrote an enthusiastic letter to Wood about his impressions. Although he still urged that work be sped along on aerial minefields for the first time he praised the air defence research programme and expressed some hope that Britain might survive a German aerial assault. He finished his letter to Wood: 'Finally, let me congratulate you upon the progress that has been made. We are on the threshold of immense security for our island. Unfortunately we want to go further and the threshold and time is short.'[58]

What finally converted Churchill was that in the last eighteen months of peace scientists at Bawdsey working in close cooperation with Fighter Command had turned the promise of early radar research into an almost fully functional daytime air defence system. The failures of non-radar research were in large measure irrelevant because radar combined with radio direction finding and the new Spitfire and Hurricane fighters made it more than likely that the Luftwaffe could be defeated if it launched a daytime strategic air campaign against Britain. Airborne Interception and Gun Laying radar appeared ready to solve the night air defence problem. In order to understand how this miracle was accomplished it is now necessary to return to Bawdsey.

10

Forging the Radar Chain

In the last months of peace finishing the radar chain became of paramount importance, as no other practical new anti-bomber technology had been found. In fact, once the government authorized the building of the radar chain in the summer of 1937, Bawdsey's scientists increasingly devoted themselves to the task of completing it and integrating the scheme into the air defence system by the late summer of 1939.

The tremendous effort of at Bawdsey, however, was only one aspect of the story of forging the radar chain. Great efforts were also being made by others involved in developing systems and techniques, and in training the necessary personnel to perfect the new air defence system. It was a concerted effort wherein air force officers and scientists had to work together to an unprecedented degree to make the system operational before British cities were subjected to German aerial bombardment. Because other histories have overlooked this crucial period of development, the story of the frantic efforts to safeguard Britain's skies has never been told before.

The Directorate of Communications Development

By the end of 1937, just three years after it had been invented, radar could no longer be confined within the Bawdsey estate, nor could the radar laboratory continue to be run in such a haphazard manner. The days of Watson-Watt's 'hands off' approach at Bawdsey were numbered as soon as the radar chain was approved. The first sign of this occurred in the autumn of 1937, when Rowe was transferred to Bawdsey to act as the Deputy Superintendent. Within eight months of his arrival at Bawdsey, Rowe would find himself the Superintendent, responsible for the day-to-day running of the growing research, development, construction and training establishment.

When discussion began in late 1937 on how to restructure the radar programme, it was immediately apparent that a completely new level of senior management was required at the Air Ministry. The number of scientists and staff at Bawdsey, around 180 by the spring of 1938, was now as large as all of the combined research establishments under the control of the Director of Scientific Research. The DSR was responsible for research only and was not supposed to be involved in supervising the construction of weapon systems or of bases. The building of the radar chain required a massive increase in planning and policy making. The DSR and his staff had been stretched to the limit supervising Bawdsey when it was only engaged in research. The development of the air defence system also required that someone coordinate radar work with research in other related technologies, including radio direction finding, aircraft radio-telephones, and information management. Since the creation of the position, the DSR had been specifically excluded from supervising radio related research, a responsibility given to the Director of Technical Development. Moreover, the War Office and the Admiralty were also developing radar for their own purposes and there was a need to create mechanisms so that inter-service competition over scarce components, such as radio valves, did not occur. The decision was made by the Air Ministry in the early spring of 1938 that an entirely new directorate, responsible for managing the radar chain, should be established.

Staffing the new position would be difficult as there were very few suitable candidates. In November 1937, Pye argued that 'the magnitude and variety of the problems to be dealt with justifies the status of Director for the man responsible, and it is further to be considered that the scientific and technical [knowledge] required in such a man is so highly specialized as to make it unlikely that anyone could fill the post adequately who had not made it his life's work.'[1] The only possible person who fulfilled all these requirements was Watson-Watt, since most of the other possible candidates – Appleton and Tizard – were outside the ranks of the civil service and unlikely to give up their academic positions, which paid much more and carried far greater prestige, to work full-time for the Air Ministry. Rowe, the only other civil servant who might have been suitable, was needed at Bawdsey to replace Watson-Watt and was far too junior for such a senior position. In fact, when it was proposed that Rowe replace Watson-Watt as the superintendent of Bawdsey, the Treasury initially objected on the grounds that this would require promoting him over several deputy superintendents at other establishments. Only forceful intervention by Pye, citing the national importance of radar, overcame the Treasury's objections.[2]

The Treasury also raised strenuous objections to appointing Watson-Watt at the director level, arguing that a new deputy director in charge of radio under the control of the DSR was more in keeping with established organizational models of government scientific research. When, in early April 1938 Watson-Watt learned of the Treasury's position, he was furious and immediately wrote to Tizard:

> I admit that a sufficiently noble character would fix attention only on the national interest, and forget alike this gossip and those 'preventative men' in secretariats and treasury, but even after a cheerless and nearly sleepless weekend I doubted whether I can claim such nobility! In any case the Deputyship is no solution; Director, superintendent or nothing are the possibilities.[3]

The Air Ministry agreed with Watson-Watt and pressed the Treasury to reconsider because of the great importance of the programme. It took several months, but in July 1938 the Treasury finally capitulated and Watson-Watt received his long coveted directorship. The new position was titled the Director of Communications Development (DCD), a name chosen deliberately to hide the true nature of Watson-Watt's responsibilities.

Watson-Watt, while at Bawdsey, had already demonstrated his lack of interest in the mundane details of day-to-day administration. Bawdsey had flourished under his leadership because he inspired the staff by his ability to pitch in and work beside them in the laboratory. Watson-Watt's unbridled exuberance and verbal karate made him a highly effective advocate for the new technology at meetings. Yet, the transformation of the radar programme from research to development meant that the time for this laissez-faire approach to management had to come to an end, and there was a need not just for someone to advocate for new policies, but who also had the ability to follow through with the implementation of these policies. This can be seen in the rapid growth in responsibilities given to the DCD.

Initially, the DCD was responsible for research and applications of radar and related technologies for the Air Ministry. By the summer of 1939, Watson-Watt's duties had expanded so that he

> acted in effect as scientific advisor to the Air Staff and to Commands on applications of RDF, as advisor to Naval and General staffs in corresponding fields, as director of research and development on communications for the Air Ministry, as director of production of RDF equipment for all departments, as chief engineer for design and installation of RDF stations at home and overseas, as chief maintenance engineer for these stations. He had also been responsible for research development and

initial provision of operations room equipment for the communication and display of RDF and other operational data in Commands, Group and Sector Headquarters. He has also, in effect been RDF advisor to the French Government.[4]

By the start of the war, as we shall see, Watson-Watt had proved unable to rise to the challenge of this increasingly demanding job.

Rowe's appointment to Bawdsey had mixed results. He was a very skilled manager and a good planner, but he did not inspire some of the scientific staff. Rowe had increasingly tenuous relations with these scientists, most noticeably Bowen and Wilkins, who chaffed under the superintendent's imposition of control over some of their more exuberant behaviour and his apparent unwillingness to accept their sage advice. For instance, Rowe banned such activities as the indiscriminate slaughter of the estate's rabbit population, which left him open to the charges of being a martinet. Bowen recalled in his memoirs:

> [Rowe] was the complete Civil Servant who knew how to work the Air Ministry machine and keep the filing system in order. Unfortunately his speciality was bomb ballistics and his knowledge of electronics rather minimal ... There is no doubt that he was highly motivated about air defence in the abstract sense, but the details were beyond him. If the truth be told, Rowe never came to terms with people like Wilkins and myself who had done the original work at Orfordness and Bawdsey. Unlike Watson-Watt, he seemed to resent any views we might express about the technology of air defence and never, in subsequent years, did he reach any kind of rapport with the pioneers.[5]

Bowen's criticisms were, on the whole, unfounded and in large measure reflected his later fallout with Rowe, which would see Bowen removed from the radar research programme in the summer of 1940. These serious personal clashes, however, lay ahead. Other Bawdsey staffers appreciated Rowe's management skills. R. V. 'Polly' Perkins, who headed the machine shop, thought that Rowe 'was the best and most efficient director I have ever known.'[6] Dr E. H. Putley, another radar scientist, commented that Rowe 'always seemed to know exactly what he wanted and more often than not he was right. He had the ability to persuade everyone most of the time that he was right, but when he failed he still persisted regardless.'[7] Rowe was on the whole successful in guiding Bawdsey through the critical period of creating the radar chain and bringing order to the chaos left by his predecessor.

Testing and Trials

Even before Watson-Watt took up his position as DCD, testing was under-way in which radar tracks of target aircraft were sent to Biggin Hill via the filter room at Bawdsey. At Biggin Hill the information received was utilized in attempting interceptions. During much of 1937, widespread intercep-tion experiments were curtailed while waiting for the completion of a new experimental inland radar station at Dunkirk, located in the Biggin Hill area. The inland station was of a mobile type using retractable masts. It was hoped that the station could take up tracking of aircraft after the chain radar at Dover lost contact when it passed the coast. This was necessary because Dover was too far from Biggin Hill to allow for successful intercep-tions at the coast. A series of tests, using a regularly scheduled KLM airline service and RAF Anson aircraft as targets, commenced in December.

By early March 1938, it was apparent that there were still major dif-ficulties to be overcome before successful interceptions could be achieved. All radar operations were hampered by aircraft not involved in the experi-ment flying into the test area. The problem of identification was so great for the inland radar that it proved to be next to useless. The Dunkirk sta-tion was dismantled in mid-February, soon after the Canewdon Chain radar became fully operational. The Canewdon station was close enough to Biggin Hill to allow for squadrons operating from the air station to intercept bombers at or near the coast. Interference from aircraft not par-ticipating in the exercises continued, however, to hamper operation. The need to find a simple and quick method to identify 'friendly' interlopers into the area scanned by radar became an increasingly urgent matter as war approached.

Another problem that emerged was that airliners proved to be an elu-sive target, with few successful interceptions being made. Some scientists believed this was caused by the sometimes erratic courses flown by the airline pilots to maintain their scheduled arrival time. Another school of thought, however, maintained that the problem was that the airliners oper-ated at low altitude and began their descent into London well before the coast. Rowe tried to play down the significance of this failure, but it was a serious issue. It was the first clear indications that chain radar had an Achilles heel, in that it had a limited ability to detect aircraft flying below 5,000 feet within 30 miles of the coast.

Still, when the chain radar stations did not suffer interference, they were quite successful in detecting RAF target aircraft and providing sufficient information to allow fighters operating from Biggin Hill to intercept them. On 31 March, Tizard recounted to Freeman a successful test he had wit-nessed from the filter room at Bawdsey.

altitude is a problem.

Yesterday, being Wednesday, there was little or no flying at Manston towards the middle of the day, and I watched an exercise in which an Anson was sent out to the coast of Belgium and came in towards Kent. It was detected with great regularity and accuracy by Dover and was intercepted by Biggin Hill right on the coast. The whole experiment was very satisfactory, particularly as although it was a fine day there were a good many clouds about and I should not have thought it was a very easy day as far as visibility was concerned.[8]

In total, there were twenty-nine interception experiments using chain radar and RAF target aircraft. Of these, fourteen resulted in interceptions in which the fighters were brought to within 2 miles or less of the target. Six others were also successful, but at ranges of up to 6 miles. There were eight failures, of which only three were caused by insufficient or inaccurate radar plots. These were impressive results, 'although,' as the official report of the tests explained, 'it is important to note that the experiments were of the simplest form and that the target maintained a steady course, which usually enabled extrapolation of its track to be made when the interception ceased.' The radar operators also proved unable to provide accurate height readings, and all these tests were conducted with the aircraft flying at a prearranged consistent altitude.[9]

The most significant aspect of these experiments was that two vital components of the air defence system were extensively tested for the first time with radar and the Biggin Hill techniques – Radio Direction Finding (DF) and the improved fighter radio-telephone set, the TR 9B. In the summer of 1937, Dowding asked the Air Ministry to equip all Fighter Command sectors with three DF sets, considered to be the ideal number to provide for radio controlled interception and navigational homing. By the end of the year, however, just five sectors had been equipped with DF sets, considered the minimum required. The Air Staff dithered on providing more sets because there was a shortage of cathode ray tube display apparatus and uncertainties about the value of DF to interception. The results of the experiments in early 1938 provided an unambiguous answer – DF was essential for successful interception. On 14 April 1938 the Air Ministry ordered twenty-nine DF sets, sufficient to equip each sector. The new sets would be purchased without a CRT display, using a less effective goniometer to display the signals. However, the DF sets were designed to be retrofitted with CRT's when available.[10]

DF worked in large measure because of improved fighter radio-telephones (R/T) and radio procedures. The new radios were provided with a quartz crystal to ensure that the sets stayed tuned to specific frequency settings and provided greater ranges. In October 1937, the RAF approved the use of a simplified set of code words first developed at Biggin Hill during the interception experiments in late 1936. These code words were designed to be easy to remember because they were dramatic and 'they bore some relation to their meaning.' Pilots and ground controller could easily understand each other using words such as, 'angels' for directing fighters to a specific height, 'orbit' for circling and searching, 'bandit' for hostile aircraft, 'tally ho' to signal the beginning of an attack, and 'pancake' to return and land. The improved radio worked well in these exercises, but there was clearly room for improvement. It was recognized that in order to maintain voice communication and contact with DF stations, the fighters' R/T sets had to be able to transmit on two separate frequencies without requiring any manual switching by the pilot.

Throughout 1938 scientists at the Royal Aircraft Establishment worked on perfecting the new radio-telephone known as the TR 9D. The set was a brilliant piece of radio engineering which allowed for each sector DF to control up to four separate formations while permitting pilots to continue to use their R/T sets for voice transmissions. The TR 9D was built to take into account the fact that it took fourteen seconds to take DF readings. Given a one second interval between each fix, it was possible for each sector DF to make four readings per minute. The sets of the aircraft of the squadron or flight leader of up to four formations could be set to change frequencies and transmit to the DF stations for a specific fourteen second period each minute. The timing would be calibrated using precision stop watches started on a signal from the flight controllers. An automated time-switch using a clockwork, known as a master contractor, was developed to switch the radio set to transmit at the DF frequency during the appropriate time period. The master controller became known by pilots as the '*pipsqueak*', a term derived from the 1,000 cycle note which sounded in the pilot's headphones during DF transmission. When not transmitting to the DF stations, the radio-telephone switched back to the frequency being used by the rest of the squadron for voice transmissions.

Each DF station was connected to a DF plotting room, where a rapid tri-angulation of results provided the accurate location of a group of fighters. The DF plotting room became another integral feature of the interception system.

[In each DF plotting room] a small circular map table was used for plot-ting with a thin cord, kept in tension by weight or elastic, drawn through a small hole at the position of each DF station. Three airmen sat around the table, each connected by telephone to one DF station. They plotted the bearings received by drawing the cord across the table from each DF station to the appropriate degree marking of the marginal protractor which ran around the edge of the table. A fourth airman estimated the point of intersection of the three cords, which usually formed a small triangle or cocked-hat, and told the position by telephone to a DF plotter in the operations room, using fighter map grid procedure.[11]

While a major improvement, it was realized long before the TR 9D entered service that even this set would not fulfill the needs of Fighter Command. On 14 March 1938, Dowding reported to the Air Ministry that the maximum range of the TR 9B radio was already inadequate for long-range interceptions. The TR 9 R/T sets operated at high frequency wavelengths and Dowding urged that research proceed as quickly as possible on a new generation of much more powerful R/T sets transmitting in the much shorter wavelengths of the Very High Frequency (VHF) band. Work on VHF R/T commenced in parallel with the development of the TR 9D.[12]

The mixed results of the interception experiments in early 1938 pressed home the necessity to continue with a rigorous series of tests to perfect each aspect of the air defence system. Scientists at Bawdsey played a diminishing direct role in these tests, operational control of the chain radar stations being assumed by RAF personnel who had passed through Raymond Hart's or Biggin Hill's training programmes. Hart, as senior RAF officer at Bawdsey, took over the running of the radar filter room at Bawdsey. Overall command of the experiments was given to the air officer in command at Biggin Hill.

While no longer involved directly in operational testing, the scientists assumed a new and increasingly vital role in improving the air defence system. On 27 May, Pye informed Rowe that Bawdsey's responsibility for these experiments was 'confined to acting in an advisory capacity, in an endeavour to assist the Service personnel concerned to make the best use of the technical equipment with which they are provided. To this end, and also with a view to improving the performance of the equipment, [the scientists were] also analysing the results obtained in the experiments.' The scientists were thus transformed into investigative observers of every test. These observations were used to provide detailed analyses to identify areas which required improvement, to provide explanations of why the present technology or techniques were not effective, and to suggest possible solutions to rectify the situation. Every aspect of the system was subject to scrutiny, from the way information was recorded in operations rooms, to the technical working of radar stations. This scientific analysis of the operations of the air defence system marks the origin of the science of operational or operations research (OR), which would prove to be an invaluable tool in improving combat performance of Allied air forces, navies, and armies during the Second World War. While certain aspects of

OR had been developed during the First World War, most notably by Hill's anti-aircraft researcher group, these lessons had been almost forgotten by the 1930s. After radar itself, OR is the single most important legacy of the years at Bawdsey.[13]

On 12 April 1938, Edmund Dixon, the former Post Office engineer, who by the spring of 1938 was in charge of communications research at Bawdsey, wrote a paper titled 'Suggested Tactical Analysis of Large Scale Air Defence Operations in Relation to RDF.' This proved to be the seminal work on OR. Dixon outlined how the lack of clearly defined policies in regard to major aspects of the radar air defence system was hampering communications research. Dixon required information on the area to be covered by each chain station, whether filter and control rooms would be located at command headquarters, group or sector levels, and how intelligence information would be distributed throughout Fighter Command. He argued that Dowding 'could be assisted by a detailed theoretical analysis of a number of alternative enemy attack operations with the assistance of data about the present performance of RDF location, RDF reporting by telephone and plotting by hand, Fighter Command operational intelligence, Sector to Fighter aircraft intelligence and Fighter Aircraft performance.' Dixon believed his group could supply information on a number of important technological, tactical and strategic issues. These included: the maximum number of raids that radar receivers should be designed to track at any one time; at what point in the command structure should filtering of information be undertaken; whether command, groups or sectors should control tactical operations; and a determination of the point at which the air defence system would become saturated and unable to cope with any more enemy raids.[14]

Approving Dixon's proposal for a wide ranging investigation of all aspects of the air defence system became one of the first decisions taken by Watson-Watt as DCD. In early July, Dixon was authorized to submit a detailed questionnaire to RAF home commands – Fighter, Bomber and Coastal – about their expectations concerning operations in times of war, what types of facilities they had established for command and control of their aircraft and how they communicated information.[15]

While OR was being established, the radar interception testing programme continued whenever weather and local air traffic conditions permitted. A great deal of trouble was experienced with other RAF commands to get them to curtail their own training activity in the vicinity of the chain stations. This brought home the need to find a workable solution to the problem of identification. With each passing test, more ways to improve performance were learned. One of the first discoveries of Bawdsey observers was that the difficulty that radar had in determining the height of aircraft was caused in large measure by the fact that height readings varied with the direction or azimuth of the target from each individual station. Signals were affected by the contour of the land surrounding radar sites, and the only solution was that each station required individual calibration in order that proper azimuth-height tables could be compiled. This would make the task of getting stations ready for operational use far more difficult.[16]

The 1938 series of tests of the radar system culminated in a major air defence exercise held from 5 to 7 August. Dowding designed the exercise to be the first one since the war in which 'an atmosphere of realism was introduced.' For the first time all five stations of the interim chain – Bawdsey, Dover, Canewdon, Great Bromley and Dunkirk – were used together. Tracks from the radar stations were transmitted to the filter room at Bawdsey by telephone and teleprinter. After filtering, Bawdsey simultaneously sent the plots of aircraft formations to Fighter Command Headquarters at Stanmore, No. 11 Group Headquarters at Uxbridge and to the sector air stations at Hornchurch, North Weald and Biggin Hill. These sectors stations were fully equipped with DF and operations rooms to conduct interceptions. In addition to the radar stations, Observer Corps posts were manned in order to test how raids tracked by radar would be tracked once they had passed the coast. For the first time in an exercise, Bomber Command 'was given comparative freedom of action with regards to the routing of the raids and the objectives to be attacked.'

Although Hart would command the radar system during the exercise, scientists manned the Bawdsey radar, while teams were sent out to observe the RAF personnel at the other four stations. Arrangements were made so that for one three-hour period on 6 August, the air force operators were replaced by 'experienced' Bawdsey operators at all stations. The scientists were instructed to record 'all data received from the chain stations, for collecting, by rough sketches and notes, a history of the Exercise' and to stand by all receivers, transmitters and communications equipment to deal with any breakdown. Dixon led a team of three to closely observe and record every aspect of the filter rooms operations. Joining Dixon was Harold Lardner, soon to emerge as the leading figure in operational research.

The exercise proved that the radar chain was a vital component of the air defence system. The filter room was able to use radar data to record 141 aircraft tracks. Many new technical components of the system were given widespread successful trials for the first time. These included teleprinters to transmit information. Much to the surprise of the Bawdsey staff, they found that RAF personnel trained by Hart were as proficient in operating radar as the scientists.

However, the exercise also revealed just how much more work was required before the chain radar system could live up to its potential. The radar operators were frequently overwhelmed by the large number of echoes appearing on their screens. The problem of identification greatly hampered operations. For the first time the radar operators were observing the echoes of large formations, and they found it difficult to accurately count the number of aircraft. Height readings, while better than in earlier tests, continued to be erratic. The chain stations' inability to track low flying aircraft was confirmed. On one occasion an undetected low-flying raid passed directly over the Dover station. The filter room staff also found itself inundated by the amount of information it was receiving from the radar stations. These problems were compounded at Fighter Command operations rooms when they attempted, for the first time, to combine the information received from the filter room with the visual tracking of aircraft over land by the observer corps. Bad weather also hampered the exercise.

All of these difficulties combined to deprive sector operation rooms of 'the information essential for the anticipation of the enemy courses and accurate interception.' As a result, the Biggin Hill method could not be used, and fighters were forced to revert back to the old and 'uneconomical' tactic of flying standing patrols. Dowding, however, was not discouraged by these results. In his report to the Air Ministry on the exercise, Dowding concluded that the basic system was sound and that only further realistic air defence exercises would make it function effectively.

In assessing radar's performance in the exercise, Rowe essentially agreed with Dowding. He believed that radar had demonstrated its immediate value if a war should break out, but concluded that the shortcomings of the system 'emphasized the need for operational research, as distinct from operational training.' The forthcoming completion of a Fighter Command filter room at Stanmore would make the Bawdsey facility available so that scientists would be 'free to concentrate upon research on how best to use the data provided by this still almost unbelievable method of locating aircraft.'[17]

However, there would be little time for further practice, research and fine tuning, however; just four weeks after the completion of the summer air exercise, radar was mobilized for the first time for war.

The Desperate Race

In early September, the Chamberlain government ordered 'the emergency deployment of a great part of the home defences.' The diplomatic controversy over Germany's demands to annex the Sudetenland region of Czechoslovakia, which had been simmering since the spring, became a full blown crisis. It seemed highly likely that war could start at any moment. The air defences were at the centre of the partial mobilization. Twenty-nine fighter squadrons, only five of which were equipped with modern Hurricanes, were brought to war readiness. Joining them were nearly 50,000 territorials to man the guns, searchlights, balloon barrages and coastal defences. Called into service as an integral part of the air defence system for the first time were all five chain stations and the Bawdsey staff.[18]

The chain immediately went 'to continuous watch, twenty-four hours of every day, save for brief maintenance periods carefully staggered among adjacent stations.' The stations worked well, but they covered the coast only from Suffolk to the south of Kent. On 10 September Bawdsey was asked what other stations could be set up on an emergency basis. The only equipment available were three small experimental sets known as type MB (Mobile Base) being built for use in the colonies. It was decided to use whatever material was available to set up these sets as 'Advanced Stations' to give some early warning of aircraft threatening the Forth-Clyde, Tyne and Wash approaches. On 16 September authorization was given to proceed with the three stations.

The first of the emergency stations, at Drone Hill, was up and running just thirteen days later. Using a MB receiver and transmitter and four 70-foot masts to support the aerial array, the station provided up to twenty minutes warning of aircraft approaching Edinburgh. The other stations proved less successful. One was damaged before it became operational when a storm knocked down its masts; the other was poorly sited and was only capable of detecting aircraft at 30 miles range, providing just a few minutes advanced warning. None of the emergency sets were ready by the time that Chamberlain returned from his infamous meeting at Munich.[19]

Despite the comparative lack of success of the emergency sets, the stations gave a good account of themselves. They tracked Chamberlain's flights to and from Germany while looking for any signs of the Luftwaffe attack that did not come. Even before the end of the crisis, the RAF's highest priority became the completion of the chain and ancillary command and control installations as rapidly as possible. On 23 September the building of a permanent underground operations block, including a filter room, was approved for Fighter Command Headquarters at Bentley Priory. Building the underground facility would take time. In order to ensure that Headquarters could control operations, in late September a team from Bawdsey was sent to Fighter Command Headquarters to establish an interim filter room in the basement of the mansion. By mid-December, the Stanmore filter room had become operational, and the Bawdsey room became a training and research centre. On 17 October, in his report to the

Air Ministry on the crisis, Dowding stated that 'the emergency emphasised the shortage of RDF stations to cover the vital area especially in the South and North East of England. The completion of the RDF chain of stations should be pressed on unremittingly.' Even before receiving Dowding's report, the Air Ministry had acted. On 6 October, at a meeting under the chairmanship of the Deputy Chief of the Air Staff, it was resolved 'that the RDF chain be hastened as to be complete by 1 April 1939.'[20]

On 12 October Watson-Watt was asked what special action would be necessary to meet the new target deadline for the completion of the chain. He replied that it was possible to have all stations in operation, but only if work on acquisition of sites, provision of towers, buildings and power supplies, and the production programme for transmitters and receivers was carried out twenty-four hours a day, seven days a week. Even with this emergency acceleration of work, the stations would be nothing like the final design agreed to the year before. Watson-Watt informed the Director of Signals that:

> The emergency provision will be limited to working on a single wave-length without duplication of transmitter and receiver units, and with very

Chain Home Radar Transmitter Tower.

restricted spares. It is probable that not less than six of the stations will have full power transmitters type CH, made by Metropolitan-Vickers; the remainder will have mobile transmitters, Type MB, manufactured by Metropolitan-Vickers, or emergency transmitters manufactured by Bawdsey. The stations will be incompletely calibrated in respect of direction-finding and height-finding.

Also most of the stations would not have their full complement of towers; the majority would only have two, and a few would have four wooden ones when completed. Only after 1 April would it be possible to go back to each station and upgrade them with two Chain Home (CH) receiver and transmitter units and the additional towers necessary to support the full antenna array.

On 14 November the Treasury approved the emergency intermediate chain programme. The desperate race was now on to have an air defence early warning system in place before the nation again faced the prospect of war. The race to complete the intermediate chain had all the hallmarks of an emergency wartime programme. The Treasury agreed that the Air Ministry could dispense with the normal tendering process and directly award contracts to companies judged most capable of finishing the work on schedule, regardless of cost.

Removing bureaucratic hurdles had the desired effect of pushing the pace of construction forward, but it remained doubtful that contractors could finish the erecting of building and towers on time. Many of the radar sites selected by Watson-Watt, in consultation with the Air Ministry's Works Department, proved less than ideal. At Ottercops Moss, the highest and most exposed station site, severe weather made winter construction extremely difficult. The site was described by the contractor as being 'a useless bog in which no satisfactory foundations could be found to a depth of 25 feet.' The Rye station was located in a marsh exposed to high wind, and the whole of the Pevensey station site was actually underwater.[21]

Completing the physical structures of the stations was only part of the work required. Transmitters, receivers, generators, cables, antennae, telephones and other related equipment had to be put into place, tested, and calibrated before a station could become operational. This was the responsibility of Bawdsey, and Rowe was doubtful that the research establishment had sufficient staff to undertake the work required and continue with research.

The shortage of personnel had been brought home during the September mobilization, when there had been barely enough trained civilian and military staff to man the radar stations and the filter room. At the end of September, when Watson-Watt ordered Rowe to begin preparations to provide another five emergency stations in case war should break out, all research at Bawdsey was suspended for the duration of the crisis.

The decision to dramatically accelerate the completion of chain stations also meant that there was a significant increase in the demands placed upon Bawdsey's scientific staff. Rowe's most immediate problem lay in securing the personnel necessary to build the radar chain and continue research. The rapid growth of the Bawdsey staff continued, increasing more than 50 per cent during 1938, from forty to sixty-one scientific personnel, but this did not keep pace with demand and research began to suffer. Rowe grew increasingly worried about the numbers and the quality of Bawdsey's scientific recruits. The choice virtually to exclude private manufacturers from setting up the stations resulted in a significant reduction of experimentation and research into new types of radar and related technologies as the Bawdsey staff were diverted to completing the chain.

Rowe reported to the Air Ministry in February 1939 that 60 per cent of the scientific staff, was engaged in setting up the early warning system. This left him with only thirty scientific personnel to spread between thirteen research projects, all of which the Ministry had assigned a top priority rating. Eight of these projects were directly related to improving the performance of the chain radar stations, including anti-jamming, and communication and operations room research. Most of the other projects were part of the airborne radar programme.[22] 'There can be no controlled interception,' Rowe confided to Tizard, 'without a chain and the chain should therefore come first.' Rowe felt that all concerned realized 'in a general sort of way that the chain [was] squeezing out research' and that Bawdsey was simply reaping what they had sown.

Rowe understood that preoccupation with the chain system had consequences to the RAF similar to the repercussions of the political decision to concentrate on the manufacture of fighters over large strategic bombers. Only in charge of radar research for eight months, Rowe already was charting the course of work at Bawdsey for the foreseeable future. The next major step in the development of radar would be to apply the technology to offensive operations which, Rowe believed, would 'revolutionize bombing, as RDF has done with defence.' Unfortunately, Rowe lamented, the shortage of staff and the emphasis on the chain meant that no research could be undertaken in offensive applications of radar.[23]

On 13 March 1939, Rowe wrote again to the Air Ministry to express his deep concerns about staffing problems. He reported that of all the new members of the technical and scientific staff recruited since January 1938, only four were appointed at the level of junior scientific officer or above. During a recent visit to Bawdsey of university physicists Rowe had made inquiries about why their students were not applying for government scientific positions. All the physicists admitted 'that posts in Government Research Establishments were unattractive to the best scientific men' because of the low initial salary, the lack of prospect for rapid promotion and the uncompetitive salaries once one reached the higher ranks.[24]

If Rowe hoped for a bureaucratic solution, he was sadly mistaken. Instead, the answer to the scientific staffing problems would come from a completely different direction. Tizard always believed that if war should come then Britain must fully and effectively mobilize its scientific resources if it were to survive. Tizard was one of a group of senior scientists who were at the forefront of a strategy to lay the groundwork for bringing university scholars into the national war effort before the war. One aspect of the plan was to brief senior academic scientists about the ongoing programme of secret research. These scientists would then be asked to assist in gathering their best junior staff and students for a preliminary training course during their summer vacations. Given the stature of the work at Bawdsey, Tizard naturally started with mobilizing physicists for the radar programme.[25]

Tizard received permission from the Air Ministry to reveal the radar secret to a select group of physicists sometime in 1938. Among the first he approached was John Cockcroft, a former pupil of Rutherford, member of the Cavendish Laboratory, and a pioneer in subatomic research. After Munich, Tizard decided to move ahead with letting additional scientists into the secret.

On 1 February 1939, Watson-Watt gave a briefing on radar to a small group of Cavendish physicists. The Cavendish scientists then drew up a list of the leading young physicists at major research centres in Britain. A plan was then implemented whereby the university scientists would be given an opportunity to be briefed on Bawdsey research programmes and then

Bawdsey Research Building.

to spend a month during the summer vacation at a chain station to gain experience with the new technology. In July, after being sworn to secrecy, the physicists visited Bawdsey, followed by a month-long stay at radar stations beginning on 1 September. Although this meant that little help for Bawdsey would be forthcoming in the few remaining months of peace, it did ensure that the nation's physicists were already learning about radar when war was declared.[26]

Rowe, therefore, was forced to balance his limited resources carefully between the completing the Chain Home stations, while solving nagging technical shortcomings of the system and continuing research into airborne radar. The only significant assistance provided to Bawdsey came from the RAF's No. 2 Installation Unit. The unit was established in June 1938 and under scientific supervision undertook much of the heavy physical work of laying out major equipment and cables, and affixing antenna arrays on the masts. Bawdsey personnel were then responsible for testing all electrical equipment, lining-up the aerials and calibration. Despite the continuing personnel problems, by the end of February 1939, Rowe was confident that the intermediate chain would be ready in time. He confided to Tizard: 'Considering the small staff and the time lost in travelling – as far as Scotland – I shall never quite understand how it has been done.'[27]

Rowe was right; on the whole Bawdsey was able to meet the Air Staff's deadline of having the intermediate chain up and running by 1 April. By the deadline, all but two of the stations were ready. Not surprisingly, Ottercops Moss and Rye were the two exceptions and they were fully fitted out, except that construction delays on the towers prevented the completion of the antennae until the end of May. All finished intermediate stations, eleven in total, were turned over to the RAF by early April. The chain now provided early warning along the south coast as far west as Portland and up the east coast as far north as Dundee. The extended early warning system was immediately placed into continuous operation, an integral component of the air defences which had been brought up to an enhanced state of readiness as a result of the German annexation of the rest of Czechoslovakia in March.

The accelerated completion of the intermediate chain caused a crisis in the training of radar station personnel. By the beginning of October 1938, Hart's Bawdsey instruction programme had trained just over sixty radar personnel out of the 260 it was then believed were required for the full chain system. The normal course took from four to six weeks to complete, depending on the technical skill required, and there was room for twenty trainees at any one time. This was completely inadequate and in November it was decided to establish a much larger radar training facility at RAF base Tangmere. The new radar school would not be ready before

June 1939, however, and as an interim measure, time spent at the existing Bawdsey facility was reduced and operators completed their training at operational radar stations.

The increased training programmes, however, could not keep pace with the anticipated rate of growth of demand for manpower. By the end of November Sqn Ldr J. A. Tester, Hart's second-in-command, calculated the need was actually between 450 and 600 radar operators and mechanics. By April 1939, it was realized that all earlier estimates of personnel requirements were grossly inadequate, and that more than 1,200 would be required to man radar stations at home and overseas. An additional 600 radar personnel would be required if Bowen's airborne radar programme proved successful. It was becoming apparent that simply further expanding the training programme would prove inadequate, because there were not enough suitable candidates available.[28]

Watson-Watt claimed to have originated the idea of training women to operate radar in a memorandum written in February 1937. In November 1938, Sqn Ldr Tester also asserted that he had conceived the concept. Whoever initiated the idea, it was far from original. Women had played an important role in the defence system in the First World War, and the Air Ministry had approved the training of women operators for the Thames Estuary mirror system in 1932. What is certain is that by June 1938, Watson-Watt had implemented a plan to test the suitability of women for the demanding task of operating a radar set.

The story of the first group of women radar trainees reflects the views of gender in British society in the late 1930s. According to Tester there were six main grounds for considering women:

1. Higher power of sustained concentration on a limited field of observation devoid of 'entertainment value' (as in the RDF indicator in the absence of targets) and lower liability to boredom in an unchanging routine.

2. Higher average finesse in relatively delicate setting of light moving parts involved in making and communicating the observations.

3. Higher scale of general conscientiousness.

4. Lower average tendency to magnify individual importance by partial disclosure of secrets specifically confided.

5. Longer period of individual availability for specialized duty in comparison with normal period of posting of service personnel.

6. Release of man-power for duties for which women are not yet suitable.

In order to test the suitability of women, it was decided to choose three Bawdsey shorthand-typists, since it was believed they had the necessary skills and character traits and had the extra advantage of being already familiar with the radar secret. The women selected were Miss Hilda J. Brooker, Miss Boyce and Miss Girdlestone.

The women's training was delayed by the Munich Crisis, but they were able to complete the course at Bawdsey by the end of 1938. Like male students, from Bawdsey the women were to be sent out to a chain station for further operational training and evaluation. This part of the programme was delayed, however, because of reluctance to send them to a station not

Women Plotters at Fighter Command's Operations Room.

equipped with separate sleeping and toilet facilities. After futile efforts to use the Bawdsey radar, which was mainly being used for research, arrangements were finally made to send the three to the Dover Station in mid-February 1939.

Brooker reported that she and her colleagues' experiences at Dover convinced them that they could carry out radar observing, but that a great deal of further training and preparation would be required before they could accurately give bearings, count number of aircraft and sort out different raids. The male Warrant Officer in charge of training at the station and a scientific observer from Bawdsey were far less critical. They considered that the women had 'performed excellently' and 'they seemed to be superior to equally experienced Service personnel of the type employed on this work; they certainly "picked up" the job more rapidly.'

Fighter Command headquarters was pleased with the results, although they had some reservations. Air Commodore Keith Park, writing for Dowding, expressed the opinion that it was likely that only a very small percentage of women would have sufficient technical skills 'to provide for adjustment of receiving apparatus' and, therefore, 'few would reach the standard of corporal operators from this point of view.' This would mean that station staff would be of mixed gender, which would 'introduce certain administrative complications.' Park also asked whether women were 'to be intentionally placed in stations in war which [were] likely to be objects of air attack.'

Despite these concerns, Brooker, Boyce, and Girdlestone proved that women could play a crucial role in the air defence system. The advantages of women radar operators, as well as operation and filter room staff, far outweighed any disadvantages. In the summer of 1939, a group of some of the earliest recruits for the Women's Auxiliary Air Force were sent to Bawdsey for training. The first WAAF watch crew went operational at Poling radar station in October 1939, just four weeks after the war began.[29]

Identifying Friend from Foe and Chain Home Low Radar

The completion of the intermediate chain was only the first step in the creation of an effective radar air defence system. The intermediate stations could only provide early warning of approaching aircraft; they could not supply the information on height and location which was required for interception. Even simple early-warning coverage was incomplete because the gap filler antennae were not fitted. Since the intermediate stations could operate only on a single frequency, they were susceptible to jamming. All the stations were also highly vulnerable to enemy air attack. The temporary buildings which housed the equipment had no protection against bombs, nor were there any antiaircraft guns supplied for local defence. The stations would have to be upgraded to the final design, which was itself constantly being improved. In October 1938, the final design was modified to include duplicate receivers and transmitters housed in separate buildings. Each building would be made proof against direct hits by small bombs, and protected by revetments against the effects of blast and splinters. The stations would have to be properly calibrated as quickly as possible in order to provide some indication of the location of the target.[30]

The state of the radar chain in the spring of 1939 reflected the general incomplete state of the entire defensive scheme. For instance, at the end of March, out of seventeen Fighter Command sectors, only one had the required three operational radio direction finding (DF) stations.[31] With the threat of war looming ever closer, the remaining five months of peace saw no slackening of the furious pace of building, testing and training which had commenced with the Munich Crisis. Added to the work of upgrading the intermediate chain stations was an expansion of the chain which resulted from the first strategic review of the system. On 19 January 1939, the Air Staff decided that in light of the increase in the number of fighter squadrons to fifty, it was now possible to offer protection to the entire country, including far northern and western regions previously overlooked. Four new stations were authorized, two to provide protection for the great northern naval base at Scapa Flow, and the others to extend the chain westward and provide early warning of raids approaching Bristol and South Wales. In May, a further two stations, on the Isle of Man and at Stranraer, were added to the scheme in order to provide early warning of aircraft approaching Belfast, the Forth-Clyde area, Liverpool and Birmingham from the Irish Sea.

The majority of Bawdsey's staff remained committed to upgrading the existing stations and expanding the chain; there remained many more hurdles to overcome than the manufacturing of equipment and the erection of building and towers before a fully functional air defence system would come into being. There were a host of technical issues that had to be addressed, the most crucial of which were the identification of targets and the detection of aircraft flying at low altitude.

The need to find a reliable system of identifying friendly and enemy aircraft went back to the First World War. Without a means to determine the status of approaching aircraft, any air defence system could be overwhelmed by the need to intercept each approaching formation. In war misidentification could even result in fighters accidentally shooting down their own patrol planes and returning bombers.[32]

For the next two years alternative means of using radios and radio direction finding proved unworkable. The best radio identification solution was dubbed the challenge and reply system. In this procedure aircraft approaching the coast were sent a radio challenge on a prearranged frequency. The aircraft would then send a coded reply which would be picked up by a ground based medium frequency DF station. The system proved to be overly complex, slow, and inaccurate; it could take as much as 14½ minutes, or enough for a bomber to travel between 60 and 75 miles, before a friendly formation could be identified.[33]

By April 1938 it was apparent that a different approach to identification was necessary. Air Cdre C. W. Nutting, the Director of Signals, summed up the general belief of the Air Ministry and Fighter Command that 'the most satisfactory system would be for the friendly aircraft themselves to either possess a recognition circuit or some other means of direct identification with the RDF Stations themselves.' Bawdsey researchers turned their attention back to an idea first proposed by Watson-Watt in his February 1935 memorandum for an aircraft mounted radio transponder. Watson-Watt suggested that fighter aircraft could be fitted with 'a keyed resonating array so that they are readily located by the same methods as those used by enemy bombers, but discriminated and identified by the intermissions in their "reflected" field.'[34]

Work began on an aircraft-mounted radio transponder in the spring of 1938. Watson-Watt states that Rowe assigned two separate two-man teams of scientists to develop a simple device that could automatically detect radar signals and then transmit a coded signal which would be clearly visible on the radar screen of a chain station. The designers used a simple radio receiver which contained a super-regenerative circuit, thus the receiver also would act as a transmitter sending back to the radar much stronger signals, particularly noticeable to radar operators at long ranges.

By the end of 1938, Rowe had assigned Dr F. C. Williams, a recent recruit to Bawdsey, to combine the best of both designs. On 4 March 1939, Williams' prototype was demonstrated to Dowding. Dowding enthusiastically reported to the Air Ministry that 'the outstanding exhibit was the automatic identification signal, which gives a very definite characteristic to the indications on the RDF tube of aircraft fitted with an apparatus, which is stated to be quite cheap and light.' Dowding wanted the new device rushed into service, but both Tizard and Rowe cautioned that it was still very much in the experimental stage.[35]

The rapid development of the new device became Fighter Command's top priority after a minor air defence exercise held at the end of March revealed that the inability to identify targets caused most of the problems for the defenders. 'The perfection of this device [was] of such vital importance' that Dowding requested that Bawdsey be authorized to provide full specifications to a private manufacturer which would hand-build thirty copies of the prototype in time for the next major air exercise in August. Dowding subsequently explained to Tizard that the risks of misidentification in war were so great that until a target could be ascertained as 'friend or foe' fighters would not be scrambled to intercept, even if this meant nullifying one of the principle advantages of radar.[36]

Bawdsey was able to have a final design of the device ready by mid-June, but only by cutting corners. For instance, it had been the original intention of Bawdsey to design the transponder so that it would search the frequencies likely to be used by all radar under development, such as the War Office's gun laying sets or Bowen's AI and ASV equipment. In the first version the device was limited to scanning only those frequencies used by the chain radar system. The thirty hand-built sets were ready in early August, but there was insufficient time to test them properly once fitted into aircraft. As a result, there was only seven confirmed sightings of aircraft using the new equipment; most sets failed due to a series of mechanical, electrical, and operational failures. Despite the poor showing, an initial production batch of 1,000 was ordered into emergency production by the end of August. It was intended that the device, now officially dubbed IFF, for identify friend or foe, would be a centrepiece of the increasingly complex air defence system, but it would not enter into general service until well into 1940.[37]

At the same time that IFF was being developed, Bawdsey was also forced to find a solution to another serious shortcoming in the air defence system. When Watson-Watt had first designed chain radar he had made several compromises as a result of the limitations of radio technology in 1935 and the need to develop quickly a set that could search as much of the sky as possible. By 1937 it was agreed that for the chain stations, the best wavelength from both technical and operational viewpoints would be around 13 metres. As the chain was developed it was realized that in the final design it would be necessary to provide each station with four different wavelengths. This was essential in order to limit jamming and to avoid interference from neighbouring stations, which actually required the chain stations to use wavelengths from 6½ to 13 metres. This was far shorter than any radio wavelengths available before 1935, and provided excellent cover of much of the sky while avoiding interference from the ionosphere and civilian radio. The one serious technical drawback was that, while by the standards of time these were very short wavelengths, they were not small enough to allow for serious gaps in the coverage to appear at low altitude, something made very apparent by the results of the summer 1938 air exercises.

Watson-Watt requested in September 1938 that Bawdsey take steps to adapt the chain to detect aircraft flying below 8,000 feet at 80 miles. On 14 September Rowe 'reminded' Watson-Watt that, due to the curvature of the earth, the range at which low flying aircraft could be detected was determined by their height and the height of the antenna of the radar set. At the 80 miles range requested, an aircraft travelling at 3,000 feet could be detected by an antenna located just 120 feet above sea level; but if the plane was at 1,000 feet, the array would have to be 1,150 high, and at 200 feet the height of the antenna would be a staggering 2,600 feet! What Rowe did not explain was that these figures were determined by a mathematical calculation in which the wavelength was the crucial factor; the shorter the wavelength the lower the necessary antenna.

The only solution suggested by Rowe was to explore suspending antennae from tethered balloons.[38] These experiments completely failed. Fortunately, the serendipity which seemed to bless the radar program produced an unexpected solution. In the spring, a War Office research team at Bawdsey began testing of the prototype of their new Coast Defence (CD) radar, designed to provide early warning of ships approaching the coast and, if possible, target information for coast defence artillery. One of the first observations made with the new set was that it was detecting low flying aircraft at far greater ranges than possible with Chain equipment. When Watson-Watt heard of the results, he immediately asked Rowe if the set could be modified for aircraft detection.[39]

The CD set's success in detecting low flying aircraft was the result of the design parameters adopted by the War Office researchers when the CD programme began in 1937. They decided to use Watson-Watt's concept of a radar which used a rotating antenna which would sweep the sky in a way akin to a single searchlight or the lamp of a lighthouse, enabling them to achieve far greater accuracy in determining the range and location of a target than the chain radar. The liability of this arrangement was that the area covered by a single radar pulse was far smaller, yet this was not as significant because there was far less area to cover along the surface of the water. Nor could the set measure an aircraft's altitude. In order to increase accuracy and reduce the size of the rotating antenna, the War Office team decided very early on in development that CD would operate at the shortest wavelength available. They quickly embraced 1½-metre-wavelength technology developed by Bowen's airborne radar group.

It was the use of the much shorter wavelength that made CD far more effective than chain radar in detecting low flying aircraft. Preliminary testing completed in late June, using the prototype set with the antenna mounted just 60 feet above sea level showed that CD could detect aircraft at an altitude of 50 feet at up to 15 miles and at 1,000 feet at 34 miles.[40]

CD was immediately ordered to be incorporated into the air defences, with new stations being built at or near the sites of existing chain radars. When used for aircraft detection, the CD sets would be renamed by the Air Ministry as Chain Home Low (CHL) radar. The final pre-production set would not be ready for testing until October, with deliveries of twenty-four operational CHL stations commencing in early 1940.[41]

The Approach to War

The introduction of these two new major components into the air defence system during the summer of 1939 illustrates that it was very much a work in progress. While researchers developed new technologies, construction proceeded on upgrading or installing chain stations, DF sites, group and sector control rooms, communication lines and all of the other ancillary parts of the increasingly complex air defence system. Training of new operators and the other skilled trades personnel necessary was increasing as rapidly as possible, with much of the schooling taking place at chain stations. From the Munich crisis onwards, however, the system was much more than a research and development project, for it was also pressed into operational service. Dowding decided that whatever the deficiencies in the technology, something was better than nothing, and that service personnel required as much practice as possible to improve the overall performance of the air defences. In early 1939, Dowding, acting upon the advice of Watson-Watt and Tizard, ordered that more frequent monthly minor air defence exercises complement the annual major summer programme. After the completion of most of the interim chain in April, Dowding directed that the radar early warning system go into full twenty-four hour a day operations. Thus the system was simultaneously undergoing construction, upgrading, expansion, experimentation, training, operational testing and operations. It was a delicate balancing act to satisfy all these demands and it caused much anxiety during the spring and summer for all those involved.[42]

Conflict between competing interests became particularly acute after 28 June when Fighter Command asked Bawdsey and No. 2 Installation Unit to upgrade '19 stations with gap fillers, height aerials and anti-jamming devices' by 7 August. This was required in order that these stations could be used for controlled interceptions in case war should break out and, if not, for use in the major summer air exercises. Rowe believed this goal was achievable, but only if his installation teams could be the sole users of the chain in the interim. Rowe realized that the international situation rendered this highly unlikely, but he remained hopeful that the target date could still be achieved if his crews worked round the clock despite the risks of making errors in the complex installations. Remarkably, on the whole, Rowe was correct and most Chain stations were upgraded by the agreed date.[43]

Dowding was not deterred by these problems, as he was well aware that only in actual operations and exercises could all the flaws in the system be detected and solutions worked out. In order to ensure that results of each exercise were carefully monitored, operational research continued as an integral part of the programme. OR grew in importance as the fine tuning of the system became increasingly crucial. This was recognized at Bawdsey in late June when Group I, which was originally in charge of developing transmitters, receivers and radio measurements, and later became responsible for OR, was split into two new Groups. The new Group I confined 'its activities to operational research and no longer carrie[d] out work directly related to radio.' The head of the OR group was Harold Lardner. The increasing importance of OR was also recognized by Fighter Command, which began after the July air exercise to circulate reports routinely from Lardner's team throughout the command.

The OR researchers made numerous observations on means to improve the entire system, including ways to increase the efficiency of filtering, better methods of displaying information on control room maps, the value of teleprinters for transmitting information between different levels of the systems and detailed analyses of every interception experiment and air exercise. Many of the recommendations made by Group I were immediately implemented by Fighter Command, but here too a shortage of resources and time precluded action on many of the suggestions.[44]

Dowding, as well, closely supervised each of the operational exercises. He also organized several major conferences in which the ideas about the workings of the air defence system were exchanged between senior officers at Fighter Command, the Air Ministry and War Office, along with scientists including Tizard, Watson-Watt and Hill. Dowding also maintained close personal contact with the scientists, including Rowe. As a result of these observations and discussions, Dowding took it upon himself to initiate two major reforms of the reporting system. The first stemmed from problems which developed when the Observer Corps was incorporated into the system. The inexperienced corps members were reporting large numbers of raids which turned out to be friendly coastal patrol planes, fighters or even civilian aircraft. This led to an utterly chaotic display on the control room maps. In June, Dowding ordered that any report of aircraft stemming from the Observer Corp should be disregarded unless it represented a raid already identified by radar. This cleared away much of the clutter on the maps and allowed controllers to focus on actual raiders. It also implied a complete trust in radar. The second innovation was the creation of a 'Lost Property Office' which would keep track of raids that might accidentally disappear from control room maps, particularly if the raid transited group and sector boundaries or was not sighted by the Observer Corps on a regular basis:[45]

The first significant test of the newly transformed air defence system was the major summer exercise held from 8 to 11 August. It would also prove to be the last opportunity to improve the system in peacetime. It was by far the largest and most realistic of all peacetime air exercises. Hundreds of raids were launched by Bomber Command, which was given nearly complete tactical freedom to launch the attacks. The exercise was also a giant experiment. Like the recently upgraded chain stations and new IFF sets, many other parts of the air defences were being used for the first time. Only days before the commencement of the air exercise, the last of the sector DF stations were completed, No. 11 Group's new underground command centre was only opened on 5 August, and the Lost Property Office was manned for the first time.

Given the experimental nature of much of the equipment and bad weather that hampered operations during the first two days, on the whole the system performed very well. The commander of No. 11 Group reported that 'RDF information and plotting throughout the Exercise was consistently first-rate, and enabled interceptions to be effected on the Coast.' Dowding echoed these findings, and emphasized that unlike the 1938 exercise, the interception system did not break down, and that particularly on the final day of the test, when the weather improved, 'daylight raids were normally tracked and intercepted with ease and regularity.' They recognized that there were many shortcomings with the system. Radar operators were usually incorrect in counting the number of aircraft in a raid, and their altitude readings were often inconsistent and erroneous. The efficiency of DF and R/T at sectors varied greatly. Fighter group controllers often delayed too long in passing information on raids to sector controllers. Still, Dowding informed the Air Ministry: 'While I am far from implying that the present system is perfect, I feel that very great advances have been made during the past year.'[46]

Bawdsey OR researchers who observed all aspects of the exercise were less optimistic than Dowding. Their report stated that when they compared the results of the 1938 and 1939 exercises, 'the Bawdsey personnel gained the impression that there was very little, if any, improvement between the performance of the Estuary stations in the two air exercises.' Yet, like Dowding, the OR scientists believed that the major faults in the air defence system were correctable. They concluded that performance was primarily hampered by the shortage of trained personnel.[47]

As many RAF officers had anticipated, the August 1939 air exercise was the last peacetime opportunity to test the air defence system. Whatever its shortcomings, remarkable progress had been made in expanding and improving the performance of the system since the beginning of 1938. While the nation was by no means invulnerable to daytime aerial assault, it was far more secure than many had thought possible just one year before. The hard work of all those associated with the development of the defensive system would soon be put to the test, for at dawn on 1 September Germany invaded Poland. Two days later Britain and France declared war on Germany and the British people braced themselves for the expected aerial onslaught to begin.

11

The Phoney War

On 24 August 1939 the Air Ministry sent the code word 'Afidock' to Fighter Command Headquarters, bringing the entire air defence system onto a war footing. Just eight days later Britain was at war with Germany. Providing early warning protection from the anticipated attack from the Luftwaffe were twenty Chain Home radar stations, none of which had been upgraded to the final design. Most of the stations were poorly calibrated, which made it difficult, if not impossible, for the generally inexperienced operators to provide accurate fixes on approaching aircraft. Particularly troublesome was the measurement of altitude; very low and, as soon would be discovered, very high flying aircraft could easily escape detection altogether. The maximum range of detection was limited by the lack of Metropolitan-Vickers CH transmitters, which operated with much greater power than those hand-built at Bawdsey and adopted from smaller mobile radar sets then in use in most stations. Not a single CHL set had been produced.

The command and control apparatus was also still very much under development. Filtering of information remained a skill that few had mastered. There was great uncertainty of when headquarters should transmit information to groups and when groups should assign individual sectors to launch an interception. The lack of IFF sets and the complete failure from the first day of the war of the 'challenge and reply' system used as an alternative, made mistakes common, and in war, such errors often had tragic results. On 6 September, a Chain radar station tracked an aircraft approaching from France. It was a friendly aircraft, but it had failed to file a flight plan. Fighters were scrambled to intercept. What then followed was a series of errors which ultimately led to tragedy. Inexperienced radar operators failed to detect the difference between the echoes given off by aircraft in front of or behind the station. Intercepting fighters were mis-identified as coming towards the coast from out to sea. More fighters were

Entrance to a Chain Home radar station.

sent up and these too were misidentified. The problem was compounded by further by Observer Corps errors. This led the Stanmore filter room to plot twenty different tracks of unidentified aircraft operating over the Thames Estuary. Soon every fighter to the east of London was sent up. Failure to coordinate these formations caused one Spitfire section accidently to attack a Hurricane squadron. Two planes were lost, and a third was shot down by friendly anti-aircraft fire. Later investigation revealed that it was unlikely any enemy aircraft were present. Thus ended the so called Battle of Barking Creek, an inauspicious baptism of fire for the air defence system.[1]

The rest of Fighter Command was in a similar state of preparedness. There were thirty-nine fighter squadrons, although only twenty-six were equipped with Hurricanes and Spitfires. The Air Staff had decided that a minimum of fifty-three squadrons were required for a fully effective defence. There was a critical shortage of anti-aircraft guns and only enough barrage balloons to provide coverage for London. Twenty-eight out of thirty-two proposed Observer Corps districts were organized.[2] While the defences were far from complete, events in September 1939 revealed just how much the situation had improved since the Munich Crisis of only a year before.

After just six days of war, with the anticipated German aerial assault nowhere in sight, Dowding informed his Group commanders that 'since the Germans haven't attacked, we must stand down from this high state of alert or risk staleness and ill health.' Operations rooms would now need to be capable of becoming fully operational in two minutes, while radar and Coastal Observer lines were to remain fully operational. Air bases with three fighter squadrons were to have one squadron on standby; in bases with two squadrons, one flight was to be held ready.

As Fighter Command stood down into a 'normal' wartime routine, Dowding confronted three major problems which had to be resolved if the air defences were to be able to defeat the German aerial onslaught which most still believed would occur in the near future. The first two have been extensively studied in previous accounts of the Battle of Britain; the third, which relates directly to the radar air defence system has been virtually ignored.

The first was strictly political and need not concern us here in any detail. Shortly after the commencement of hostilities, four of the precious squadrons of Hurricanes were dispatched to France as part of the Air Component of the British Expeditionary Force. In the months ahead, Dowding would do everything in his power to restrict the number of fighters sent to France until he had sufficient quantity of aircraft and trained personnel to man the defences. This issue was only resolved eight days into the German attack on France in May 1940. Dowding, ably supported by Air Chief Marshall Sir Cyril Newall, the Chief of Air Staff, was able to convince the Chiefs of Staff and the government that no further fighter squadrons could be sent to France.[3]

The second issue, like the first, involved the quantity and quality of the defence forces. Even without the drain of resources to the continent, Fighter Command was desperately short of aircraft and trained pilots, and had little in the way of an effective reserve. Even during the summer of 1940, when Fighter Command was given absolute priority, the shortage of fighters and pilots remained one of Dowding greatest worries and almost led to Britain's defeat.[4]

The need to complete the radar air defence system as quickly as possibly, while at the same time improving its operational performance, was the third crucial issue. It continued to bedevil Dowding throughout the rest of his tenure at Fighter Command. No one predicted that the ultimate test lay ten months in the future, just enough time for most of the key components of the daytime defences to be made ready.

Reorganization of Research, Development and Operations

The commencement of the war brought about a series of rapid changes in almost every aspect of the way that air defence research and development was managed. By the start of the war the political dimension of air defence research had virtually ceased. Winston Churchill was brought into the Cabinet shortly after the war began, returning as First Lord of the Admiralty. Now part of the government, with responsibilities for a different service, Churchill, along with Lindemann, had little to say about air defence research for the time being.[5]

Just as the ADRC quietly disappeared, so did the committee for the Scientific Survey of Air Defence. The last recorded meeting of the Tizard Committee took place on 16 August 1939. The minutes of this meeting do not indicate that it was to be its concluding session. Undoubtedly, the start of the war was the deciding factor. There was no longer the need for this informal advisory body, since the barriers, which prevented academic scientists from working full time for the armed forces, had been swept away by the exigencies of war.

Even before the war, the Air Ministry took steps to insure that it would not be deprived of the invaluable services of Tizard. By the end of August, Newall arranged for Tizard to be appointed as the scientific advisor to the Chief of Air Staff. It was at the time a unique appointment, giving Tizard the ability to provide scientific and technical advice on almost all aspects of the air war. Tizard maintained his close contacts with Dowding. Dowding was pleased that his close friend was now working at the Air Ministry. On 15 September, he replied to a request by Tizard for the loan of two Blenheim heavy fighter squadrons for experimenting with the photoelectric bomb:

> At the moment, however, I am in desperate straits in regard to the Fighter situation, which has turned out to be much worse even than when I spoke to you. I shall welcome your advice and assistance in this matter if you will help me.
>
> What sticks out a mile is that you are not going to get two Blenheim Squadrons for this purpose in two months, or in any period I can foresee.
>
> 'Come over into Macedonia and help us.'[6]

For the time being Watson-Watt remained as DCD, responsible for all aspects of the radar chain. The vastly expanded burdens placed upon him and his tiny staff, and Watson-Watt's lack of organizational skills would soon cause a major reorganization of the management of radar in the RAF. In part, the changes were inevitable, as radar ceased to be simply a research programme and became an operational weapon system. One of the first signs of difficulties at DCD was that many stations proved unable to maintain the twenty-four hour a day coverage demanded by Dowding. Watson-Watt, inexperienced with operational conditions, had failed to arrange for an adequate organization of supplies. As a result there were insufficient spare parts immediately available to the chain stations. All too often, in the first few months of the war, radar sets were made inoperable by the lack of replacement parts. The equipment staff for the entire radar chain consisted of one inexperienced officer and two civilian store-keepers. According to one contemporary account, 'their whole time was absorbed on the physical obtaining and dispatching of stores,' leaving no opportunity to establish a systematic organization of supplies. There were no reference numbers on components, nor was there a set inventory. This made ordering parts very difficult, since station personnel had difficulty in communicating the identity of any component that needed replacing.[7]

The spares shortages and continuing operational deficiencies of the chain were brought home during a Luftwaffe raid on the fleet anchored at Rosyth on 16 October. Given prior notice that a raid was forthcoming by the sighting of reconnaissance aircraft and by radio intelligence, the Germans' attack should have been easily broken up before reaching the fleet. At the critical moment, however, the local radar station broke down, and the force of about a dozen bombers was able to start their attack unmolested. Only after several ships were heavily damaged did two squadrons of Hurricanes appear and shoot down two He 111s and chase the others away.[8]

Four days later Dowding requested that the Air Ministry appoint a committee under the chairmanship of Tizard to investigate the problems with the radar chain. Joining Tizard were several senior officers, including Air Marshal Joubert, recently appointed by Newall 'to investigate the RDF chain from the point of view of service control,' Raymond Hart, then stationed at Fighter Command Headquarters, and Watson-Watt. It took only a month for the committee to make wide ranging recommendations for reform. The committee suggested that a new radar command be established, to take over responsibility for radar operations from Fighter Command, and radar maintenance, construction, development and training from the DCD. Dowding successfully appealed to Newall that the operations of the chain must remain in his purview, but most of Watson-Watt's responsibilities for the air defences, except for basic research and development, were to be stripped from him. As soon as possible, these responsibilities were given to a new RAF group to be set up under the command of a senior air force officer. No. 60 Group began functioning in February, 1940.[9]

Research continued on the radar chain, here too, however, the war brought some significant changes in organization. On the first day of the war, most of the scientific staff of Bawdsey headed north to Dundee. Once located at Dundee, the radar laboratory was renamed rather innocuously the Air Ministry Research Establishment (AMRE). The departure from Bawdsey occurred because Watson-Watt and Rowe feared that the research centre was vulnerable to a German attack. While the transfer to Dundee did remove the invaluable staff and equipment from harm's way, it led to the virtually complete collapse of research into new equipment. This collapse of research had limited effect on the chain because most of the basic components of the system for daytime defence had already been completed or designed. Crucial work on radar for defences against night bombing was, however, crippled. In early 1940, this would lead to Watson-Watt's removal as DCD. The full story of the move to Dundee will be explored later.

Operational Research

A very positive development from the Bawdsey move must be considered here. In the summer of 1939, Hart had left Bawdsey for Fighter Command Headquarters in order to work on improving the command and control of operations. Hart realized that one of the key ways to improve the entire air defence system was to expand the programme of operational research and to allow these scientists to observe first-hand the workings of the system on a daily basis. The great value of OR had been brought home during the various air exercises in 1939, culminating in an insightful analysis of the August programme. Air Cmdre Keith Park, Senior Staff Officer Fighter Command, and Dowding supported Hart's plan for retaining Lardner and his OR group at Stanmore on a permanent basis.

The key role of Lardner's OR team in fine-tuning the air defences during the Phoney War period has been generally overlooked, yet without their analyses of tests and of actual interceptions of Luftwaffe raids, which became increasingly frequent in the winter of 1939-40, the system would have been much less effective during the Battle of Britain. Lardner, a Canadian engineer trained at Dalhousie University, would continue to lead the team. By early 1940, the OR team was responsible for examining the operations of the radar chain; the performance of Fighter Command's operational system, including filtering and the working of operation rooms; and the techniques by which information was displayed. By the spring of

1940 Lardner was in charge of twenty scientists and technical assistants, supported by a dozen WAAF clerical staff. E. C. Williams headed investigations into the radar chain's performance; G. A. Roberts led research into operational systems. Both were Scientific Officers who had been working on OR at Bawdsey since the spring of 1939.[10]

The OR group's eyes and ears at the radar stations were the Chain Scientific Observers, a new team of researchers established by Rowe from the large pool of university scientists who entered government service at the start of the war. These scientists were sent out to half a dozen chain stations to observe their workings and to make recommendations on ways to improve training and operational procedures. At first the stations commanding officers viewed the scientists as headquarter spies, but the scientists gradually won their trust by proving themselves invaluable in improving observing and in maintaining equipment. Stations fortunate enough to have a scientist present were always among the top performers of the chain.[11]

The first report issued by the Stanmore RDF Research Section, as the OR group became known, was an examination of the 'Lessons learned as a result of the differences between "operating conditions" of the "RDF Chain" in "War" as compared with "Peace".' Written by Lardner on 11 September, just two days after Dowding ordered Fighter Command to stand down from its initial mobilization, the report is a model of crisp and clear prose which powerfully presents an important message. It also is an example of the freedom that Dowding allowed the OR staff in assessing the performance of Fighter Command.

Lardner began his report by warning of the dangers of misidentification, given that the vast majority of aircraft that were detected by radar were friendly. It was better, he argued to allow the occasional single raider through than to risk accidentally shooting down friendly aircraft or issuing a false air raid warning 'which probably causes several civilian heart failures and wastes terrific amount of time and money generally.' The single most important fault in the system, therefore, was the lack of IFF. Other problems included inconsistent height measurements by chain stations. He outlined one incident as a result of poor height measurements that occurred on 8 September:

A single aircraft was observed to be approaching, according to Bawdsey, at 7,000 ft. and according to Dunkirk and Great Bromely, at 18,000 ft. The Coastal Liaison Officer said his aircraft would be at not over 5,000 ft. Mainly because of test flights on Tuesday, it was judged that the aircraft was 10,000 ft. or over, so it was called an enemy and ordered to be intercepted. It was intercepted and proved to be a single friendly Anson

– luckily it was not fired at. Perhaps if it had been at night, the story would not have had a happy ending!

Lardner concluded that, despite these problems, the air defence system was working well and that 'there seemed very little reason to doubt that it will prove to the greatest value when active enemy attacks against this country commence.'[12]

Lardner's group was instrumental in introducing the first reform to the system brought about by wartime experience. On 10 September, the Filter Room at Stanmore issued 'Instruction No. 1', concerning reforms in the raid tracking system. Each aircraft track was assigned a letter and number code for use on the filter and operation rooms' maps. The letter codes indicated whether the formation was civilian, friendly or enemy. Prior to the war no designation was given to tracks of unidentified formations. OR observers noted that this caused much of the confusion when multiple tracks were being plotted, tracks of uncertain identity were invariably listed as enemy. The letter X was designated to denote 'Doubtful raids,' a simple but effective measure to ease confusion.[13]

Throughout the Phoney War there was a very active aerial campaign of raids, patrols, minelaying and attacks on shipping launched by a small force of Luftwaffe bombers. While a far cry from the scope or type of operations anticipated from the German air force, these small scale attacks provided ample opportunity to improve the performance of the defences. Because most Luftwaffe aircraft operated well out to sea and frequently, at low altitude, they proved a difficult challenge to the system which was designed to meet raids as, or just after, they crossed the coast. Still each Luftwaffe mission provided an opportunity for the OR researchers to assess the performance of the system and to recommend improvements.

Reports on specific raids began in November as a direct result of the failure to prevent the attack on the fleet at Rosyth. Lardner wrote the first report which assessed the track of a reconnaissance mission by a lone German aircraft flying up the Thames Estuary. The Filter Room had first identified a single target as an expected friendly civilian airliner. Only after this aircraft landed at Shoreham was a second track detected. This track was assumed to be an enemy aircraft and Hurricanes were scrambled to intercept. The Hurricanes damaged the German aircraft, and ultimately forced it to ditch at sea. The disturbing part of this raid was that the German aircraft closely followed the path of the civilian aircraft, so Lardner recommended that Filter Room staff keep close watch of a possible German ruse to trick the defenders.

By late January 1940, the analysis of raids had become a regular routine of Roberts' section. Any raid that appeared to illustrate weaknesses of the system was examined in detail and suggestions were offered to correct the problem. For example, on 19 January, a lone He 111 approached the Scottish coast near Peterhead. The raider was tracked by radar, but when it crossed the coast the Observer Corps spotted the aircraft some 10 miles north-west of the plot provided by the chain station. As a result the Observer Corps controller listed this flight as a separate raid. This caused the flight of Hurricanes sent out to intercept the initial raid to receive no information on its position for a crucial ten minutes, while a second section was scrambled to intercept the imaginary second raid. By the time No. 13 Group controllers realized that there was only one raid and instructed both sections to intercept, the Observer Corps had lost sight of Heinkel, which was by then proceeding out to sea. After crossing the coast the bomber was promptly picked up by radar, but once again this new track was not properly connected to the previous raid. Apparently there existed no procedures for plotting raids going back out to sea. The Hurricanes looked in vain for the German aircraft, which had made a leisurely escape, disappearing from the radar screens over thirty minutes after it was first detected.

Roberts assessed that the main cause of the confusion was that there was no nearby radar guarding that part of the coast, and also, that the two stations that had tracked the bomber were operating at an extreme range of around 120 miles. These factors caused their plots to be 10 miles out, which led the Observer Corps to misidentify the raid. Roberts recommended that radar plots of less than usual accuracy be used only in an emergency, and that controllers were informed of any potential problems. He also stressed the need to have continuous raid numbers, and to develop a system whereby raid numbers were carried over from Observer Corps tracks to outgoing radar tracks.[15]

Another example from the morning of 30 May illustrates the importance of OR and the continuing weakness of the defences at this late date. This report examined another raid by a single aircraft which escaped interception even though it was detected some 65 miles out to sea and tracked for an hour and twenty minutes as it made its way along the coasts of Norfolk and Lincolnshire to the Humber River estuary. The main reason for the failure was that the Humber marked the boundaries of No. 12 Group, responsible for defending the Midlands, and No. 13 Group, which protected the north of England and Scotland. Roberts discovered that standard procedures meant that No. 13 Group only began to receiving information of the raid when it was 20 miles south of it boundary or '4 minutes at 300 mph, part of which is swallowed up in the time lag of RDF filtered plots, thus giving the defences practically no warning.' Roberts also determined that once No. 13 Group was aware of the situation the delayed interception failed when radar operators confused the track of the German aircraft with that of the fighters as they both headed out to sea.

Roberts recommended that filterers begin passing plots to Groups when a raid was within 40 miles of its boundary. This would give sufficient time for a Group to scramble fighters and alert the ground defences. The reform was quickly implemented. As for the problem of misidentification, there seemed no easy solution. All that Roberts could recommend was that ' a more satisfactory method of identifying our own fighters in the RDF area is required, as it is noticed that cases of delay in this matter are very frequent and usually result in the enemy getting away while the fighters chase their own tracks.'[16]

When not analyzing raids the OR researchers conducted detailed investigations into various aspects of the air defence system. Among the most important OR studies were those that ventured into the area of sociology and psychology. Much of the work of Lardner's team was in finding ways to ease confusion and speed the transmission of information. This inevitably led them to consider what personnel were suitable for particular positions within the system. On 11 January 1940, Williams completed a landmark study on Filter Room organization and technique. He outlined the evolving role of the Filter Room which was 'originally envisaged to sort out and "filter" [radar] information and to pass on to Fighter Command the most probable answer.' By early 1940 the room 'had gradually receded from its original function as a purely technical room' until it had become 'more of an operations room than a filter room, and contained all sorts of people and devices [that had] nothing whatever to do with RDF.' Key tactical decisions were now made in the Filter Room and the correct application of filtering procedures was necessary if the system was to function at all. Yet, Williams observed that filtering had been corrupted by a breakdown of the system developed at Bawdsey. Under procedures worked out by Hart, the Filter Control Officer allotted stations to raids and prevented unnecessary duplication. The filterer was then assigned the task of getting the necessary information from stations to determine the true path of a raid. At Fighter Command Headquarters, however, 'the "Filter Officer" attempts to interfere far too much in the detail of the filtering; he has become an established pseudo-filterer, and much valuable time is wasted by his asking for information which has been requested already by the filterer via the plotter.'

The cause of this breakdown was 'the appallingly low standards of filtering' which had cursed Fighter Command since the completion of the interim chain in March 1939. Williams argued that it was no exaggeration to say that the efficiency of the radar system was actually less in January 1940 than during the Munich Crisis. Since radar instruments, while still

Aerial view of CHL Radar Station.

far from perfect, had undoubtedly greatly improved over this period 'the degeneration must, ipso facto, be due to personnel and methods of handling the information.'

The problem in the Filter Room was that most of the filterers were poorly educated, non-commissioned officers. The decision to use NCOs had been made during prewar exercises with the Bawdsey Filter Room. The shortage of personnel had led Hart to select two of his more intelligent NCOs as filterers. 'The fact that they both succeeded well,' wrote Williams, 'is unfortunate, in a way, in that it may have given a false impression that any NCO was capable of becoming a filterer.' Most NCOs, however, were simply not up to the task. Filtering was 'simply and solely an assessment of probability; a weighing of the accuracy of various sources of information or the estimation of the accuracy of any source of information.' Filterers 'should therefore be of the best possible type; people with technical and scientific training to assess probabilites entirely unbiased.' Williams concluded with a frightening warning:

> The writer of this memorandum is quite convinced that a complete breakdown of the Filter Room will occur if the enemy attempts a large scale air attack of even half the magnitude of that tried during the Air Exercises of 1939 ... It is hoped that in the few months of comparative inactivity which may yet remain to us, the reorganization outlined above, will be carried out. If not, then the consequences may be serious.[17]

What Williams did not explain, but was well understood by those reading the report, was that filtering was becoming more difficult, in large measure, because of the new sources of information available. This included the introduction of the first CHL radar station on 1 November 1939. CHL tracks were much more accurate in bearing but, at least initially, far less accurate in range then those provided by CH. Additionally, CHL could give no indication of altitude. Filterers had to weight the relative merit of the information from various sources, now had another factor to consider.

Both Fighter Command and the Air Ministry agreed with Williams' assessment and decided to approach the Treasury about securing financial authority for the appointment of Flying and Pilot Officers to replace the NCO filterers. The memorandum to the Treasury extensively quoted Williams' report, the best proof available of the significant role played by OR researchers. The Treasury approved the new appointments on 19 February, and the first fifteen trained officer filterers, mainly university graduates in the sciences or mathematics, reported for duty on 10 June, just in time to improve the overall quality of filtering as the Battle of Britain began.[18]

The success of OR in improving the overall performance of Fighter Command can best be judged by the increasing trust in and, therefore, growing responsibilities given to Lardner's group. On the morning of 15 May, Lardner was summoned to a meeting with Dowding. The Air Marshal outlined his growing concern about increasing demands for fighters to be sent to France. He asked Lardner: 'I know you have been studying my operations; is there anything you scientists can suggest bearing on this matter?' Dowding then concluded by saying that anything that could be done had to be done before he left for the Cabinet meeting two hours later. Lardner described what followed when he brought Dowding's request to his scientists:

> So, at the suggestion of E. C. Williams, a rapid study was carried out, based on current daily losses and replacement rates, to show how rapidly the Command's strength was being sapped and how much more rapid this would become if its losses were to be doubled while the replacement rate remained constant. For ease of display and understanding the findings were presented into a graph form. The next morning Lardner asked if the findings had been of any help. Dowding replied, 'They did the trick.'

While the graphs provided no new information, they clearly illustrated Dowding's arguments against a further weakening of his fighter force. Lardner considered this incident a 'turning point and a hint of the wider scope that lies ahead for Operational Research – the prediction of the outcome of future operations with the objective of influencing policy.'[19]

Expansion and Re-Equipment

While the OR personnel worked on improving overall performance of the air defences, the building up of the infrastructure continued as quickly as possible. Freed from most prewar concerns about money, material expansion and re-equipment continued at an ever-increasing rate. The pace of this work was, to say the least, frantic. The minutes of one meeting of radar scientists held in January 1940 stated bluntly: 'The conference was concerned with panic measures for the quick provision of CHL equipments.'[20] Each RAF command was supposed to keep a daily record of events in their Operations Record Book. No. 60 Group, which was responsible for radar maintenance, training, and, for much of 1940, the fitting of stations, was unable to keep even this simple daily record. Instead, well after the fact, some time in 1941, an officer explained this breach of standing orders in No. 60 Group's Operations Record Book:

> Owing to the extreme pressure of work which fell on the very meagre Equipment Staff existing at the formation of Headquarters No. 60 Group, it was impossible to keep a daily record of events of historical and general interest, and this task was subjugated to the paramount necessity of keeping the Stations, which played such an important part in the Battle of Britain, on the air.[21]

In certain areas this frenetic pace led to significant improvements in the battle-worthiness of the defence system. On 9 March 1940, the new underground Filter and Operations Room at Stanmore was completed and put into operation. Every effort was made to make this most vital of command centres invulnerable to bombing. Three separate cable feeds carried all outgoing telephone lines; each was buried 6 feet below the surface and encased in steel pipes. Similar underground operations rooms were established for the fighter groups by the end of May. By the time the German onslaught did commence, however, most sector control rooms and many radar stations remained vulnerable to air attack.[22]

On 2 January 1940, Prime Minister Chamberlain, after being briefed by the Air Staff on the importance of radar, accorded the highest order of priority to complete the air defence system.[23] The first phase of upgrading the Chain Home stations was by this time well underway. This involved the fitting of Metropolitan-Vickers CH transmitters, only two of which were in service at the start of the war. By the end of May 1940, all thirty CH stations had one of them.

The performance of CH stations were also being improved the introduction of a variety of new equipment designed to assist operators in speeding up the process of accurately determining the location of aircraft and improving the ability to resists both deliberate and unintentional jamming. Perhaps the most interesting of this new equipment was the only 'computer' actually part of the air defence system. Radar stations were required to pass to the Filter Room the plots of aircraft as grid reference points on a standard map. This required the converting of observed results by the manual plotting of ranges and bearings onto a grid map. Similarly, observed angles of elevations had to be converted to altitude using a height convertor chart. Each station had to take into account the peculiar characteristics of its location. These manual methods worked reasonably well when only a few target were being tracked, but it was realized well before the war that an alternative method was needed or station staff could easily be overwhelmed by a large number of raids. In the summer 1938 a number of mechanical calculator devices were tested at four chain stations.

G. A. Roberts continued work on a calculator at Bawdsey. By the summer of 1939 Roberts was collaborating with a post office engineer named Mr Marchant to develop a device that could automatically transform:

1.   Polar coordinates (the bearing indicated by the goniometer and the range) into the grid reference.
2.   Goniometer (height position) reading into angle of elevation.
3.   Angle of elevation into range and altitude.[24]

The basic design of the CH Electrical Calculator was completed in the late autumn of 1939, with the GPO's Circuit Laboratory undertaking the final design and development. The postal engineers heavily based their prototype on electro-mechanical telephone switching equipment. The prototype of the calculator or 'fruit machine' as it became affectionately known, was completed on 20 November. By the end of June the twenty-first machine was ready for installation, completely equipping all the East Coast CH stations just in time for the battle. Although criticized as being 'unnecessarily complicated and expensive' the calculator proved an invaluable addition to CH radar as it gave 'higher plotting speed and greater accuracy than any alternatives.' The calculators also proved extremely tough, one survived a 'very near bomb', and none were put out of action by the Luftwaffe.[25]

The new companion set to the CH began entering service in late 1939. On 1 November 1939, the first CHL station at Fifeness went operational. Another eight stations were raced into service on an emergency basis by 11 February 1940, primarily to combat low flying mine-laying aircraft. The rush to deal with this severe threat was so great that a number of stations were set up before enough CHL equipment was available. Instead, the stations were initially equipped with War Office gun-laying radar. In early January, the growing threat of anti-shipping strikes led to the approval of a second CHL emergency programme of seven stations before the first group was complete. All of these stations were in service by the end of February. By the end of May, thirty-six CHL transmitters had been completed.[26] The crucial role played by CHL was almost immediately apparent. By early April they were detecting almost as many raids as CH stations.[27]

At the same time as CHL was being introduced, a major expansion of the entire radar chain was underway. In late 1939, Luftwaffe aircraft began on a small scale to attack shipping in areas that were not fully scanned by radar, such as the Irish Sea. On 12 December Joubert recommended that a further eleven CH and fourteen CHL stations be rushed into service in order to provide all around protection to Britain. An addi-

Typical East Coast Radar Chain Station.

tional radar of each type was required to defend Belfast. This programme was approved in January and was on-going throughout the remainder of the year. In order to speed completion of the West Coast Chain, these stations were designed to operate with only two frequencies, therefore requiring only half the towers and antennae array of final design East Coast stations.[28]

Hand-in-hand with the growth of the chain was the introduction into widespread service of IFF. Dowding informed Tizard in November 1939 that the absence of IFF was one of three 'chief troubles' facing Fighter Command.[29] At the beginning of the war 1,000 IFF sets were under emergency production. One hundred sets were due to be completed by the end of November, with fittings on Bomber and Coastal Commands aircraft, the most likely to interfere with the defences, receiving top priority. Production was delayed while various devices were tried which would guarantee the destruction of the IFF set if an aircraft was shot down. This was necessary in order to prevent the German's from capturing IFF and learning how to duplicate its responses to radar signals. The device had to be able to destroy the set without damaging the aircraft, since pilots of badly damaged planes would be ordered to detonate it before bailing out. A safeguard also was required to automatically set off the charge in the case the IFF was not destroyed before crashing. Tests determined that the detonation of two electric detonators was sufficient to destroy the set, while the aircraft would be protected from the explosion by a steel protective band around the unit.

By the end of February 1940, only 258 aircraft, mainly bombers, were fitted with IFF. This was sufficient for Lardner's OR group to begin monitoring the performance of the equipment. They found that radar operators correctly identified IFF equipped aircraft only about 50 per cent of the time. There were several causes of this high failure rate, including faulty initial settings, and the failure of pilots to switch on the equipment, but, above all, it was the difficulty inexperienced radar operators had in identifying IFF signals. The chain station personnel found it hard to distinguish between the brief response generated by an IFF set and the indication given off by a flight of several aircraft. It was noted that in stations where scientific observers were present, the rate of successful IFF identification was greatly increased. Despite these difficulties, mounting of IFF continued at an accelerated pace. By 17 April, Fighter Command reported that a large proportion of bombers and coastal patrol aircraft were fitted with the identification device. By the end of May, improvements in the performance of radar identification led Lardner to recommend to Fighter Command 'that the only positive form of identification for all aircraft operating over the sea' should be IFF.[30]

In March, the shortcoming of IFF caused a great deal of concern, because a new improved version of the device, the IFF Mark II, was just about to go into mass production. The new set was to have a marginally shorter signal then the previous model and might be more difficult to detect. IFF Mark II was the set that Bawdsey researchers had originally wanted to design in the spring of 1938. The principle improvement was that the set was able to respond to a wider range of frequencies, particularly those used by anti-aircraft gun-laying radar. Acceptance tests of the Mark II commenced in March 1940, and while there were concerns about its pulse duration of a tenth of a second, most radar operators were able to identify the signal under normal operating conditions. The set, however, had a serious shortcoming – it could not respond to the 1½-metre wavelength used by CHL and airborne radar. There was no time for redesign, however, and on 11 March the immediate mass production of the set was ordered to go forward. By October 1940 virtually every RAF aircraft was fitted with IFF. A total of 21,000 Mark II sets were manufactured.[31]

These tremendous efforts to expand and develop the defensive system, like the introduction of IFF, certainly led to major improvements in the defences, however time and resources were not enough in every case. Nor was there sufficient prioritizing of requirements; this led to a dispersal of efforts which resulted in the defences being less effective than they could otherwise have been in the summer of 1940. One of the most serious shortfalls was in the failure to introduce Very High Frequency (VHF) air-to-ground radio-telephones before the Battle of Britain commenced. Reliable communications between fighters and ground controllers and direction finding was central to the interception system. The introduction of the High Frequency (HF) TR 9D radio in the late 1930s had brought about a great improvement in communications, but as early as March 1938, Fighter Command was aware that these sets were critically deficient in terms of range and signal clarity, as well being liable to jamming. Research on a VHF radio, which would offer much better performance while requiring the same amount of space and using a similar amount of power, had been under way at the Royal Aircraft Establishment since 1935. In July 1938, RAE reported that it would take at least four more years of research to perfect VHF equipment. Alarmed at this, the Air Staff took steps to improve HF performance by building a series of forward radio stations for each sector, to ensure that fighters could continue to receive messages from control rooms as they approached the coast. RAE was also asked to cut corners and to produce an interim VHF radio which, while not perfect, would be ready well before 1942. On 9 January 1939, Watson-Watt and the Director of Signals reported to the Air Ministry that RAE could produce such a set and equip eight sectors by September.

Proceeding with the VHF programme at this date was, admitted the Assistant Chief of the Air Staff, a bit of a gamble because VHF and HF were not compatible. Still, the benefits were felt to outweigh the risks, and on 16 January, Newall approved the introduction of the new TR 1130 VHF radio. In order to minimize the risks of getting caught partially equipped with the new technology, the TR 1130 was cleverly designed to be similar in shape and size, and have identical means of attachment as the TR 9D; this made it possible for an aircraft radio to be switched from one type of set to the other in little over a hour and a half.

As with many programmes Watson-Watt's forecast for the completion date of the first stage of fitting of VHF proved to be overly optimistic. Only in early October was the first aircraft fitted, and full scale trials with six Spitfires did not begin until the end of the month. The TR 1130s had more than twice the range of the early sets, air-to-ground ranges of up to 140 miles and air-to-air ranges of 100 miles. According to the official historian: 'Speech was clearer, pilots' controls simpler and quicker to operate, direction finding was sharper; in every way the TR 1133 was beyond comparison with the TR 9.' Within days Newall ordered the VHF radio into general service.

Production of the TR 1133, however, was 'tardy,' and by May 1940, only the initial production run of 240 sets had been delivered. Further deliveries would not be forthcoming until the summer. When British-based aircraft were called into providing protection for the Dunkirk evacuations at the end of the month, Fighter Command was faced with operating two incompatible radio systems. The decision was very reluctantly made to remove, for the time being, all VHF radios from service. The TR 1133 would not reenter service until mid-August when supplies of the sets improved.[32]

Even more serious was the failure of the training organization to keep pace with the expansion of the air defences. The maximum pre-war estimates called for a total of 1,800 radar personnel. By early October 1939, the Air Ministry had set a target of training 3,600 men and women per year to meet home and overseas requirements. This target was unreachable in the first year of the war and the training programme could barely be expanded to accommodate the minimum number required to man the home defences. Despite prewar commitments to set up another school, Bawdsey remained the only such establishment until January 1940, when the training of radar mechanics was transferred to Yatesbury. Fortunately, the removal of the scientists from Bawdsey left plenty of room to expand. The security of the training school was not considered either sufficiently imperiled or important to require a move to a safer location, despite the fact that many of the same top secret devices in use by the researchers were also in operation for training purposes.

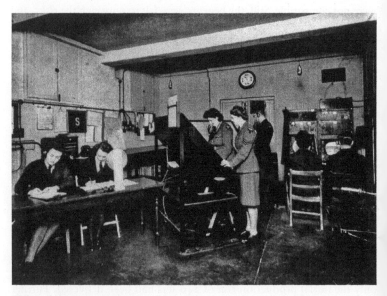

CHL Radar Receiver Room.

On 9 October J. A. Tester, who was placed in charge of Bawdsey training programme after Hart's transfer to Fighter Command in August, reported that even after expansion his school could only train thirty-six radar operators per fortnight. This rate could only be achieved by operating all equipment twenty-four hours a day six days a week. The main training equipment consisted of four mobile radar receivers which were fitted with a device to simulate a single track of an aircraft on the cathode ray tube. The two week programme was, Tester admitted, inadequate. Trainees graduated with no knowledge of CH receivers, how to track multiple raids, anti-jamming devices, IFF, height determination, counting of aircraft, and plotting and filtering. Radar mechanics were even less well prepared, having no knowledge of any transmitters, most receivers, and 'any work culminating in the equipment actually going "on the air."'[33]

While improvements in the training programme were gradually introduced, most graduates were completely unprepared for operations. It was left to the radar stations themselves to finish the training, but expansion diluted the pool of experienced talent available at the chain, and the standards of performance steadily dropped. Virtually every OR report and many Fighter Command documents make reference to the diminishing skills of the average radar station personnel throughout the first year of war. The only exception to this trend was in the half dozen stations which were fortunate to have a scientific liaison officer present.

As well as the decline in operators' performance was a continuing catalogue of technical shortcomings with the radar stations. Expanding the coverage of the chain was a necessary strategic decision, but it precluded the completion of any of the stations to the final design until early 1941. In one exhaustive and exhausting sentence, Watson-Watt, in his inimitable style, recalled the state of the air defences at the start of the battle:

The CH stations entered the Battle with good and dependable indoor equipment; aerials not yet nearly as good as we would have liked; calibration seldom complete, completion of calibration greatly hampered by necessary restrictions due to battle conditions; range-finding good; direction-finding fairly good; height finding, as always a delicate and difficult operation; height-reporting sometimes made more misleading by our very success, because first height reports might be made on formations which had not yet completed their climb to operational height; counting an art still in the learning; continuity in individual tracking impaired by still half-developed procedures in heavy traffic conditions; filtering and display still being learned in the only available school, that of full-scale utilization; supply of suitably minded and adequately experienced observers, reporters and plotters very acutely limited.[34]

The stations particularly suffered from their inability to accurately measure height. Throughout the Phoney War, perhaps the largest efforts to develop new tactics were devoted to methods to use a direct radio-telephone link between radar stations and aircraft to intercept Luftwaffe aircraft operating off the coast. The position of sub-controller was created, responsible for directing the attack by direct observation of the radar screen. Experiments with the 'all RDF methods of interception' began in the summer of 1939 using CH radar and were greatly expanded in the first nine months of the war, particularly after CHL became available in early 1940. There were a few notable successes, such as in the morning of 12 May, when a Bristol Blenheim was able to intercept and shoot down a Heinkel bomber patrolling some 55 miles off the coast at Bawdsey, yet on the whole, the scheme was beyond the capabilities of the system. On 27 March, Dowding lamented to the Air Ministry the very limited ability of radar to provide accurate height readings and its repercussion on interceptions.[35]

In early June, two extensive OR reports were produced concerning the failure of these experiments and on radar's problems with measuring height. They concluded that interception over the sea usually failed because CH was not accurate enough and CHL, owing to it being originally designed as a Coast Defence set, had no capability to measure height.

As designed, the CH stations were to be able to measure the angle of elevation from the station between 1½ and 15 degrees, however, Williams reported that, 'at the moment, no single RDF CH station has all its equipment on even one wavelength. And it is not considered likely that the whole Chain will be finally equipped for some time to come.' Most stations could only measure aircraft at from 1½ and 6 degrees. Since at whatever the angle from the station aircraft could be detected if flying between 5,000 and 25,000 feet, the results were that operators received false readings when trying to measure height if the actual degree of elevation was either below or above those capable of being accurately measured. This led to often wildly inaccurate height readings, particularly by inexperienced operators.[36]

The height problem dilemma was caused by more than just the slowness of the physical completion of the stations. The main piece of equipment missing was sufficient numbers of height antennae in the masts, which would allow switching between them so that the angle of elevation could be determined. It was not sufficient to simply mount the necessary antennae, because each time modifications were made to the antennae array, the entire station had to be re-calibrated. Calibration of a CH final design station, of which none existed in the summer of 1940, would require a total of 400 different observations. Fighter Command and the Air Ministry allocated far too little resources for calibration prior to the Battle of Britain. There were only two autogyros, two ships equipped with balloons, and a single flight of aircraft which was also responsible for aiding in the training of ground controllers at Bawdsey. As a result of these limited resources and bad weather, which made accurate calibration impossible for much of the winter, most CH and CHL stations were, at best, only partially calibrated by the summer.[37]

While there were numerous technical shortcomings, perhaps the greatest uncertainty about the air defences rested not with the equipment and personnel but with the fundamental nature of its command and control organization. Dowding was so intimately connected to the research, development, and deployment of virtually all aspects of the system that his name would ultimately become synonymous with the defensive scheme; it would later be dubbed the 'Dowding System.' As much as he relished being the master of all of the technical details, he also insisted that he retain control over most important strategic decisions. Dowding was what we would call today a micro manager, unwilling or unable to divorce himself from the details and the responsibilities of command. The Filter Room at Fighter Command Headquarters controlled information from all radar and other long-range sources of intelligence, except for a handful of stations in the far north of Scotland. At the start of the war the filtered information was sent from the Filter Room to the Operations Room next door. Only after Dowding or the senior watch officer could assess the situation on the Operations Room map would the appropriate group headquarters be informed. Group would then have to analyze the situation and allocate which sector should be assigned to intercept approaching aircraft. Dowding believed that this centralized approach was necessary in order that he could give orders for air raid warnings to be given to major population centres, and to ensure that the raids were allocated to the right group.

The limitations of Dowding's approach were apparent at the start of the war, and as the defences grew in size and complexity, it became increasingly difficult for all filtering to be done at one centre. Dowding's insistence that he retain control over all aspects of the radar chain was supported by the Air Staff in late 1939 when they rejected the recommendation of the Tizard investigation into the chain to turn over radar tracking to a separate command. This idea had been sold to the Tizard's Chain Committee by A. M. Joubert. Joubert, who had been in charge of the fighter forces immediately prior to the creation of Fighter Command, and then became the first Commander-in-Chief of Coastal Command, believed he shared with Dowding the credit for the successful promotion of radar as a key component of the air defence system. Joubert was appointed by Newall in the autumn of 1939 to act as the Air Staff's senior officer responsible for radar. When Joubert failed to persuade the Air Staff of the need for a separate radar command, he continued to watch with growing anxiety the way Dowding was managing the dissemination of information in Fighter Command.

In early January 1940, Joubert convinced Newall that there was a need for a meeting between senior Air Staff officers and Dowding in order to discuss the failure of Fighter Command to intercept enemy aircraft operating off the East Coast. Joubert recommended to Newall that Dowding must be forced to accept changes in procedures which would lead to 'the reduction or removal of any time lag which occurred between the receipt of the RDF information and the dispatch of the fighters.' Joubert also worried that the single filter room at Fighter Command Headquarters might be overwhelmed by the sheer amount of information it would receive during a major attack and that it was also highly vulnerable to enemy action. The creation of similar rooms at the group level, he believed, would both speed the flow of information and provide extra security.

The minutes of the meeting held on 12 January are brief, which belies the fiery confrontation that must have taken place between Dowding, Newall, Joubert and other senior staff officers. It appears that Dowding was ambushed, having been given no advanced warning about the subject of the meeting. Dowding was forced to agree to change filtering procedures so that raid plots would henceforth be passed to groups 'at a stage not later than when sufficient information has been received to enable one directional arrow to be placed on the Fighter Command filter room table.' It was also agreed that plots from CHL stations should be transmitted directly to nearby sectors which could immediately take scramble fighters to intercept low flying aircraft. Dowding, however, was able to hold his ground over the question of establishing filter rooms at group headquarters. He agreed only to 'consider' the matter.[38]

In late January, the Air Staff felt it necessary to take the unusual step of following up with a letter to Dowding confirming the points agreed to at the meeting. Joubert wrote the draft of the letter carefully, trying to ensure that Dowding had no room to wriggle out of the Air Staff's demands.[39]

On 31 January, Dowding submitted a fierce rebuttal to the Air Staff, refusing to implement most of their demands. He informed them that while he had already ordered that plots be transmitted to groups as soon as the Filter Room had labelled a raid, he refused to consider any alterations in the way information was disseminated. He argued that a sector could not be informed about a possible raid until it was clearly determined that its area was threatened. If a sector was designated too early, the enemy might change course forcing other sectors to also scramble interceptors, and resulting in 'a great waste of effort.' The Air Staff's concerns over the security of the Headquarters Filter Room were 'easily disposed of' by the nearly £100,000 spent on placing the Filter Room underground, the existence of a backup filtering centre at Leighton Buzzard, and the ability of chain stations to communicate directly with sectors in an emergency. Dowding doubted the Filter Room staff could be overwhelmed by the massive amounts of information flowing in as the result of a major attack. He concluded with a direct challenge to the Air Staff's authority to make tactical demands on his command. He wrote:

> My contention is that the Air Council have the right to tell me what to do but should not insist on telling me how to do it so long as I retain their confidence.
>
> I have spent a great deal of time on this subject, starting from the proposal to form a Communication Command, I calculate I have devoted about 30 hours on the subject, and a great deal of nervous energy which might have otherwise been expended. I therefore deprecate the suggestion that a further meeting should be held to discuss the subject, since there are several matters of great importance which I have had to put aside pending the discussion of this issue.[40]

Dowding won this showdown with the Air Staff, since the only answer to his blunt refusal was to demand his resignation, something Newall was unwilling to do. The issue of centralization, however, would not go away. Several OR reports, and reports from technical officers in the first half of 1940, mention the need for group filtering and greater independence of sectors.[41] Joubert did not revisit the issue until the Battle was just beginning, when this was far too late to affect any changes to the system. On 11 July, Joubert wrote a response to a letter he had received from Archibald Sinclair, the Minister of State for Air in the new government of Winston Churchill. Sinclair asked about the need for introducing a less centralized command and control system. Joubert began by cautioning Sinclair that: 'I know that the C-in-C Fighter Command feels most strongly on this subject, and to raise it again would be tantamount to moving a vote of want of confidence in him.' While he would not recommend any action be taken at this time, he explained to Sinclair that his own personal viewpoint was that, when he commanded the fighter force and was planning the communication system for Fighter Command, groups and sectors were supposed to have much more direct access to information and freedom to act independently. He concluded:

> Probably the answer is that no two men work any system alike. I believe in decentralization, particularly when speed is required. C-in-C Fighter

Radar

Command believes in retaining things under his own hand. So far nobody has been brave enough, except myself, to tell him that he is wrong, but I am exceptionally rash![42]

Joubert tried to soothe Sinclair's worries about the system by reducing his concerns about the basic command and control structure into a mere difference of opinion. Joubert was unquestionably right about taking this approach. There was no need to worry the politicians about the air defence system, since whatever its flaws, it was far better than anything that anyone could have imagined in 1934, when Rowe and Wimperis had began the scientific investigation into a means to stop the bomber from always getting through. The chain was the backbone of the air defence system; if it worked sufficiently well enough and the dissemination of information flowed as Dowding hoped, then Britain would have a fighting chance to survive. The ultimate test of the air defences lay just days after Joubert wrote to Sinclair, for 15 July 1940 is the day most British sources consider the Battle of Britain to have begun.

12

The Battle of Britain

The Phoney War provided Fighter Command and its supporting scientists an opportunity to expand and fine tune the air defence system. Major gaps and inadequacies in the system remained despite desperate measures to fix them. Luftwaffe activities tested the defences, but never seriously threatened Britain's security. The air war was not unfolding as pre-war pundits had predicted. At the start of the war, anxious civilians were issued gas masks and thousands were evacuated to the countryside or even overseas to seek safety from the expected all-out bomber offensive. When these attacks failed to materialize, complacency gradually returned. In the summer of 1940, however, Britain suddenly faced the prospect for an all-out air offensive and the fear of the Luftwaffe, mixed with a fair bit of British defiance, returned. The remarkable courage and pluck of the average citizen was startling because none knew of the miracle of radar which now shielded Britain from attack. All evidence of war so far, suggested that Baldwin was right, the bomber would always get through.

On 1 May 1940, Germany launched its invasion of France and the Low Countries. Throughout the campaign, as in Poland and Norway before, the Luftwaffe proved to be a vital component in the Blitzkreig. Flying modern twin-engined bombers such as the Junker 88, the Dornier 17, the Heinkel 111 and the deadly accurate Junker 87 Stuka dive bomber, the German Air Force was able to provide close air support to the army in the field, devastate air bases and other military installations well behind the lines, and spread panic among both soldiers and civilians. The German fighter force, which consisted of the superlative single engine Me 109 and the overrated twin-engine Me 110, proved more than a match for the British and French air forces.

The Allied air forces were crippled by the lack of an air defence system equivalent to that which defended Britain. The French were only informed of radar in the summer of 1939 and were only just coming to appreciate the complexities of the command and control apparatus of Fighter Command. The British, desperately trying to finish their own air defences, provided little material help. The French air rearmament program was well behind the British and German, and they only had few fully modern aircraft. The RAF's own army co-operation squadrons, which had been given lower priority then either the fighters or strategic bomber forces, were no better then the French. Flying Fairy Battles and Bristol Blenheims, much of the BEF's air component was wiped out in desperate but unsuccessful efforts to destroy German bridges across the Meuse. The Hurricanes sent to protect the bombers did their best, but they were overwhelmed by superior numbers of German aircraft, and their bases were repeatedly attacked with little warning. Nearly 200 Hurricanes and, more importantly, a large number of pilots, were lost. This represented about 25 per cent of Dowding's fighter force. It was this huge drain on resources that led Dowding and Newall to convince the War Cabinet that no more fighter squadrons could be sent to France.

On 26 May, the BEF and a large number of French Divisions were surrounded at the French port of Dunkirk, and the famous evacuation of troops began. Hitler allowed Reichsmarschall Hermann Goering, the Commander-in-Chief of the Luftwaffe, the honour of finishing off the trapped troops and preventing their escape. For the first time the Luftwaffe was confronted by the Fighter Command aircraft operating from British airfields. Unable to utilize the radar early warning system because of the short range of their HF radios and the distance from their bases, the fighters were forced to fly standing patrols over the beaches and approach routes to them. The need to maintain cover meant that the RAF was often heavily outnumbered by large scale German attacks. The RAF won the battle over Dunkirk, since the Luftwaffe was unable to stop the evacuation of some 338,000 Allied soldiers. It was, however a costly victory; some 106 RAF aircraft and fifty two pilots were lost during the nine day evacuation. The British shot down 132 German aircraft, including those shot down by the Royal Navy.

Fighting would continue in France until 21 June when a cease-fire was declared. Germany had conquered France in little over a month. When Britain, under the leadership of Winston Churchill, who was made Prime Minister on 10 May, decided to reject German overtures for peace and fight on, the stage was set for the greatest air battle fought up to that time. If Hitler wanted to defeat Britain then the German army had to invade across the English Channel. This could only be accomplished with absolute air superiority, which was required to counter both the RAF and the still great might of the Royal Navy. Goering and other senior Luftwaffe officers were confident that once their air force was ready for full scale military operations in early August it would take them just four days to destroy the fighter defences in Southern England and a further four weeks to eliminate the rest of the RAF.

The Luftwaffe had good reason to be confident. It had destroyed the air forces of Poland, Holland, Belgium, and France, as well as the RAF aircraft supporting the BEF. It had been instrumental in countering the superiority of the Royal Navy during the Norwegian campaign. Once regrouped and re-equipped after the defeat of France, the three Luftwaffen or air fleets that confronted Fighter Command numbered some 2,800 aircraft. This dwarfed the approximately 700 Hurricanes and Spitfires available to Dowding by mid-July. That the fighter pilots, those that Churchill's immortalized as the 'few', were able to hold their own has often been viewed as a miracle and a tribute to the fighting power of the Hurricanes and Spitfires, and the heroism and skill of their pilots. John Ray, in a recent re-examination of the Battle, has challenged this view that the RAF struggled against tremendous odds, arguing that the number of British fighters should be compared to the quantity of Me 109s available – just 760 – which slightly outnumbered Fighter Command.[1]

Ray and earlier historians of the battle, however, have not placed enough emphasis on the radar based air defence system. No matter how one compares the numerical strengths of the respective forces, these comparisons are meaningless without equal emphasis on examining the strategy and tactics of the opposing air forces. There can be little doubt that the extraordinary bravery of Fighter Command's pilots and the mechanical excellence of their planes would have been irrelevant, if not for the early warning and tracking system and the sophisticated command and control apparatus created to disseminate this information quickly and efficiently from the CRT tubes of the radar screens to the cockpits of individual fighter pilots. If this system had suffered a complete failure, either through its own inadequacies or enemy action, then surely the RAF would have been defeated.

There have been so many operational histories of the Battle of Britain that there will be no effort made here to provide yet another blow-by-blow account. Instead a detailed examination of the effectiveness of the air defence system, as opposed to aerial operations, will be provided. However, a brief summary of the campaign is useful as a reminder. British histories of the Battle of Britain generally divide the campaign into several distinct phases. The first began around 15 July, when the Luftwaffe began a series of escalating attacks on coastal convoys, ports, and, occasionally, coastal airfields along the south and east coasts. German fighter escorts were used to wear down the resistance of Fighter Command. The second and most crucial stage of the battle began in the second week of August with the opening of Goering's four-day offensive to destroy the RAF – code named

Alder (Eagle). The Luftwaffe launched large-scale strikes against RAF installations including airbases, communications centres and, to a limited extent, radar stations. This was the decisive phase of the battle, in which the RAF was able to inflict severe casualties on the Luftwaffe while, just barely, maintaining the air defences in operation. The third stage commenced on 7 September, when the Luftwaffe switched it strategic priorities to attacks on London and other British cities. Daylight attacks were designed to draw the RAF into the final decisive battle which never came. Day and increasingly frequent night attacks on urban areas were also designed to break British civilian morale. The final stage of the battle commenced in early October, after the Germans gave up their plans to invade Britain. This was the period of the Blitz when the Luftwaffe ceased large-scale daytime attacks and concentrated instead on night attacks against British cities. The Luftwaffe continued to attack specific targets by day using fast fighter-bomber formations flying either very high or very low. By this stage of the battle, the threat of invasion was gone and the RAF had won.

The Radar Air Defence System

The crucial role of the radar air defence system to the RAF's victory in the Battle of Britain has been appreciated since 1940. On 21 December 1940, Watson-Watt wrote to Archibald Sinclair, the Secretary of State for Air, on the future of radar construction. He claimed: 'It is, I believe, agreed that, so far as concerns material, the "First Battle of Britain" was won by RDF and the 8-gun fighter. Our old statement that RDF would multiply by three, and perhaps by five, the value of our fighter force, has been justified.'[2] This is one occasion when Watson-Watt's exuberant claims are not exaggerated.

Virtually all sources on the Battle of Britain agree that radar early warning and interception were key components of British victory. This view was held at the time of the battle by less-biased observers than Watson-Watt. During the first stage of the battle, Keith Park, who had been promoted to the rank of Air Vice Marshal and placed in command of No. 11 Group which bore the brunt of Luftwaffe attacks, reported that 'RDF at its worst is most valuable to my Fighter Group, and at its best, it is quite invaluable and a great boon.'[3] In November 1940, Stanmore OR researchers concluded that 'the fundamental part played by RDF in destroying any hopes which the enemy had placed on the results of the mass raids of August and September can hardly be contested.'[4]

Dowding wrote extensively about the role of radar and the raid reporting system in his official report on the battle, composed in 1941 and published after the war. After outlining the various deficiencies that existed with the chain in July 1940, Dowding stated: 'In spite of these handicaps, however, the system operated effectively, and it is not too much to say that the warning which it gave could have been obtained by no other means, and constituted a vital factor in the Air Defence of Great Britain.'[5]

The Air Ministry's historical section completed a number of secret official studies of radar, the air defence system, and the battle. All of these studies emphasized the crucial role played by radar in the battle and provide a more thorough analysis of the benefits provided by the system. The study of radar in the air defence system pointed out that in large measure, 'the air struggle was fought without any large deviation from the technique of raid reporting, and fighter control organization evolved for defence in air exercises before the war.' The system provided Fighter Command with a 'tremendous advantage' which in large measure counteracted the superior numbers of the enemy.[6]

Historians have, for good reason, generally followed these assessments. No major study of the battle has failed to cite the central role played by radar and the Dowding system, although none, until this study, has examined in any detail the history of the creation of the air defence system. While historical revisionism has challenged some of the standard arguments about the battle, the centrality of radar has withstood intense scholarly scrutiny. The only modification of this aspect of the story has been the inclusion of information regarding the role that intelligence played in providing tactical information to Fighter Command. Unquestionably the Air Intelligence Branch of the Air Ministry proved an invaluable source to Dowding on German operations before, during, and after the fact. Intelligence used high level decryptions of Enigma-coded radio traffic and low level signal intelligence (Sigint) of Luftwaffe air-to-air and other low-grade radio traffic. The issue of the tactical value of this intelligence to daily operations remains, however, a 'vexed question.' This is particularly true of the Enigma messages, since historians still remain uncertain if, and when, Dowding was informed of the existence of the ultra-secret. The official historian of intelligence viewed Enigma as having only strategic value.

According to the best study of intelligence in the battle, however, 'the contribution of low-grade Sigint to the difficult task of assessing the enemy's intention from the confused and conflicting radar on the operations room table was obviously of great importance.' Sigint worked as an adjunct to information generated by radar. It was often able to detect the radio traffic of raids taking off from their bases in France before radar could detect them and sometimes provided more accurate indications of height than the chain stations. Moreover, it was occasionally possible to determine

the makeup of the raids and their targets on the basis of the Germans'
radio chatter; this was something that radar could not do. Sigint had significant limitations, however. It could not direct a successful interception,
nor could it always track the various changes in course and altitude taken
by particular formations. In this, radar was the only useful source until
the Luftwaffe aircraft crossed the coast.[7] Comparing the relative tactical
value of radar and radio intelligence is not particularly productive. In May
1941, Wg/Cdr Thomas Lang, a controller in the operations room at No.
11 Group during the battle, provided an account of his experiences. He
viewed both sources as being indispensable in providing Park with a clear
tactical appreciation of the developing German attacks.[8]

While the great value of the radar-based air defence system cannot be
questioned, it is important to examine just how well it actually performed
in operations. Overall, the performance of the air defence system was far
better than might have been expected given the circumstances. We have
already examined in detail the shortcomings of the air defence system
as the battle began. Moreover, the system faced a strategic and tactical
dilemma it had never been designed to meet. The situation that confronted
Fighter Command in the summer of 1940 was far different than what had
been considered in prewar planning and air exercises which had molded
the system. When France was defeated, the Luftwaffe was able to take
over airbases far closer to Britain than anticipated. This caused several
major challenges to the air defences. Many of the air bases in France were
within the maximum range of CH stations. This tended to decrease warning times and make height readings far more difficult, since formations
could be observed climbing into position. Particularly in the earlier part of
the battle, when much of the fighting took place over the English Channel
and North Sea, radar was frequently deceived and gave inaccurate readings as to both height and numbers. The technically complexity of height
finding and number counting in these circumstances was exaggerated
by the inexperience and poor training of many of the operators and the
incompleteness of the radar stations.

The second surprising challenge faced by the defenders was that German
bombers would be escorted by high performance single seat fighters. It is
a tribute to the overall strengths of the system, Park's handling of No. 11
Group, the skill of his fighter pilots and the excellence of their aircraft
that such an extraordinary challenge could be met. Only once was the
air defence system able to function as it was designed. On 15 August, the
day the Luftwaffe exerted its greatest effort to overwhelm the defenders,
Luftflotte 5, which was based in Norway, made its only major daytime
attack of the campaign. Three separate groups approached the coast from
Flamborough Head in the south to the Firth of Forth in the north. Radar

Control Room indicators.

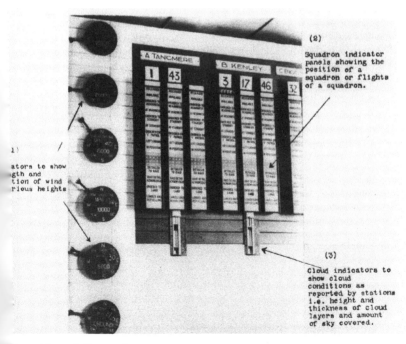

Control Room Indicators close up.

provided over an hour's warning allowing Nos 12 and 13 Groups more than adequate time to get fighters into the air for successful interceptions. The Me 109 was incapable of flying such a long-range mission and the only fighters available were thirty-four Me 110s, which escorted two of the three groups. All three formations were met, and only one was able to press through to their primary target. All the formations suffered heavy casualties; in total some sixteen out of 123 bombers, and seven out of thirty-four fighters were shot down. Many other aircraft were damaged, and others abandoned their mission on the approach of the first Hurricanes and Spitfires. It was a rate of loss that no force could sustain, and Luftflotte 5 played no further role in the battle until its aircraft were transferred to support the forces operating from France and the Low Countries.[9]

Given the circumstances the overall performance of the system was excellent. As one Air Ministry study pointed out, 'while the RDF stations occasionally failed to locate an enemy formation, they never failed, on a single occasion, to locate a major attack.'[10] From a strictly mechanical perspective, despite the limitations of the stations and their crews, radar proved to be a highly reliable piece of equipment. Throughout the period of the most intense air attacks, from early August to early September, the vast majority of stations, over 97 per cent, remained on the air twenty-four hours a day. Moreover, throughout the battle the efforts to complete the system were on-going, and by October, when the daylight campaign began to peter out, the raid reporting system was generally far superior than it had been in July. On 1 July there were fifty-four stations in operation; by 30 September this had grown to seventy-six. The rate of growth in radar installations accelerated in the winter of 1940-41. By mid-April 1941, after just twelve months, there was a fourfold increase in the number of CH and CHL stations in service.[11] As well, stations were in generally better technical shape. By the end of 1940, many of the CH stations were at last completed to final design standards and important technical improvements were being gradually introduced into CHL radar. These improvements included a motorized rotation mechanism for the antenna, which allowed for far greater rates of search, and a Planned Position Indicator much more accurate and swifter interpretation of radar (PPI), which presented information in plan like a map and allowed for much more accurate and swifter interpretation of radar information. *Developments*

Shortcomings in the Air Defence System

The raw statistics on radar performance, however, do not mean that radar and the command and control system always worked well. There were serious problems throughout the campaign caused by the weaknesses of the system. In the first stage of the battle, during which the Luftwaffe focused on attacking coastal convoys to draw the RAF into a battle of attrition, radar and filtering errors often left squadrons heavily outnumbered and flying at the wrong altitude. On 24 July Park reported to Rowe that 'the RDF chain and CHL stations on the South Coast have been behaving just like a young lady – as fickle as can be – during the past month or so.' He outlined the problems that radar's fickle nature posed to his command:

> The other day, a raid variously reported as 3 plus, 6 plus and finally 9 plus, turned out to be about 80 aircraft; so the fighter pilots reported after being shot up rather badly by vastly superior numbers. We dispatched 18 fighters to intercept this raid, because our controller sensed that it must be more than nine. This is not typical, but there was another instance almost as bad this afternoon, when two and a half squadrons were surprised to find that they were engaging between 45 and 50 enemy fighters and bombers in the Dover Straits. This sort of thing makes economical fighting very difficult, and I am particularly anxious that the RDF should not be discredited.[12]

Rowe agreed with Park, believing that 'the question of under counting [was] an extremely serious one because the service [was] apt to oscillate from complete disbelief in RDF to a pathetic belief in its infallibility.' Rowe asked W. B. Lewis to investigate the situation immediately. He reported back the next day that counting 'suffers most from attempts to simplify it, e.g. it has been said "count the number of independent peaks (on the CRT display), multiply by 3 and add a plus to the answer."' Unfortunately, 'this could only possibly be right for a certain formation viewed from a certain direction, and with a certain tuning and bandwidth on the receiver.' The underlying cause of this simplified approach to counting was lack of training and experience.[13]

Lewis' conclusion was supported by two other well informed sources. On 17 June, as an emergency measure, the Air Ministry ordered that the radar operators' course of training be shortened from four to two weeks. On 29 July Wing Commander D. R. W. Thompson, the commander of the radio training school at Yatesbury, bluntly informed his superiors 'that this drastic reduction in time has resulted in an extremely poor performance on the part of the trainees.' If this shortened course of training was adhered to, it was 'certain' that the performance of radar stations would be 'seriously reduced, and information passed to Fighter Command [would] be unreliable.' Thompson was so adamant that those graduating from the training programme were not ready for their new duties that he refused to sign his name on papers certifying the graduates as radar operators.[14] On 6 August, E. C. Williams informed the head of the signals directorate that 'there is no doubt that at the moment the average standard of operators on the RDF Chain is lower than it ever has been.' The shortened course at the radio school 'should never had been allowed', but even worse was that in desperation there had been 'direct intake of completely untrained personnel to the Chain.' The latter was 'very dangerous and should never had been undertaken or authorized.'[15]

Harold Lardner, from his vantage point at Fighter Command, was not convinced that the undercounting and poor height measurement were completely the fault of the radar stations, but 'that there was some evidence to show that stations [were] not always successful in getting Filter Room to display on the raid plaques the number of aircraft as estimated by the stations.' He attributed the confusion to Luftwaffe tactics in which formation were spaced vertically one above the other. This was presumably how radar observed formations of fighter aircraft flying escort above the bombers. While radar tracked two or three separate formations closely stacked together and provided numbers for each, the filters were discounting the existence of these multiple formations and indicating the count for only one group of aircraft.[16]

Whether it was radar operators or Filter Room personnel, and it was likely both, who were responsible for undercounting and poor height readings, the problems persisted throughout the battle. It remained a serious cause for complaint from Park, although gradually the counting of the number of aircraft in raids became better as operators gained experience. As for height readings, improvement here was more difficult owing to the technical shortcomings of the stations, and changes in German tactics which will be discussed below.

As attacks intensified in August and moved inland to strike at No. 11 Group's ground installations, the air defence system began to suffer from other deficiencies. Perhaps the most significant was that at peak periods, radar stations, the Filter Room, and communications between them became fully saturated and unable to process all the data on Luftwaffe activity. In effect the system began to suffer from what today we call information overload. In an OR report completed in 5 August 1940, Lardner predicted that the system would begin to break down. He pointed out that there was a fundamental flaw in the air defences that would lead to saturation if Luftwaffe attacks continued to increase. The system, Lardner concluded, was trying to serve two separate functions, long-range early warning and interception, which required different types of information and equipment. One could not improve one without detracting from the other. For instance, CHL stations could track a single target with great accuracy, except in determining altitude, by ceasing to sweep its antenna around in a search pattern and instead concentrating on the single target. Early warning of other approaching aircraft, however, was sacrificed in this process.

However, the overloading of the system did not lead to its collapse, in large measure because the crucial elements of the reporting system were human and not mechanical or electronic. Radar station and Filter Room, staff very quickly improvised what OR scientists would later dub the macroscopic reporting system. Rather than trying to report details on each track to the Filter Room it was found necessary to 'report the characteristics of the activity as a whole.' Only the property of a mass raid could be sent to the Filter Room, indicating 'its probable strength (e.g. 100+), the mean track followed by the raid as a whole, its range spread and the mean height.'[17] This macroscopic reporting method allowed the system to continue to function, but just barely. Sufficient information was passed through to groups in order that the majority of raids were intercepted. Perhaps as important, the Luftwaffe was generally deprived of its favorite tactic of destroying enemy aircraft on the ground. As Park, reported for the period between 8 and 18 August, 'group controllers had to be very vigilant to prevent forward squadrons from being attacked on the ground at forward aerodromes, and on only one occasion was a squadron caught on the ground, and this was because they failed to maintain a patrol overhead while the rest of the squadron refuelled.'[18]

This improvised system, however, did not always function and there were a number of cases where a radar station was ordered to cease transmitting information to the Filter Room because they were 'unable to handle the amount of information available.'[19] In mid-September, Dowding grew concerned with the growing number of cases during periods of 'saturation' in which 'important information regarding enemy tracks either failing entirely to appear on Group or Command Operations Tables, or only in part and after undue delay.' The problem appeared to Dowding to rest with the tellers. The teller were responsible for transmitted information by voice from radar stations to Fighter Command Headquarters, from Command Headquarters to Group and then from Group to sectors. Dowding informed No. 11 Group and the Commandant of the Observer Corps that:

> The general conclusion that has been reached is that the telling of information in these circumstances must be selective, that is, the teller must understand the point of view of the Operations Room to which he is telling, and the type of information, i.e. the plots, which are most essential, and also the purpose for which information is required.

Dowding established an ad hoc research team of OR scientists and officers, under the command of Roberts, to investigate.[20]

Only after the crucial period of the battle was it possible to scientifi-
cally assess what caused the over-saturation of the air defences. In late
November the OR group published its exhaustive investigation into the
capacity of the system during the battle. They found a variety of reasons
why detailed reporting had been suspended, including telling, but the principle cause was the inadequacies of the Filter Room and the failure to
decentralize filtering to group headquarters. They concluded:

> At present 5 filterers and 3 tellers tend to areas covered by 11, 12 and
> 13 Groups, whereas, according to the analysis, 7 filterers and 3 tellers
> would be needed to handle all the information that could be supplied,
> given adequate intelligence links, by the RDF stations serving 11 Group
> only. Since the output of information from RDF stations can certainly
> be doubled, it follows that the Filter Room as constituted at present is
> likely to be totally inadequate for its task for the major portion of each
> working day.[21]

In September, another major technical shortcoming of the system was
revealed which forced Park to revise an old idea in order to meet a new
German challenge. When the Luftwaffe switched from bombing No. 11
Groups ground installations to targeting London and other urban centres, the large formations proved much easier to intercept than the smaller
forces which had attacked airfields and other installations in August. As a
defensive measure, the Luftwaffe gradually began to fly much higher, often
at heights over 15,000 feet. CH stations had increased difficulty accurately
tracking targets at this altitude, and radar 'warning deteriorated considerably.' Even if radar provided early warning, the height of the attackers
often precluded interception until after they had crossed the coast. Park
was forced to withdraw his fighters from air bases along the coast to those
further inland. This gave the fighters sufficient time to climb into position
and allowed them to avoid meeting high flying Me 109s while still trying
to reach combat altitude. This left it to the Observer Corps to provide the
information needed to complete an interception, but they found it almost
impossible to see the high flying aircraft, a problem compounded by the
increasingly cloudy September skies.[22] Park was forced to revert to a form
of standing patrols which would take off at the earliest indication of a
large Luftwaffe attack. Controllers would try to keep the fighters on their
patrol lines and feed them whatever information was available about the
enemy's location. This information was scanty, and the long term solution
was to develop an inland radar chain. In the meantime more desperate

measures were required.[23] Park 'decided to employ single VHF Spitfire to shadow enemy raids.'

The final period of the Battle proved in many ways the most challenging to the air defences, but paradoxically, it was strategically the least important. When the Luftwaffe withdrew its bomber force from daytime operations in October, they were replaced by very high flying formations of fighter-bombers, often operating at over 20,000 feet. Fighter Command appreciated that these attacks were a mere nuisance, since the Germans had abandoned the notion of launching a cross-channel assault in 1940, and the fighters carried a very small bomb load. They did cause enough damage to warrant interception, but CH radar proved highly inaccurate and was generally incapable of detecting aircraft flying at over 25,000 feet. Many formations were missed altogether as they accidently approached the coast in a gap in local radar coverage. Gaps between lobes of the radar signals increased in size with altitude. Emergency measures to improve gap filling were undertaken and completed by the end of October. Still, CH radar remained incapable of detecting aircraft flying at over 30,000 feet, a height at which many German formations operated at the end of the daytime campaign.[24]

Even when radar could provide early warning the high flying raids still slipped through. The new German formations were moving much faster than earlier ones and, unlike the bombers, made a direct approach to their targets, thereby further reducing interceptions. Standing patrols of fighters, flying just below oxygen height, about 15,000 feet were introduced in order to increase the rate of interceptions.[25] No. 11 Group also expanded its experiment with high flying Spitfire aircraft; a special unit, 421 Fighter Reconnaissance Flight was formed in late September. According to No. 11 Group's summary of operations, the unit 'provided on numerous occasions invaluable reports of the approach of enemy raids; information as to height and strength that was not provided by either RDF or the Observer Corps, whose work has been greatly hampered by the enemy raids flying above 20,000 feet over 10/10 clouds.'[26]

The Luftwaffe and the Air Defence System

By far the most serious threat to the air defence system, however, did not come from any of its technical failings or the inexperience of its personnel. The system had to deal with Luftwaffe attacks on various components of the system, as well as efforts to deceive and jam the early warning radar. Fortunately for Britain, Luftwaffe intelligence did not develop an accurate understanding of how the system functioned or its crucial importance to

Fighter Command. The Germans lack of comprehension of the air defences was one of the keys to Fighter Command's victory.

The existence of radar was one of Britain's most closely guarded secrets in the years leading up to the war, but there was no way that the masts of chain stations going up along the coast could remain hidden. Efforts to disguise the purpose of the construction met with limited success, particularly after the rapid completion of the interim chain stations in the winter of 1938-39. The Germans had begun research into their own radar systems as early as 1929, well before Watson-Watt's 1935 memorandum. Perhaps as early as 1937, Luftwaffe intelligence speculated that the towers on the coast might be some form of an electronic early warning device. Ironically, British intelligence ignored ample information which indicated that the Germans also had radar. It was only in mid-May 1940 that the Air Ministry officially recognized that the Germans had their own version of the device.[27]

In the spring of 1939, General Wolfgang Martini, head of signals for the Luftwaffe, proposed that airships be used to conduct radio intelligence missions in order to determine the function of the large radio towers that had sprung up along the coast of Britain in the last 12 months. He was convinced that they were radar stations. Martini wanted to use a dirigible and to fit it with the large amount of radio detectors and direction finders needed to determine the type of signal emanating from the stations.

Before the war began, Martini was able to conduct two intelligence missions along the British coast. The first took place in late May, and the second in early August. Both flights were closely followed by chain stations which could not miss such a massive target as a Zeppelin. A vivid account of the observations made by one young radar operator has survived. On the afternoon of 3 August 1939, radar operator Len Dobson had just come on duty at the chain station Douglas Wood, located some 10 miles north of Dundee, when he observed a positive response of a large target located some 120 miles from the station. Dobson passed on the plot to Fighter Command which ordered him to keep it under observation. The target gradually approached the coast lumbering along at 60 miles an hour. As it came closer the response on the CRT display grew larger and larger and soon exceeded anything Dobson had experienced before. He wondered if it could be an invasion fleet, but he told the Filter Room that the response was big enough for a formation of from fifty to a hundred aircraft. Dobson followed the target for several more hours before losing it as it made its leisurely way along the coast off Aberdeenshire. He saw the large signal once more some six hours later and again reported the information to Stanmore.[28]

Despite being carefully tracked by a number of CH stations, Martini's flying laboratory failed to detect the British radar signals. The technical dissimilarities between the two countries' radar was the problem. German radar was much closer akin to CHL than CH, in that they used a much shorter wavelength and a lighthouse-type signal. CHL was not in service in the summer of 1939. The Luftwaffe's equipment detected the huge amount of radiation of the chain in the 10 to 13-metre band, but failed to interpret the signals correctly. The German's concluded that the signals were a form of radio interference, possibly in part caused by German ionospheric research experiments which continued despite a request that they be stopped during the intelligence missions.

Martini remained suspicious that the RAF had an operational early warning radar system, but the byzantine structure of the German intelligence apparatus prevented the signals intelligence group from being informed of other sources of information, such as interrogation of prisoners of war, which might have confirmed the existence of the chain. The Germans, therefore, started the campaign with no appreciation of radar, and as little understanding of the command and control system of Fighter Command. A detailed comparative survey of the Luftwaffe and Fighter Command's striking power, completed on 16 July 1940 by German air force intelligence, makes no mention of a British early warning system. Only in the first stage of the Battle in late July did intercepts of RAF ground-to-air radiotelephone traffic confirm that British fighters were being directed towards interceptions. The only way Fighter Command could detect the Luftwaffe aircraft operating off the coast was by radar. Martini urged Goering to destroy the radar stations before launching the first of his major strikes at Fighter Command airbases. Goering reluctantly agreed to begin his great *Alder* offensive by trying to blind Fighter Command by destroying every radar station from Portland to the Thames Estuary.[29]

On the morning of 12 August, a specialized group of sixteen Me 110s fighter-bombers, Erpro 210 under the command of Hauptmann Walter Rubensdoerffer, took off and headed west down the Channel. The fast-moving aircraft dropped down to near wave top level. They then turned suddenly north towards Eastbourne and Pevensy. At 9:00 a.m., just as they reached the coast, the formation broke up into four sections to attack the Pevensey, Rye, Dunkirk and Dover CH stations. The attacks were a complete surprise. Pevensy was hit at 9:32 by eight 500 kg bombs, which destroyed many buildings, including the NAAFI or canteen, and cut the main power mains. Rye was the next station to be hit some fifteen minutes later. Dover was attacked at almost the same time as Rye. Bombs exploded near the masts rattling, but not severely damaging the structures. Every hut was destroyed, but a strong cross wind spoiled the bombers' aim and they missed the main buildings. At Dunkirk all eight bombs hit their target, but no vital damage was sustained by the station.[30]

Although all were caught unaware and heavily hit, the stations proved remarkably resilient to bomb damage. When properly shielded, transmitters and receivers remained unscathed except by direct bomb hits, and the lattice work of the towers proved capable of absorbing tremendous damage. Most vulnerable were power lines and other ancillary cables such as those leading up to the towers, which had not been properly protected against bombing. Dunkirk continued to operate without a break despite the bomb damage. Dover operated with minimal interruption using emergency backup equipment. Rye was back on the air by noon using an auxiliary generator. Pevensy was in service again a few hours later. The rapid repair of these stations was a tribute to their crews, many of whom stayed on duty throughout the attacks.

Far more successful was an attack later that morning on the Ventnor chain station on the Isle of Wight. A group of fifteen Ju 88 dive bombers broke away from a larger formation that was about to bomb Portsmouth and headed for Ventnor. They were intercepted by two fighter squadrons which managed to interrupt the long shallow dives of some of the large bombers. The leading Junker was shot down by the station's own defensive Bofors guns, but not before the plane had dropped its bombs with pinpoint accuracy. Despite the valiant defence, the German bombers were deadly accurate. The majority of the surface buildings on the site burned down when a lack of water hampered the fire fighting efforts. Delayed-action bombs on the site made it impossible to immediately begin repairs.

Although these raids had only managed to knock one station off the air for more than a day, inexplicably the Luftwaffe did not systematically attack the radar stations until they could no longer function. On 15 August, Goering, who remained unconvinced about the importance of radar to Fighter Command, ordered that little more time was to be spent on attacking such secondary targets. It was a good thing for Fighter Command, as two later attacks on chain stations demonstrated the Luftwaffe could knock chain stations off the air for a considerable period of time.

On 16 August, five Stukas, shielded by nearby larger raids on Gosport and Lee-on-Solent, again attacked Ventnor. The station was still being repaired at the time. Seven high explosive bombs hit the compound, new fires broke out and destroyed most of the remaining structures. The intense heat of the fires actually damaged the towers. Ventnor was declared effectively destroyed and had to be replaced by a reserve mobile station, which was not up and running until 23 August.

Two days later, at 2.00 p.m., the Poling CH station tracked three enemy formations of eighty plus north of Cherbourg, twenty plus east of Cherbourg and ten plus north-west of Le Havre. The formation crossed the channel and reached the coast twenty-five minutes later. To the amazement of the radar operators one of the formations, which turned out to be thirty bombers, headed directly for the station. Over ninety bombs were dropped and extensive damage was done. The receiver block was demolished by a direct hit and a receiver aerial was destroyed. Again an emergency mobile station was called in to replace the station, and it was operational within forty-eight hours. The mobile stations were, however, useless for anything but early warning and were, even in this limited role, far less accurate than the CH radar they were replacing.[31]

The damage done in even these few raids severely effected Park's ability to manage his group's operations. He warned his operations room controllers on 25 August: 'Owing to the damage caused by enemy bombing, Ventnor, Poling, and Rye RDF stations are working on lash-up equipment, which gives less reliable results than previously. Controllers must accept with reserve the strength reported for enemy formations approaching the South coast West of Beachy Head.'[32] Writing in mid-September, Park considered that during the attack on his airfields and other group installations prior to 7 September, 'the main problem [he faced] was to know what was the diversionary attack and to hold sufficient fighter squadrons in readiness to meet the main attack when this could be discerned from the very unreliable information received from RDF, after they had been very heavily bombed.'[33]

Direct assault on the radar chain was only one way the Luftwaffe could disrupt early warning. To the radar, Filter Room, and operations rooms personnel, the Luftwaffe appeared to deliberately engage in tactics that were designed to confuse the air defences. We have already talked about certain of these tactics, including the stacking of formations and the gradual drift to higher altitude raids which began in September. Many other techniques were also employed by the German to hide their true intentions. This included the regular use of diversionary raids which broke off from main formations to attack targets of a secondary nature. As a rule the bombers never approached their target directly, engaging in frequent alterations in course and altitude. What was behind these maneuvering awaits a study of Luftwaffe tactics, but it is unlikely they were specifically designed to throw off radar tracking. Whatever the reason, the early warning system was ready for these types of deceptions, as the Biggin Hill system of interception had been predicated on bombers trying to throw off the defending fighters. Fighter Command controllers were generally able to have a reasonably accurate tactical picture of German intentions.

Deliberate radio jamming, which had been greatly feared by radar scientists prior to the war, was hardly attempted by the Germans and then only begun in late September when the Dover CH and CHL station experienced deliberate radio interference. The stations were able to use a variety of methods and devices to limit the effectiveness of the jamming, including frequency changing and coloured filters on the CRT screen which masked signals generated by radio interference. Like the direct attack on the stations, the German radio jamming programme was carried out too sporadically to greatly hamper the chain radar stations.[34]

German misconceptions about Fighter Command extended well beyond radar and included a fundamental intelligence failure to grasp the true nature of the command and control apparatus. The 16 July Luftwaffe intelligence comparative analysis of the two air forces also made no reference to the fighter control organization, even though it had existed in a far more primitive form in the First World War.[35]

Even when signals intercepts should have provided the Luftwaffe with a comprehensive picture of the Fighter Command, the Germans clung to the notion of an inflexible and easily defeated system of air defence. An intelligence summary completed on 7 August recognized that sectors controlled fighters by the use of radio-telephones, but insisted that this was a tactical weakness. The Luftwaffe intelligence service believed that fighter squadrons were 'tied to their respective ground stations and thereby restricted in mobility.' This would prevent the RAF from concentrating forces against large scale attacks, which would invariably cause 'considerable confusion' and reduce the effectiveness of the defences.[36] No such inflexibility existed, however, as squadrons could be transferred between groups and sectors relatively painlessly.

Ironically, only once a did the Luftwaffe come close to severely disrupting the command and control apparatus of No. 11 Group and this was apparently by accident. During the last week of August and the first week of September the Luftwaffe began a series of concentrated attacks on major sector airfields located well inland. Although Fighter Command was able to intercept many of these raids, the Germans pressed through to their targets and did considerable damage to the air stations. Badly damaged in a number of these attacks were three of seven sector operations rooms. These vital command centres had been foolishly placed in the airfields complexes, when there was in fact no reason why they could not have been located in a more inconspicuous location. Although most of the sector control rooms were shielded by earthen revetments, little effort had been made to protect the power and communications cables that made possible the working of these rooms. Emergency operations rooms proved too small and ill-equipped to replace the main centres. Only the sudden decision by the Germans to discontinue the systematic pounding of the sector airfields in favour of attacking London saved the command and control structure from complete collapse.

Even after the Battle senior Luftwaffe officers persisted in their fundamental misunderstanding of how the British air defences functioned and the crucial role it had played in their defeat. Shortly after the war two senior officers, Erhard Milch, the inspector general of the Luftwaffe, and Adolph Galland, who commanded a group of fighters during the battle, were interviewed by Air Ministry historians. The historians recorded this revealing summary of their conversations:

> As for RDF stations, the prisoners agreed that they thought serious damage had been done to one or two – which was so – but that in general they were considered difficult targets to effectively damage. This was also true: but the fact remains that neither seemed to realize how important were the RDF stations to Fighter Command technique of interception or how embarrassing sustained attacks upon them would have been.[38]

The Air Defence System in Battle – A Final Analysis

For Dowding and Park, radar and its integration into the air defences was the key tool in their successful campaign to preserve Fighter Command and defeat the Luftwaffe. Despite its shortcomings, the system had performed admirably, proving once and for all that the bomber would not always get through. The men and women who manned the defences were as much responsible for saving Britain as the fighter pilots. Although there were faults in the technology and in the training of the personnel, it was only when the Luftwaffe attacked radar stations and sector operations rooms that Park and Dowding worried that the system might not safeguard Britain's skies. Whatever the weaknesses in the air defence early warning and command and control apparatus, they were by no means the most serious deficiencies in Fighter Command. Certain aircraft, most noticeably the Bolton Paul Defiant, a single-engine fighter with a 4-gun power-operated rear turret, and the Bristol Blenheim, a converted medium bomber, proved completely unsuitable for daytime operations. Fighter tactics using three-plane V formations were found to be inferior to German fighter formations. There was a constant worry about the supply of Hurricanes and Spitfires, and, far more critically, the number of fighter pilots available to fly them. The fighter's armament of eight .303 calibre machine guns did not provide the firepower needed to bring down armoured German bombers swiftly. Efforts to introduce cannon-armed fighters were hampered by a persistent problem with jamming. Finally, as the Luftwaffe climbed higher, aircraft performance, like radar, suffered. Unlike their German counterparts, the British fighters were not equipped with a twin supercharged engine, which put them at a distinct disadvantage when tangling with high flying Me 109s.

Perhaps the greatest tribute to the overall performance of the air defence system was that only one aspect of the system came under persistent criticism and demands for immediate reform. As we have seen since well before the battle, Dowding was criticized for not decentralizing filtering to group level. This became an issue again in late September. On 25 September the Air Council, which included the Secretary of State for Air as well as the Air Staff, met and decided to press Dowding to accept that filtering should be done at the Group level. Also in attendance was the Marshal of the Royal Air Force, Sir John Salmond, retired Chief of Air Staff who had been asked to investigate the night air defences. Two days later, Dowding again rejected the proposal. This time the Air Ministry would not back down. Dowding was summoned to meet with the Air Staff on 1 October. At the meeting he was finally forced to agree to establish Group filter rooms; the only concession he was able to obtain was that the new centres would not become operational until all fighters were provided with IFF. This would enable radar operators to identify all RAF aircraft, without having to go through the screening procedures that could only exist at a centralized filter room.[39]

Dowding was furious about the way the system was being interfered with and took it upon himself to write personally to Churchill asking that he stop the Air Ministry from building group filter rooms. Churchill, who after the great victory over the Luftwaffe was a firm supporter of both the air defence system and Dowding, took up Dowding's case and wrote to Sinclair to question the wisdom of the Air Council's decision on filtering.[40]

Sinclair asked Joubert to write a draft response to Churchill. Joubert's letter contained some unveiled criticism of Dowding, stating for instance that the Air Council was 'alarmed at the slowing-up of the air raid reporting system.' He bluntly warned that if the Luftwaffe offensive recommenced in 1941, filtering in one room would be 'an impossible task' because of the growth in the number of German and British aircraft. The Air Minister's private secretary substantially toned down Jouber's comments and the letter was sent by Sinclair to Churchill on 18 October. Sinclair informed the Prime Minister that 'the question which faced the Air Staff might have been summed up as one between the merits of a system which gives the simplest dissemination of air raid warnings and one which, at a moderate expense of money, labour and material gives us a vital addition to the time available to the squadron (in which) to attempt to intercept the enemy.'[41]

Churchill was unconvinced and continued to support Dowding. On 27 October he wrote to Sinclair asking if it was really true that time would be saved if filtering was done at groups and, if so, how much? Joubert again wrote a draft reply to Churchill's inquiry, stating that filtering at groups removed one step from the process before information could go to operational units, and that the time saved would amount to thirty seconds in periods of light enemy activity to several minutes during heavy attacks. Joubert's letter did not sit well with several other members of the Air Staff, who questioned the validity of the notion that group filtering removed a step from the process, since filtering was filtering no matter where it was done. The advantage was not in removing a step, but in creating more filter rooms which would handle less total information, thus reducing the chance for the system from to become over-saturated. After much discussion Joubert's letter was edited to omit the controversial paragraph and was only finally sent to Churchill some two weeks later.[42]

Here the issue rested, at least for the time being. While Churchill still supported Dowding, his confidence in him was swiftly being eroded by another far more serious issue. Although the daytime air defence system had saved the nation in the summer of 1940, during the autumn nights the Luftwaffe was ruthlessly bombing Britain's cities, and Fighter Command appeared unable even to hinder them. The triumph of the great work at Bawdsey would be marred by the failure to provide an effective night defence in time. It is to this sad tale we now must turn.

13

The Failure to Stop the Night-time Blitz

On 13 November Dowding met with Sinclair in the minister's London office. It is uncertain if Dowding was aware of the subject of the meeting, certainly he did not anticipate that he was to lose his command. Dowding was informed that he had been selected to head a mission of senior RAF officers being sent to the United States to assist with equipment purchases and to establish closer liaison with the American armed forces. Dowding's replacement as Commander-in-Chief of Fighter Command was Douglas, the Deputy Chief of the Air Staff. The next day Dowding appealed the Air Ministry's orders to Churchill. However, this time Sinclair ensured that he had the Prime Minister's full approval in advance, and the Air Marshal received nothing but confirmation of his removal from command. The subject of Dowding's dismissal just as the daytime air defences he had done so much to create had succeeded in saving Britain, along with the subsequent removal of Park from command of No. 11 Group, remains one of the most controversial acts of the British government during the Second World War.

Several explanations for Dowding's departure have been raised in studies of the Battle of Britain. The most often stated is that Dowding failed to mediate a dispute between Park and AVM T. L. Leigh-Mallory, the commander of No. 12 Group, over fighter tactics. The former advocated the use of a small number of squadrons to engage in timely interceptions of Luftwaffe formations. The latter argued that the correct approach was to assemble 'Big Wings' of five or more squadrons to overwhelm some of the German strike forces, even if it took so much time that the enemy were able to reach their targets before being intercepted. This controversy was combined with growing animosity between Dowding and the Air Staff. Dowding was seen as being too imperious and, as demonstrated in the dispute over filtering, outright disobedient. This animosity was compounded by jealousy over Dowding's use of his considerable political influence to retain his command far longer than was normally the case; he had been due for replacement in 1939, and had extended his service well beyond the normal age for retirement. Dowding, however, was able to withstand all criticisms on these issues because, until early November, he had received the unwavering support of Churchill.

There can be no doubt, as John Ray has shown in his recent study of the Battle of Britain, that it was the poor performance of the night-time defences in the autumn of 1940 that led to the collapse of Churchill's backing and delivered the *coup de grâce* to Dowding's controversial tenure as the Commander-in-Chief of Fighter Command.[1] As Britain's first great hero of the war, Dowding would have remained at Fighter Command, but when the Luftwaffe began to attack British cities at night, during the late summer and autumn of 1940, the Royal Air Force's aircraft defence system, so successful in thwarting daytime raids, was almost useless. In September, the Luftwaffe flew 6,135 night sorties over Britain, and Fighter Command managed to shoot down only four aircraft. In the last two weeks of September alone, some 3,200 civilians were killed by bombing. The situation grew increasingly worse throughout the rest of 1940.[2]

The greatest irony of Dowding's dismissal was that while Fighter Command proved unable to defend Britian from the Blitz, it was Dowding, and not his critics, who understood how to defeat the German night bombers. Dowding's position began to be seriously undermined after the Marshal of the Royal Air Force and former Chief of Air Staff Sir John Salmond was appointed on 14 September to head a committee to investigate the situation. Salmond was no friend of Dowding, and there is evidence that he was predisposed to criticize him rather than to present a balanced overview of the situation. Salmond moved briskly, interviewing senior officers and other informed parties over a three-day period beginning on 16 September. Dowding was only called to testify on the final day, and only after the committee had written the bulk of its recommendations. The Salmond Committee report contained little that was new to Dowding, and much that was simply incorrect. For instance, the committee had heeded the advice of Joubert that filtering at Group level was one of the key ways to improve the rate of successful interceptions of night bombers. Yet, Dowding had long since realized that the key to shooting down raiders after dark was through localized control of night fighters at the sector level.

The attacks on Dowding continued throughout the rest of his tenure in command. Most of his critics were as uninformed as the Salmond Committee. On 16 October, Adm. P. S. V. Phillips, the Vice Chief of the Naval Staff, produced at Churchill's request, a report summarizing the problems and shortcomings of the night air defences. Phillips, after consulting a number of unnamed RAF officers, concluded that the best way to stop the Luftwaffe was to revert back to the methods of the First World War when 'the night raiding of London was in fact stopped in about 6 weeks by the specialisation of 2 fighter squadrons in night fighting.' To emulate this, Phillips urged the creation of night fighter squadrons of Hurricanes. Phillips admitted that the problem of dealing with the night bombers was more difficult than it had been in 1918 owing to their operating at a far greater speed and altitude, but he was confident that Hurricanes flying standing patrols along the bomber stream could shoot down a significant number of the attackers.[3]

Churchill gave Dowding an opportunity to reply to Phillips' letter. The Air Marshal cut to shreds any notion that the methods of 1918 could be applied to 1940. The speed and altitude of modern bombers rendered it highly unlikely that any repeat of tactics from the earlier conflict could succeed. For example, searchlights guided by sound detectors could no longer illuminate a target long enough to guide a fighter to its prey. Dowding pointed out that Phillips suggested 'no method of employment of the fighters, but would merely revert to a Micawber-like method of ordering them to fly about and wait for something to turn up.' While he was willing to experiment with a few squadrons of night flying Hurricanes, he was dubious of their success. He told Churchill: 'In my considered opinion, the only method of approaching the problem is to devise a methodical system of utilising the two new aids which are coming forward to our assistance, viz A.I. [airborne interception radar] for the fighter and radio aids for the searchlights.'[4]

Phillips, however, had not ignored AI. Instead, he admitted that, if perfected, AI was the key to successful night interception, but he informed Churchill, 'the hard fact is that it does not deliver the goods today. At the beginning of the war, AI was stated to be a month or two ahead. After more than a year, we still hear that in a month or so it may really achieve results.'

Dowding and Phillips were both correct; AI was vital and in the autumn of 1940, AI did not work. Dowding was quite right that emergency measures including pressing single seat day fighters like the Hurricane into night operations would prove ineffective. Dowding, however, was not completely forthcoming to Churchill. What Dowding did not tell him was that the solution to thwarting the night raids was far more complex than just providing adequate AI. The technical complexity of providing for the night defences dwarfed that of daylight operations. Three additional weapons and detection systems were required. First, it was critical to have a night fighter with a two-man crew, capable of carrying a powerful armament for the often-fleeting night-time encounters. Second, to guide the fighter to within 3 to 4 miles of its prey, there had to be an entirely different type of ground based radar, GCI or Ground Control Interception. Finally, it was imperative that the night fighter be equipped with an AI radar, capable of taking up where GCI left off, steering the interceptor from 20,000 to just under 500 feet from the target. At that point, the pilot would use his eyes to guide him into firing position. The GCI and AI radar had to provide clear indications of the bomber's relative air speed, direction, heading, and height to allow the pilot to manoeuver his aircraft into an appropriate attack position. This was an extremely precise operation, since night fighter pilots soon discovered that the best position to attack from was directly astern and slightly below a bomber. Thus, in order to avoid a lengthy chase, the fighter had to be guided above the bomber so that it could dive down on the unexpecting target, picking up air speed in the process, and only end up chasing the aircraft for the last few minutes of the interception. To avoid overshooting the bomber before the fighter pilot even saw it, he had to be at approximately the same air speed, only gradually moving into gunnery range.

Of these three systems, the most vital was AI radar; without it, few night interceptions were possible. The earlier versions of AI were a dismal failure. Just two examples will suffice to illustrate this point. From 1 to 15 October 1940, Mark III AI-equipped night fighters operating from Kenley made ninety-two flights, making contact with a mere twenty-eight bombers without destroying a single one.[5] The Operations Record Book of Fighter Command recorded the typical poor results of this period. On 21 December it reported: 'On this night 320 e/ac operating, 250 on Liverpool, thirty-six fighters patrolled that night, making not one single interception, and one Blenheim [night fighter] crashed.' The results the next night were no better, even with ninety-four night fighters. Concerned about the effect on the public of the disastrous showing Fighter Command ordered that: 'If any concern is shown by the Civil population at the apparent inactivity of the defence on Fighter nights, report should be made to Headquarters.'[6]

Just as the Chain and Chain Home Low radar had made possible Dowding's triumph in the Battle of Britain, the lack of an effective AI radar resulted in his dismissal. The great triumph of Bawdsey was marred by the delay in getting an effective airborne radar into service. While the Chain was only just ready for operations in July 1940, the night defences would only begin to shoot down significant numbers of Luftwaffe bombers in March 1941, too late to save Dowding or to be a determining factor during the Blitz. In the end, it was the courage and tenacity of British civilians that defeated the Luftwaffe and proved wrong prewar prognosticators about the effectiveness of aerial bombing on morale.

Airborne Radar Research Before the War

When we last discussed Bowen's airborne radar research team, it was late 1937. They had successfully demonstrated the ability of 1½-metre airborne radar to detect ships. While a working prototype of the air-to-surface vessel detection radar (ASV) set was ready in early 1939, AI sets proved far more difficult to perfect. Unlike ASV, AI had to be able to track comparatively small targets in three dimensional space. This was done by chain radar, but AI was far more complex because of numerous factors: space and weight constraints, the stress placed on any equipment carried in an aircraft, the need for far greater accuracy, the requirement to build sets at a far smaller wavelength with greater power than any device previously attempted, and the need to shield the set from any interference from the aircraft's equipment. While the preliminary research stage was successful, there were already indications that the development of the basic prototype into an effective weapon system was beyond the capability of Bowen's small team. In large measure, the delay in developing a successful AI was the direct result of the shortage of personnel and time. With much of Bawdsey's staff devoted to completing the chain in the last eighteen months of peace, Bowen's group suffered. Nor was any outside assistance sought until it was too late to influence the course of events during the Blitz.

Like the chain radar, no electronics firm was fully briefed on the design of the new radar prior to the outbreak of the war. Instead, a few manufacturers were asked to assist only in designing and constructing certain individual components. Even these limited contacts were haphazard and amateurish. Bowen's memoirs contain what he viewed as an amusing anecdote from October 1938 about arranging for the manufacturer of an aircraft alternator to serve as a power source for airborne radar equipment. With no formal system for integrating the resources of the electronics industry into Bawdsey's work, and unwilling to go through the normal acquisition channels, Bowen made an appointment to see the Managing Director of Metropolitan-Vickers. He flew to the Metro-Vickers plant at Sheffield in one of the Fairy Battle bombers then being used to fly-test airborne radar. On arrival Bowen took a spanner, removed the aircraft's DC generator and took it to the meeting. Without explaining why it was required, Bowen said that he wanted Metro-Vickers to build 'an alternator, not a DC generator, looking exactly like the one on the table, occupying the same space and using the same fixing bolts and spline shaft.' Over lunch in the Managing Director's dining room it was agreed that the company could design and build the alternator as requested. Bowen then flew back towards Bawdsey, but his pilot was forced to land in a farmer's field after their plane hit a large goose. Bowen made his way to London the next day, consulted with Watson-Watt, by then working in the Air Ministry as the Director of Communications Development, and immediately orders were placed for eighteen pre-production units.

Fortunately, even working without written specifications, drawings, or designs, Metro-Vickers showed the high calibre of its industrial design team and had a working model of the alternator running successfully within a month of Bowen's visit. It would prove to be one of the most important components designed for airborne radar. Used throughout the war, over 133,000 of these alternators were built in Britain alone, powering many Allied airborne radar systems.[7] It was an example of what might have been achieved if industry had been brought into the complete radar picture earlier.

The absence of development expertise began to haunt the AI program in the summer of 1939. In June and July, Bowen's team demonstrated the prototype AI installed in a Fairy Battle. The results were first reported by Rowe, who found them 'most striking' with the target being followed by the radar 'at distances greater than 1,000 feet and at a maximum range equal to that of the defending aircraft above the ground.'[8] The maximum range was affected by the height of the aircraft above the ground because of ground reflections. The reflections of targets that were further away than the height of the AI-equipped aircraft above the ground merged and became indistinguishable from the ground clutter. This greatly limited maximum range, particularly below 10,000 feet. When approaching the target, the reflection was lost at 1,000 feet because of interference caused by the AI set's transmitter.

Dowding was the most important person taken aloft in the Battle for a first hand demonstration. Two remarkably contradictory accounts of Dowding's visit in early July 1939 have survived. In his memoirs, Bowen recounts that before becoming airborne, 'Stuffy' Dowding emphasized the importance of the AI's minimum range performance. The two men crammed themselves into the rear of the Battle with the radar set in front of them. For safety reasons, the test took place in daylight; in order to simulate night-time operations a black cloth covered the two men's heads. Dowding asked during the last interception, once minimum range had been reached, that he be allowed to remove the black cloth while the radar operated so that he could judge for himself just how close they could track the target aircraft. Bowen recalls: 'I whipped the cloth off and Stuffy looked straight ahead. He said: "Where is it? I can't see it." I pointed straight up; we were flying almost directly underneath the target. "My God" said Stuffy, "tell him to move away, we are too close." We broke away and headed for home. The Commander-in-Chief was clearly pleased with what he had seen.' The visit concluded with an almost two hour-long afternoon meeting between Bowen and Dowding where the Air Marshal explained to the scientist his views of night fighting. Here, Bowen states, he learned for the first time that AI had to be mounted on a twin engined, two-man fighter, and that visual identification of the target was vital to avoid the accidental shooting down of friendly aircraft.[9]

Dowding's account of the day survives in both a letter written to Rowe and in his official report to the Air Minister. Dowding told Rowe that he 'was tremendously impressed by the potentialities of the device' but that he had explained to Bowen that the minimum range of around 1000 feet would prove to be 'a serious handicap, since it would only be on exceptional nights that a pilot could pick up an unilluminated target at this range.' When Dowding returned to the air base in the afternoon Bowen informed him 'that a sensational advance had been made during [his] absence, and that the transmitter ray had been suppressed altogether, with the result that the minimum range had fallen to 220 feet.' Dowding did not question this extraordinary turn around, but attributed it to being 'typical of the manner in which difficulties disappear before the intensive and intelligent work' of the Bawdsey staff. [10]

It is impossible to explain the discrepancies in these two accounts, but what is certain is that Bowen could not duplicate either the results described in his memoirs or, if we rely on Dowding's account, his 'sensational advance.' Whichever source one believes, the ramifications of this day were serious. Assured that AI, although still in a crude prototype stage, was giving useful operational results, Dowding urged the Air Ministry to authorize Bowen's team to begin fitting twenty-one hand-built sets to Blenheim twin engine fighters. Dowding desired that top priority be given to equipping a squadron with AI in order that training and operational tactics could be worked out before the radar entered mass production; it had the extra advantage of providing a small measure of night protection if war should break out. He saw this as so important that it should take precedence over continuing efforts to make improvements to AI or of providing assistance to manufacturers who would eventually have to mass produce the sets.

Dowding showed a remarkable lack of awareness about the complexities of the development phase of the AI programme. The Air Ministry was equally ignorant. No one, not even Watson-Watt or Rowe, objected to turning Bowen's research team into what amounted to a small-scale manufacturer. On 14 July the Deputy Chief of the Air Staff concurred with the C-in-C Fighter Command, and Bowen's team was ordered to produce and fit the first twenty-one AI sets before resuming research on improving their crude prototype.[11]

What followed was the kind of development and production debacle which was a frequent occurrence in British industry during the prewar and wartime rearmament program. The major difference between the AI story and other incidents described in Correlli Barnett's *Audit of War* is that private industry was not to blame for the failure to produce an effective AI set in time for the autumn battles of 1940. Instead, private industry would salvage the AI program from the utter chaos and technical mismanagement which befell Bowen's group.

The Move from Bawdsey

Prior to Dowding's request to begin installation of AI, little had been done to prepare for development and production of the experimental set. In late 1938, Watson-Watt ordered six receivers and transmitters from Metropolitan-Vickers and A. C. Cossor Limited, respectively. The equipment was to be based on Bawdsey's drawings and specifications. Both manufacturers requested that models also be provided, but none were available of current equipment. Cossor received only 'a complete run-down on the receiver'; Metro-Vickers had to settle with taking away a two-year-old transmitter used in some of the very first experimental flights.

There was no supervision of the manufacturers. Cossor's receivers proved to be a disaster, weighing more than the entire prototype set combined and having less than one tenth the sensitivity of the Bawdsey design. These sets were replaced by a new model built by Pye Radio. Metro-Vickers, instead of relying on the specifications, copied the two year old receiver. Bowen only learned of this when the first transmitter arrived in the summer of 1939. When asked to explain the foul up, the company's representatives told Bowen that Watson-Watt had visited their plant in Manchester and insisted they copy the working model of the old receiver rather than utilize the written specifications provided by the radar scientists.

In fact, until August 1939 only one complete prototype existed, and it was almost lost along with Bowen and Frederick Lindemann in an AI demonstration flight. The pilot aborted the mission just after takeoff when all of the oil streamed out of the battle because a mechanic had failed to replace a sealing cap. Without a complete set of specifications and drawings the loss of Bowen and the prototype would have set the entire program back for months.[12] The lack of blueprints and specifications was typical of British research projects of the period and greatly hampered the transition to mass production in Britain and elsewhere for much of the war.

In his memoirs Bowen describes a litany of impediments to producing and fitting the AI Mark I. Many of the problems were the fault of the

Air Ministry, which had little appreciation of the difficulty in fitting the radar to aircraft. Replacing the Fairy Battle, an obsolete single-engine light bomber, with the Blenheim meant a complete rethinking of the mounting of the antenna arrays. When this had been worked out and the first two sets fitted, the Air Ministry sent to Bowen's group, without warning, a different version of the Blenheim with a totally dissimilar nose design, thus forcing the complete reworking of the array yet again. Other problems with the Blenheims included the failure to shield the electrical equipment of some of the earlier aircraft to avoid interfering with the radar, and the poor construction of new planes from the assembly lines. On two occasions propellers broke off aircraft in flight.

While these types of obstacles were time consuming, far more serious was the complete unpreparedness of the Bawdsey team for the development stage of fitting the first pre-production sets into the aircraft. The total staff of the airborne radar group was just twenty-three, including the typist. Not a single person was an experienced aircraft fitter. Moreover, the orders to Metro-Vickers and Pye were not for completed sets, but for the basic components. The only outside help came from a party of five experienced aircraft fitters from the Royal Aircraft Establishment. The RAE party proved to be 'pearls without price' and soon took over much of the actually fitting of the sets into the aircraft, while the scientific team focused on assembling the sets and training the first group of AI operators.

Throughout August the airborne group struggled outside, exposed to the elements, trying to fit the sets into Blenheims. The Air Ministry anticipated have a full squadron fitted by the end of the month but only six were finished by the 31 August. When war was declared on 3 September just one operational AI-equipped night fighter patrolled the skies over London. The radar operator was Hanbury Brown, one of the airborne group's senior scientists. This marked the high water mark of Bowen's airborne radar group, for a while they had managed to persevere through August, the beginning of hostilities led to great changes in radar research which would turn a bad situation into utter chaos.[14]

As we have seen, by December 1938 it was accepted that radar research would have to be moved from Bawdsey no later than the commencement of hostilities. Rowe recommended that no move take place before then because of the need not to disrupt ongoing research. He reasoned that the war would inevitably disrupt research anyway and that this would be an ideal time to transfer Bawdsey elsewhere, preferably to the south-western-most CH radar station, by the spring of 1939 likely to be Poling. In order to make the move as orderly as possible, Rowe suggested acquiring land and erecting suitable buildings in advance of the move.[15]

In some respects it was fortunate that Watson-Watt chose not to act upon Rowe's recommendation. After the fall of France, Poling became one of the most vulnerable CH station to enemy attack, and, in fact, suffered from heavy bombing raids in the summer of 1940. Ironically, Bawdsey did not experience a single major attack during the war. Yet, what Watson-Watt did instead was perhaps as destructive as the most devastating German air attack.

Watson-Watt decided that, rather than acquire a new site, the radar researchers should move to Dundee, Scotland and use the existing university college facilities. Watson-Watt had been a student at Dundee and in a visit of northern radar sites he stopped in at his old school. He introduced himself to the Vice-Chancellor. Bowen sarcastically described what he believed followed:

[Watson-Watt] must have been overcome by nostalgia and remembrance of things past because he mentioned that there was a problem – in the event of war being declared, some members of an important Defence Laboratory might need emergency accommodation. 'No problem at all', said the Vice-Chancellor gaily 'we would be only too glad to help in a case like that' – without enquiring too closely into how many staff were involved and how much space was required.[16]

Watson-Watt did not follow up on this casual conversation, even though he had not mentioned the size of the research group that he wished the college to house. When the first cars and trucks from Bawdsey arrived at Dundee they found that no one expected them; the Vice-Chancellor could barely recall the chat with Watson-Watt some months earlier. Gearing up for the new term, the college administration found they could spare just two rooms, each 20 square feet, to house the Bawdsey team. Eventually, some additional laboratory space was loaned to them by a sympathetic professor, but much of the equipment brought up from the south remained in storage, some of it in large crates in the parking lot.

Space was not the only problem with Dundee; electricity and radio sets from the town interfered with the much of the sensitive test equipment, some of the civilian staff refused to go north after learning that the Air Ministry would not compensate them for the cost of their move, and the scientists soon found themselves completely isolated from contact with operational commands and other research establishments. Research on the chain radar ground to a halt, but this was not a serious issue since most of the technology for the large ground based radar was already out of the laboratory and in mass production. The move from Bawdsey, however, crippled AI research at a critical moment and ended any chance that an effective night defence system would be in operation during 1940.

As bad as the situation was at Dundee, it was worse for Bowen's airborne radar group. When Rowe first suggested a move in late 1938 one of the requirements was for a nearby large airbase to house aircraft needed for the research and fitting program. Watson-Watt had managed to find a location about as far away from a suitable airfield as possible, the closest being at Perth, a small civilian establishment some twenty-two miles from Dundee. Once again the arrival of the Bawdsey refugees and test aircraft came as a complete surprise, this time to the civilian Station Manager. He too only vaguely remembered Watson-Watt's visit and was adamant there had been no follow up. After several calls to the Air Ministry, the Station Manager agreed to share facilities with Bowen's group until a more suitable airfield could be found. This gave the airborne team and all the aircrew exactly one small hangar and a few offices. Needless to say it was a completely unsuitable situation whether for research or continuing the fitting program.[17] So bad were the conditions, that by the end of October the airborne team was moved yet again, this time to St Athan in Wales.

The appalling conditions at Perth were only partly to blame for a sudden decline in morale of the AI researchers. This was readily apparent to young university scientist Bernard Lovell, who joined the airborne group in early October. On 14 October he wrote to his friend, P. M. S. Blackett, a remarkably candid letter about the deteriorating morale of the AI researchers and technicians. He told Blackett that much of the frustration coming from a lack of consistency in Air Ministry policy, which resulted in much work needing to be completely redone. Worse still was the performance of the AI sets.

> In 6 tests out of about 12 it has caught fire in the air, due to extremely bad design. The power packs flash over, thin flex leads break off etc. etc. The tester knows exactly how to put things right in future designs, he is *never* consulted, and has given up trying to be helpful in sheer despair. It is designed by men sitting in secluded offices here, who to be frank have no social sense and no vestige of organising ability. By the peculiar [Air Ministry] system they have attained positions for which they are in no way fitted. (Please do not think I refer explicitly to Bowen. I like him immensely). The situation is really unbelievable. Here they are shouting for hundreds of aircraft to be fitted. The fitters are working 7 days per week, and occasionally 15 hour days. In their own words "the apparatus is tripe even for a television receiver".[18]

Blackett passed Lovell's letter to Tizard. Tizard in turn, without attribution, passed on select quotations from the letter to Rowe.[19] Rowe wa willing to downplay the severity of the situation, but Tizard did not accept this view.

Lovell's letter had simply confirmed much of Tizard's worst fears that things were going terribly wrong in much of the radar program, particularly with AI. In Tizard's view Bowen and the other scientists were being wasted in assembling and fitting sets, when what was needed was a far better model. Concerns were mounting because at the end of October, despite these poor early results, 300 production versions of AI, the Mark II, were ordered.[20]

The severity of the crisis also did not go unnoticed elsewhere in the Air Ministry. The Tizard investigation into the radar chain also examined the causes of the breakdown in radar research as well as the teething technical problems with existing equipment. The enquiry began with a request that all those concerned with radar submit reports on the current situation. Two of the most influential papers focussed on the requirement to separate research from development. Rowe wrote on 30 October that it was essential that the dividing line between research and development, 'however tortuous and transient,' be established and that researchers be removed from development work.[21] AVM A. W. Tedder, the Director General Research and Development, agreed with Rowe. He assessed that the root cause of the 'difficulties' was the 'inevitable' practice of going 'straight into large scale production of equipment which is really still in the research stage.' The obstacles to successful manufacturing were tremendous: 'There are at present no fully detailed specifications (such as are required for planned production), no schedules of parts, no installation layouts, no provision for interchangeability.'[22]

It would take Tizard's committee until January 1940 before it was able make a full analysis of the situation and implement a sweeping reform of the management of radar research and development. Meanwhile there was a pressing need for more immediate action to kick start AI research. On 10 November Tizard met with Rowe and Watson-Watt at Dundee and suggested that they enlist the aide of 'GEC Research Laboratories, and perhaps some other industrial research laboratories.' For the first time it was proposed that private firms be fully briefed on all aspects of AI and be encourage to conduct primary research. Watson-Watt disparaged the capabilities of the private companies and Tizard did not disagree. But the latter told Dowding, in his report of the meeting: 'You may have guessed rightly that one of my objectives in suggesting this plan is to provide some much needed competition to DCD and his staff, which I am quite certain will act as a stimulant. They won't want to be beaten on the results or on time.'[23]

Dowding concurred with Tizard's recommendation because he saw his chief trouble at the moment as being 'the complete stagnation in the fitting of AI.' He only added Electric Musical Instruments (EMI) to Tizard's list of manufacturers to be brought into the AI picture.[24] This exchange of correspondence marks the turning point in AI research; for it would be EMI that would eventually make the improvements that would turn the radar device into an effective weapon system.

The Failure of AI

Unfortunately, the decision to bring in the electronics industry into radar research came too late for an effective AI set to be ready for widespread service in 1940. The initial deployment of AI Mark I was a disaster. The Operation Record Book of Fighter Command laconically summarized the results on 12 December 1939: 'The use of AI flights at Martlesham and Manston discontinued, equipment unsatisfactory.'[25] Even when working correctly, the sets could not achieve anything like the minimum range that Dowding knew was necessary for a successful interception, nor could they operate below 10,000 feet. Several major design faults needed to be corrected, including the inability of the AI operator to determine if the signal was coming from ahead or behind the fighter and, if the target was not directly ahead or behind, if it was to the right or left.[26] The production version, the Mark II, was no better and until research resulted in improvements in range performance, AI was operationally useless.

There was, however, one positive development from the failures of late 1939. Repeated testing demonstrated the need for there to be highly accurate radar ground control of night fighters if they were to have any chance of getting into the right position to launch an AI guided attack. There was a particularly acute problem with the Blenheim, which was, at best, only marginally faster than the bombers it was supposed to shoot down. On 24 November Hanbury Brown, who had spent much of the autumn with AI-equipped fighter squadrons, wrote a seminal paper on the need for a radar set which would provide direct information to a ground controller. The controller needed to be able to place the night fighter just behind the target.[27]

Brown's proposal fitted in very nicely with the on-going series of experiments on all radar interceptions implemented in the summer of 1939. The creation of direct radio-telephone links between fighters and sub-controllers based at radar stations was intended to guide fighters to elusive day raiders operating over the sea. The scheme had limited success, in large measure because CHL could not measure height, while CH could not give sufficiently accurate readings in any dimension for pinpoint interceptions of lone targets operating off the coast.

A new type of radar utilizing the highly accurate lighthouse sweep of CHL, while providing equally exact information indication of altitude of the target was required. In order to simplify the job of the interception controller a new type of radar display which could show the position in plan of both the fighter and bomber was necessary. In December 1939, at Dundee Lewis directed G. W. A. Drummer to develop a CRT tube in which the radar station was at the centre of the screen on which was displayed information like a map as the radar antenna swept the sky. Using techniques similar to those used by sector controllers using the Biggin Hill techniques, the GCI controller could easily direct the fighter to within AI range of its prey. The prototype plan position indicator (PPI) was finished in May 1940.[28] The first of the new radar sets, the GCI, was only ready for testing in September 1940 and went into service the next month. It would be early 1941 before sufficient numbers of the new set were available to begin the regular direction of night fighters to their targets.[29]

While GCI development work began at Dundee, the Air Ministry spent much of the period from December 1939 until February 1940 correcting the mistakes of the previous four months. The first step was to overhaul the senior management structure looking after radar. Watson-Watt's responsibilities were divided among six senior managers and advisors working in three different Air Ministry organizations. In December, Watson-Watt was given the new position of Scientific Advisor on Telecommunications, removing him from day-to-day management. Replacing him as DCD was Sir George Lee, the former Engineer-in-Chief of the Post Office and chairman of the Radio Research Board. Lee had two deputy directors, one in charge of research and the other development. A Director of Communication Equipment Production took over the responsibility of supervising the manufacturing of radar sets and components.[30] In February, the RAF established No. 60 Group to administer the setting up and maintenance of ground based radar stations.[31]

Efforts were also made to get AI research restarted in early 1940 with a two pronged program. First, the existing AI equipment was to be modified to improve its reliability and its range performance. This work was to be undertaken by AMEE, and by a competing team at EMI. The second part of the programme involved long-term research by GEC to develop a completely new AI operating at a much smaller wavelength. Smaller wavelengths theoretically allowed for a far more accurate set with range performance well within Air Ministry requirements. As fundamental research on radio valves, receivers and transmitters was required for shorter wavelengths, the GEC work was viewed as requiring at least eighteen months before it could come to fruition.[32]

Research at AMEE recommenced sometime during the first few months of the year. There are few detailed records of this work, and the memoirs of the scientists involved are highly unreliable. In part, this is because AMEE's efforts were eclipsed by EMI, something that is obscured in the memoirs. It is also the result of the coming into the open of a long simmering personality clash between Rowe and Bowen. The two had never seen eye-to-eye concerning management techniques. Bowen believed Rowe a humourless civil service martinet who was mainly responsible for the chaos of the last half of 1939. Rowe's supporters have presented Bowen as being jealous of the appointment of W. B. Lewis as Senior Scientific Officer in July 1939. Lewis had only joined Bawdsey in early 1938 and he was promoted over several more senior scientists, including Bowen. Relations between the airborne group at St Althan and the main laboratory at Dundee, now several hundred miles apart, became increasingly strained. In January, Lewis at Dundee instigated what he viewed as complementary, and Bowen saw as competitive, research on AI improvements.[33]

Throughout the winter of 1940 there was growing anxiety about the ineffectiveness of the night defences, the solution to which was viewed as being of premier importance to the Air Ministry and Fighter Command. In mid-March, in order to ensure there would be no repeat of the chaos of 1939, a new Night Interception Committee was created under the chairmanship of Douglas, the Deputy Chief of the Air Staff. The committee oversaw all research, development and production of night defences. A sense of desperation hung over the committee because of the AI problems and the knowledge that time was swiftly running out. One of the remarkable features of the minutes of this committee is the time spent examining non-radar solutions, often involving the dredging up of ideas long since abandoned as impractical. Yet, the committee never lost sight of the fact that only radar appeared to offer an effective solution, if AI could be made to perform as promised. The committee was instrumental in pushing forward as rapidly as possible research into GCI and AI. Bowen's group was finally completely freed from development and installation work by the creation of a fitting unit, first based at St Athan and later at RAE at Farnborough.[34]

One of the Night Interception Committee's first measures was to act on Tizard's recommendation for the formation of an elite Fighter Interception Unit (FIU) which would test all new night fighting equipment. Towards the end of March the FIU began trials of a newly re-engineered Mark II. It was far more reliable than earlier sets, and several of the nagging technical problems, such as back echo, were eliminated. However, both the maximum and minimum ranges remained less than adequate.[35]

Bowen's team was soon hard at work replacing the AI transmitter with a far more powerful one originally designed for ASV radar; this became the

Mark III AI. Another version of the set incorporating modifications suggested by Lewis' research at Dundee, the Mark IIIa was also constructed.
Both versions were tested at the end of May by FIU. The results remained
disappointing, with the minimum range remaining stubbornly at 1,000
feet.[36]

By late May 1940, however, time had run out as neither the Luftwaffe
nor the new Prime Minister, Winston Churchill, would wait for any further refinements of AI. With the Germans racing through France, Churchill
demanded a full briefing on the state of AI research and production, which
was to be followed up with weekly progress reports. The Air Ministry felt
that, given the situation, something was better than nothing and they ordered
the Mark III into full production. Other refinements would have to proceed
after the widespread fitting of a radar they knew to be inadequate.[37]

Nothing else is heard from Bowen's airborne radar group. It appears
that he and Rowe had a final falling out in May when all of AMRE moved
once again, this time to Worth Matravers near Swanage. Bowen found the
facilities here little better than in Scotland, and he again blamed Rowe.
His criticism is perhaps borne out by the fact that the establishment was
forced to move yet again in 1942, this time to Malvern. Bowen, however,
would never work at Malvern. No details of his final months at AMRE
survive, but in August, Bowen was the only radar scientist made available
to Tizard for his technical mission to North America. It can be surmised
that his availability was caused by the failure to find a solution to the
AI range problem and his personality clash with Rowe. Unlike the other
members of the Tizard Mission, Bowen was not recalled to Britain at the
end of the year, but remained behind to act as a radar consultant.[38]

A prototype of the first truly effective AI set, the Mark IV, was only
ready for testing in July. This set was the result of research at EMI, a clear
demonstration of the capability of private industry to contribute to basic
research design. It incorporated the existing Mark III transmitter and
receiver with a new modulator and power pack. This produced a set with
far less transmission interference at short range and greater power for
long-range tracking. The set could track targets from 19,000 to between
400 and 500 feet. While not perfect, it was so superior to the Mark III that
it was immediately ordered into production.[39]

With herculean efforts the Mark IV began to enter squadron service
in small numbers in October. It would take several months before the
numbers and training would yield positive results. By March 1941, however, large numbers of German aircraft were being shot down by Mark IV
equipped night fighters. From the 2 March until the 13 April AI equipped
fighters detected 189 German bombers, brought down thirty-two aircraft,
possibly destroyed another five, and damaged a dozen more.

Beaufighter fitted with Mark IV AI Radar.

It is impossible to fully judge the consequences of the delay in getting an effective AI into service, for AI itself was only part of the solution. Coincidentally, the introduction of the Mark IV occurred virtually simultaneously with GCI and just after the arrival into service of a new and far more effective night fighter, the Bristol Beaufighter. However, before the introduction of the Mark IV, a grand total of just six aircraft were shot down by Mark III AI equipped night fighters, many of which were Beaufighters. While the majority of kills in March and early April 1941 were by aircraft guided to the proximity of their targets by GCI, aircraft hunting with AI Mark IV alone, or 'free-lancing', 'obtained notable success in turning AI detection into combat.' Also it must be pointed out that the GCI/Mark IV combination still had grave limitations. Few kills were obtained when the moon was below the horizon, since the minimum range remained inadequate for these conditions.[40] Thus, it is certain that if a more effective AI had been available just six months earlier a far more creditable, but far from perfect defence could have been mounted against the night-time blitz.

Participants in the events had little doubt about the seriousness of the failure and of its causes. As AVM Tedder confessed to Tizard on 24 January 1940: 'I am afraid much, if not most, of the trouble is due to our fatal mistake in rushing ahead into production and installation of AI before it was ready for production, for installation, or for use. This unfortunate precipitancy necessarily wrecked research work on AI since it involved diverting the research team from research proper to installation.'[41] Added to this must be the great delay in bringing in assistance from the electronics industry, which would ultimately find the solution to the problems with AI, and the extraordinarily ill-conceived move from Bawdsey. The responsibility for most of this lies with the Air Ministry, particularly Watson-Watt, and, to a lesser extent, Bowen, Rowe and Dowding. The consequences of the deception that occurred during the demonstration to Dowding were far reaching, but the roots of the problem remain the general mismanagement of radar development.

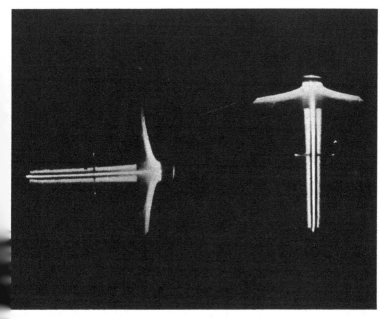

Mark IV AI Radar Display.

Conclusion

Radar, in consequence, has played a great and increasing role right from the beginning of the present war. It has more than any other single development since the airplane, changed the face of warfare; for one of the greatest weapons in any war is surprise, and surprise is usually achieved by concealment in the minutes or hours before an attack. The concealment formerly afforded by darkness or fog or cloud or artificial smoke or the glare of the sun simply does not exist in the world of radar. The tactical thinking of an attacker or a defender must take this fact into account.

Radar: A Report on Science at War, 1945

With the winding down of the Blitz in the spring of 1941, the great Luftwaffe air assault on Britain finally ended. The daylight air defence system had withstood the aerial onslaught of the summer before. The failure during the Blitz, while tragic for the victims, proved not to be a decisive factor in the outcome of the campaign. If the Germans had returned for another attempt to gain air superiority over Britain in 1941, they would have found both day and night defences far stronger than they had been in the previous year. The air defence system, with little change, would continue to successfully protect Britain's skies for the rest of war, although in the remaining five years of the conflict, there would be few serious challenges to test the defences. In 1944, however, two new types of weapons – the V-1 unmanned flying bomb and the V-2 ballistic missile – would once again bring terror to Britain's cities. V-1s pushed the defenders to the limit due to the high speed generated by their ram jet engines. The V-2 proved unstoppable; only the defeat of Germany brought about a close to this threat to Britain's cities. Both weapons proved that the air defence system was verging on obsolescence and would have to be replaced in the postwar period by far more sophisticated and complex radar and command and control apparatus.

Still the radar chain had delivered what Watson-Watt had promised in 1935 – an effective way to detect approaching bombers. Linked to an expanded command and control system, which originated in the First World War, and using the techniques developed at Biggin Hill to direct fighters to intercept the bombers, Fighter Command was just able to hold the Luftwaffe at bay. The scientists at Orfordness and Bawdsey had performed a technological miracle in just five years. A completely new type of weapon system had been taken from its primitive beginnings at Daventry and made a vital and integral part of the nation's air defence system. The scientists had overcome many bureaucratic and managerial shortcomings, which had curtailed the hiring of suitable numbers of qualified staff. Much of the credit for this goes to the two men who ran the radar research laboratory during these years. Watson-Watts' enthusiasm and hands on leadership approach proved inspirational to the mainly young men who worked long hours to push forward the development of radar. Rowe had been there at the beginning as a key player in the Air Ministry, and had come to Bawdsey at a time when firm managerial control was necessary to bring order to chaos.

Tizard, joined in the early years by Wimperis, was crucial in promoting the development of radar at the highest levels of government. He was placed in the unenviable position of having to defend the air defence research programme against the generally ill-conceived ranting of Lindemann. When faced with Lindemann's far more politically dangerous friend, Winston Churchill, Tizard had to rely on the support of cabinet ministers including Londonerry, Swinton, Wood and Inskip. The fight over the control of the agenda of air defence research degenerated into a nasty backroom struggle which did little credit to any of those involved. The only positive result of this conflict was that the radar secret remained hidden; after June 1935 Churchill and Lindemann never carried out their threats to bring the debate back into the public eye. The politics of air defence was never allowed to hamper the work at Bawdsey, although it did lead to the promotion of a variety of bizarre technological schemes that never should have seen the light of day, most notably, the aerial mine.

While Churchill and Lindemann attacked the Tizard Committee and the Air Ministry, the air defence research programme received consistent support from the governments of Baldwin and Chamberlain. As well, senior air force officers took an active role in developing the air defence system well before the invention of radar. With radar, officers such as Dowding, Joubert, and Newall saw an opportunity to provide a means of stopping the bomber. While they continued to believe in the need to concentrate efforts on building the RAF's strategic bomber force, no senior RAF officer seriously questioned the wisdom of rushing forward with the building of the radar chain even before the Bawdsey scientists could demonstrate that the technology would work as promised. It was, wrote C. P. Snow, 'an act of astonishing intellectual courage.'[1]

Service technical experts provided two of the key technological components of the air defence system – radio direction finding and radio-telephones. Servicemen worked with scientists to develop and implement the Biggin Hill techniques and filtering which allowed Fighter Command to defeat the Luftwaffe in the summer of 1940. Close collaboration between uniformed personnel and civilian scientists would continue throughout the war. Sometime after the Battle of Britain air force officers would coin the term Boffin, as an affectionate term for the military scientists who made such a huge contribution to the defeat of the Axis powers.[2]

Dowding, of course, must be singled out for special praise. His support for the Bawdsey scientists never wavered after the Daventry experiments, despite some good reason to mistrust their abilities after the failures in testing in 1936-37. He and Tizard were solely responsible for promoting the Biggin Hill Experiments. It was Dowding who oversaw virtually every aspect of the implementation of the new technologies into the air defences between 1935 and 1940. While Dowding may have been wrong about centralized filtering and misled by Bowen concerning the effectiveness of AI, his errors in judgement pale in comparison to his central role in developing the Dowding system. His dismissal from Fighter Command was caused by both political and personal factors which came together only when the scientists were unable to deliver in time the technology needed to defeat the Blitz.

The saving of Britain in 1940 was an extraordinary achievement, but it was only one of numerous developments from the Bawdsey years. Perhaps after the technology itself, the most significant product of the prewar radar programme was the transformation which occurred in the role of the scientist in senior level military and political decision making. Tizard was treated as an equal by senior RAF officers. He freely consulted with cabinet members and senior government officials. His friendship with Dowding opened the doors to scientists into the inner circle of Fighter Command. OR research gave Dowding an invaluable new tool to improve the performance of the air defence system. The ability of OR researchers to analyze critically the fundamental structures of the defences sped improvements at a crucial period when any delay might have led to disaster. Tizard's appointment at the start of the war to the position of Scientific Advisor to the Chief of Air Staff, was indicative of the growing importance of the scientists in military planning and operations. Personnel with scientific training were considered invaluable; most senior Allied commanders after the summer of 1940 would have some sort of scientific staff. OR would spread from Fighter Command, first to the War Office's Anti-Aircraft command, then to Coastal Command, and from there via Patrick Blackett to the Admiralty. After the war the growing role of science in the nation's affairs was recognized and institutionalized. In 1953, Tizard wrote to one of the official historians about the long term consequences of the scientific war.

> The ultimate result of this struggle of many of us was good. There is now a permanent scientific organization in the Ministry of Defence and so far as I can see scientific advice will now always be taken with strategy (as well as tactics). On the civilian side there is an Advisory Council on Scientific Policy the influence of which on national policy is growing.[3]

Well before the Battle of Britain, radar was being applied to all aspects of warfare. The Royal Navy was mounting their own sets on warships, and maritime patrol aircraft were searching out German submarines using ASV. It could be said that the battle represented simply the end of the pioneering phase of radar research and development. Even before France was invaded a whole new type of radar apparatus was suddenly made possible by the exploitation of far shorter wavelengths for radar applications. This stemmed from a direct spinoff over the debacle of AI radar. When private manufacturers were asked to assist in fixing AI, one solution was thought to be the development of radar operating with a wavelength measured in centimetres or in the microwave range. This would improve accuracy, allow for a much smaller antennae array, and reduce the interference from ground reflection. The best that Bowen's team could create using conventional radio tube technology was a set that operated at a wavelength of 1.5 metres. The antenna required for this set could only just be carried by larger aircraft and escort vessels. A group of scientists at Birmingham University, under the leadership of Professor Mark Oliphant, was one of several asked to investigate means of achieving shorter wavelengths.

In February 1940, two members of Olipant's laboratory, John Randall and Harry Boot, produced a revolutionary piece of technology – the cavity magnetron. It was the prototype of the first radio valve that could generate sufficient quantities of microwave radiation. At its first trial the crude prototype produced more than double the energy any other shortwave valve. More advanced models of the magnetron were soon producing 100 times more power than the original. By the beginning of August 1940 the first dozen production models of this radical new device were completed.[4]

The magnetron became the key component of an entirely new generation of radar sets that would begin to enter service in 1942. The magnetron proved perhaps more important as the key offering in one of the most extraordinary diplomatic missions in history. As part of broader efforts to improve diplomatic relations with the United States in 1940, Tizard lead a group of scientists and military technical experts across the Atlantic. Accompanying Tizard, were Bowen and John Cockcroft, the Cambridge physicist who since the beginning of the war had been heavily involved in the army radar research programme. Radar was the most important of the technologies taken by the Tizard Mission, and the Bowen's ASV radar and the magnetron proved to be the stars.

The Americans, who had radar of their own, where greatly impressed by the way the British had integrated radar into a comprehensive air defence system. They were amazed by the small size and sophistication of ASV, but were overwhelmed by the extraordinary promise of the magnetron. Within weeks American scientists, led by Vannever Bush, would turn the entire American military scientific organization on its head by orchestrating the establishment of the Radiation Laboratory at the Massachusetts Institute of Technology. The Rad. Lab. was made responsible for exploiting the potential of the magnetron. Staffed by America's best physicists and electrical engineers the Rad. Lab. soon perfected a large number of microwave radar sets. The huge American electronics industry ensured that the revolutionary radar would swiftly enter service. Bowen, unable to return to work under Rowe, remained in America to help teach them about some of the most important lessons learned at Bawdsey.[5]

In Britain, once finally ensconced in Malvern, the former Bawdsey research laboratory, renamed the Telecommunications Research Establishment (TRE), also developed a new generation of radar operating at microwave radiation. Rowe remained in charge of the research facility until the end of the war. He incorporated many managerial techniques to keep the enterprising spirit of the Bawdsey years alive and well. The most famous of these was the Sunday Soviets, which brought together laboratory staff with senior officers, service radar operators, and industrial researchers into wide-ranging discussions on all aspects of the radar war.

Ultimately, radar proved invaluable in the defeat of the Axis powers. It was vital in overcoming the U-boat in the Atlantic, in bringing accuracy to the bomber offensive over Germany, and in the destruction of the Imperial Japanese Navy and merchant marine. IFF, the secret of which was carried to America by the Tizard Mission, became a universal system by which Allied forces could identify friendly aircraft. Various radio navigation aides developed from radar technology saved countless lives both during and after the war. The radio proximity fuze, another technology acquired by the United States through the Tizard Mission, was perfected there and gave the Allies a deadly advantage in anti-aircraft artillery in the last year of the war.

After the war, the work at Bawdsey would continue to have important ramifications. Robert Buderi, in his recent study on the long-term implications of the development of radar, dubbed it, 'The Invention that Changed the World.' Wartime radar scientists would use their knowledge and techniques to conduct groundbreaking research into transistors, computers, medical imaging techniques and television. Even the purest of scientific disciplines, astronomy, would be transformed by radar. Former radar scientists, including Bowen, became pioneers in the new field of radio astronomy, revolutionizing our understanding of the universe.

Of all the accomplishments of radar, however, none stands out as important as the defeat of the Luftwaffe in 1940. Civilization stood at the abyss of what Churchill dubbed 'a new Dark Age made more sinister, and perhaps more protracted, by the lights of perverted science;' ironically only the invisible electronic shield of radar created by British scientists prevented Hitler's conquest of Britain. Despite the efforts of the radar scientists, most were poorly treated by their country after the war. Watson-Watt received a knighthood during the conflict. He took the opportunity to add a hyphen between his mother's maiden name and his father surname to become Sir Robert Watson-Watt. The rest of his war was anti-climactic and disappointing. His position as Scientific Advisor on Telecommunications, which was created for him upon his removal as DCD in early 1940, had no power and little influence. Like most of his fellow old radar hands from Orfordness and Bawdsey, he was quickly overshadowed by the first-rate university scientists that flooded into government service after the commencement of the war. There would be no return to scientific research for Watson-Watt. There was an attempt in the spring of 1941 to remove Watson-Watt from all major radar policy committees. Tizard intervened and saved his positions. Yet, even Tizard had grave reservations. He wrote Appleton on 21 July: 'We all know that in many ways he is his own worst enemy; at the same time a great deal of use can be made of him if he is handled properly.'[6]

In 1942, Watson-Watt was made Vice Controller of Communications Equipment, a middle level managerial position. After the war he resigned from government service when he discovered that he would receive no increase in pay, although other senior managers at his level received pay boosts of up to 50 per cent. In 1946, he became a part-time contract scientific advisor to the Air Ministry, the Ministries of Supply, Civil Aviation and Transport. Still Watson-Watt was an increasingly unhappy man who could never recreate his glory days at Bawdsey. After the war, when he was free to publicize his research, he found interest in the science of radar eclipsed by atomic physics. Here too, men from the Cavendish and other university laboratories dominated.

Watson-Watt and several other of his former Bawdsey staff chafed at the lack of recognition. Not one radar scientist was invited to participate in the Victory Parade. No other knighthoods were awarded in the post-war honours lists, although several received lesser honours. Finally in July 1950, a group of ten Bawdsey scientists, including Wilkins, Bowen and Watson-Watt applied to the Royal Commission on Awards to Inventors for financial compensation for the great economic and military value of their discoveries. Watson-Watt was determined to right what he saw as a grave injustice. He would devote the next eighteen months of his life to the fight for what he believed was their just rewards.

Rowe was appalled by what he perceived as the greed of these few scientists. He wrote to the Royal Commission 'that radar was evolved by many teams and by many people. To make money awards to a few of those involved would, to me, be a sorry end to a grand story of team work and my personal hope, shared by many of my war-time colleagues, is that no claimant will get a penny for doing what came his way to do.'[7]

Watson-Watt could not be deterred by Rowe's criticism or by the government's decision to fight the claims. The government position was that the development of radar was part of the normal work of its scientists and that few of the so called 'discoveries' were particularly new or original, but adaptations of the ideas of others. This simply stiffened Watson-Watt's determination to proceed. At the hearings of the commission, Watson-Watt tried his famous verbal karate. His opening speech consisted of 75,000 words and took nearly six hours to deliver over two days. His closing arguments consisted of some 200,000 words and took up four days of hearings. In total over an exhausting forty-four days of hearings Watson-Watt estimated he spoke a third of a million words. 'It would be disingenuous,' he related in his memoirs, 'to leave any impression that I did not enjoy the unusual experience.' Despite the weight of his verbal assault, the commission was only partly swayed and awarded all ten petitioners £87,950, a far cry from the £675,000 that Watson-Watt felt was their minimum due. Watson-Watt received £52,000 as his share of the award, a not-inconsequential sum in those days, particularly since it was tax free.[8]

Shortly after the commission's decision, Watson-Watt left for Canada to start a new life as a consultant. He would remain bitter about his treatment after the war, one of the few clear points made in his otherwise remarkably convoluted, distorted and error-filled memoir, *Three Steps to Victory*, published in 1957. He only returned to his beloved Scotland just before his death on 5 December 1973.

Only Rowe survived the war in a position of importance. He remained head of TRE until physical and mental exhaustion finally drove him to the breakdown he had predicted would occur before the war. Afterwards he served briefly as the Deputy Controller of research at the Admiralty. In 1948, he was appointed the Vice-Chancellor of the University of Adelaide. Rowe was almost completely forgotten by his countrymen after the conflict. Except for receiving a CBE in 1942, he received no further recognition for his vital work on radar. Yet, unlike Watson-Watt and some of the other prewar radar scientists, Rowe never complained. He died in relative obscurity in May 1976, aged seventy-eight.[9]

Tizard and Lindemann would renew their epic personal battle during the war. This time with Churchill as Prime Minister, Tizard was bound to lose. His influence at the Air Ministry was destroyed by Lindemann's intrigues with Lord Beaverbrook, the new Minister of Aircraft Production and Archibald Sinclair, Churchill's choice as Secretary of State for Air. In a dramatic confrontation in June 1940, Tizard attacked the R. V. Jones' assertion that the Luftwaffe had developed radio-navigation beams to guide their bombers to their targets at night. When Tizard's viewpoint was rejected he resigned from the Air Ministry. Tizard was wrong on this occasion, the Luftwaffe did have the beams and Jones was instrumental in ensuring that they could be jammed when the German's used them over Britain during the Blitz. Despite his resignation, Tizard's considerable influence continued, and he was appointed to head the British scientific mission to the United States just as the Battle of Britain commenced.

'The Prof' prospered under Churchill's premiership. When he once again failed to win a parliamentary seat, Churchill elevated him to the House of Lords as Lord Cherwell and made him a member of his War Cabinet. Cherwell was the senior scientific advisor to the government, although he commanded little respect from most other scientists within and outside of government service. Cherwell and Churchill continued to promote mainly useless gadgetry throughout the war. Cherwell also used his position to promote particular strategies and tactics, even if it meant distorting the scientific evidence.

On his return from America, Tizard returned to his old role as unofficial scientific advisor. He served both the Ministry of Aircraft Production and the Air Ministry in a variety of capacities, including heading the inter-service committee dealing with radar and, later on when the body was reconstituted as a cabinet level agency, The Radio Board, he served its senior member.

In the spring of 1942, Tizard and Cherwell would clash once more, this time over strategic bombing. Tizard challenged evidence presented by Cherwell that supported the Royal Air Force's belief that the war could be won by massive strategic bombing of Germany. Supported by several other scientists, including Blacket and J. D. Bernal, Tizard argued that

Cherwell's claims about the efficacy of bombing were grossly exaggerated. Again Tizard lost this dispute, in large measure because Churchill inevitably sided with his friend.[10]

In 1943, Tizard was elected to the Presidency of Magdalen College, Oxford. This took him away from some of his war work, but he continued to serve on the Radio Board and in other advisory capacities. At the end of the conflict, with Churchill and Lindemann out of power, Tizard was urged to return to government service on a more official basis. In 1946, he reluctantly accepted the Chairmanship of both the Defence Research Policy Committee and the Advisory Council on Science Policy. This made him the most politically powerful scientist in the country. He did much to ensure that science was well represented in defence and civil policy decision making. He undertook these two arduous tasks for four years before finally retiring from public life. Unlike the radar scientists, Tizard received great recognition during his lifetime, including numerous honourary degrees and fellowships. He died in 1959.[11]

Cherwell returned to the Clarendon Laboratory at Oxford after the war, while remaining an active member of the House of Lords and the shadow cabinet. In 1951, he joined Churchill's new Cabinet. Among his accomplishments was the creation of an independent atomic energy authority. Cherwell resigned from the Cabinet in 1953. He died in 1957.[12]

Neither Tizard or Cherwell spoke in public about their great falling out over air defence research. Only after their deaths did their friends and supporters begin a nasty public and private airing of the events of the 1930s and the war. Most notable or notorious, depending on your point of view, were the writings of C. P. Snow and Blackett, in defence of Tizard's leadership and, in rebuttal, the Earl of Birkenhead, Cherwell's friend and biographer, and R. V. Jones. None of the writers took the time correctly to assess the full dimension of the air defence debate, which is revealed in this study for the first time. Instead, they focused on the personalities of the two former friends, lionizing one while ripping apart the other in an effort to settle a long-simmering feud. The debate only cooled in 1965, when Ronald Clark finished his masterfully balanced biography of Tizard.[13]

The failure to recognize adequately the work of the radar scientists after the war was matched by a general absence of recognition of the historical importance of the activities that had taken place at Orfordness and Bawdsey. Both sites remained military bases until the 1980s. The former became a National Trust sanctuary, the latter a private English language school. Few traces of the radar work in 1935 survive at Orfordness, although by some remarkable quirk both of the 'temporary' First World War huts which housed the transmitter and receiver are still standing, albeit in a dilapidated condition. Although the staff of the National Trust wishes to restore these structures and provide a suitable interpretative display, there is no money to fund the project. At Bawdsey, much of the estate survives, including the grand manor, stables and the radar research and training blocks. The protected transmitter and receiver buildings of the Bawdsey Chain radar exist, but most of the equipment has long since disappeared. The last partially intact antenna mast was demolished in 2001. Just a single small plaque at the entrance to the manor marks Bawdsey's crucial part in winning the Battle of Britain.

A few other remnants of chain stations survive along the coast. A sector operations room is preserved at the Imperial War Museum's Annex at Duxford, and Fighter Command Headquarters' Operations Room has been rebuilt at the RAF Museum at Hendon. Surprisingly, there is much greater survival of the remains of Tucker's acoustical mirrors, perhaps because they were constructed of reinforced concrete. Five mirrors still stand, including all three at Denge. Alas, here too, time and neglect are overtaking these wondrous white elephants. The three at Denge, including the 200-foot mirror, have been undermined by quarrying, and stand precariously on the edge of a man-made lake.

It is hoped that, if nothing else, this study will generate sufficient public awareness of the work of the scientists and service personnel who created the air defence system and that this knowledge will result in a national tribute to them, equal to those already accorded to Fighter Command's pilots. Watson-Watt was right to feel unappreciated after 1945, since he and his team achieved one of the most remarkable scientific and technological accomplishments of the twentieth century. They provided Britain with a shield of radar which allowed Fighter Command to defeat the Luftwaffe. This enabled Britain to survive and, arguably, saved western liberal democracy.

Note on Sources

Primary Sources

As far as possible this work has been based on contemporary documents. The most important sources are the government records at the Public Records Office. These include the surviving records of the Air Ministry and the Ministry of Aircraft Production (AVIA and AIR), the War Office (WO), Cabinet (CAB) and Treasury (T).

The papers of Sir Henry Tizard at the Imperial War Museum (HTT) were a vital source. Tizard took extensive records with him when he left the Air Ministry in 1940. Of great importance in tracing the events of 1934-36 was Henry Wimperis' Diary, located at the National Archives of Canada. The personal papers of Patrick Blackett at the Royal Society provided useful information on the early research into proximity fuzes and air-to-air bombing. The Cherwell papers at Nuffield College, Oxford, were important in tracing the great feud between Lindemann and Tizard, as well as 'The Prof's' association with Churchill and Austen Chamberlain. Numerous personal paper collections at Churchill College Cambridge filled in important gaps in the story. This included the papers of Sir Winston Churchill (Chartwell Papers), E. G. Bowen (EGB) and A. V. Hill (AVHL). Other collections consulted included the Dowding Papers at the Royal Air Force Museum and the collections of various cabinet members, including Londonderry and Swinton.

Special mention must be made of the two secret Air Historical Branch histories of radar and the air defence system, *Signals: Volume IV, Radar in Raid Reporting* and *Signals Volume V: Fighter Interception and Control* (1950). Since they have never been published and use many now unavailable sources, these works are essential reading.

Memoirs

Almost all existing histories of radar in Britain rely extensively on the memoirs of scientists, most notably Watson-Watt's memoir, *Three Steps to Victory*, published in 1957. So filled with errors that Tizard refused to review it, and so convoluted that the American publishers would not print it until it was substantially re-edited and reduced in size, the memoir must be used with extreme caution. The American edition, which appeared two years later as *The Pulse of Radar*, is more readable but lacks much of the content and all of the flair of Watson-Watt's unique style. Watson-Watt's memoir can be supplemented by the interview he gave in July 1961 as part of *The New York Times*-Columbia University Oral History Program.

Rowe's account, *One Story of Radar* (1948), was written while he was in Australia and had no access to any archival material or personal papers. There is some useful information particularly around events surrounding the discovery of radar, but on the whole this work is uninformative about most of the key events that Rowe witnessed.

More recently several of the junior scientists have published their own account. Wilkins provided an brief remembrance of his experiences at Orfordness and Bawdsey in two books by Gordon Kinsey, *Orfordness – Secret Site* (1981), and *Orfordness, Bawdsey – Birth of the Beam* (1983). E. G. Bowen's informative but biased memoirs, *Radar Days* (1987) is particularly useful in tracing the atmosphere during the very early days of radar research and the development of airborne equipment. R. Hanbury Brown's *Boffin* (1991) is invaluable in providing the point of view of a member so Bowen's research group. Two collections of oral history interviews of radar scientists and personnel by Colin Latham and Anne Stobbs, *Radar a Wartime Miracle* (1996) and *Pioneers of Radar* (1999), contain much new information, although most of the interviews are about events after the Battle of the Britain. R. V. Jones' *Most Secret War* (1978) presents his own account of the events, particularly in regard to infrared research.

Most memoirs of air force officers give short shrift to the prewar period and contain little useful information on the development of the air defences. General E. B. Ashmore's *Air Defence* (1929) remains the best personal account of the air defence system in the First World War.

Secondary Sources

The strategic bombing campaign in the First World War is examined in depth by Christopher Cole and E. F. Cheesman in *The Air Defence of Great Britain* (1984). Supplementing this is an article by John Ferris, 'Airbandit:

C³ and Strategic Air Defences during the First Battle of Britain, 1915-18' (1989). The only well researched study of Tucker's acoustical mirrors is Richard N. Scarth's *Mirrors by the Sea: An Account of the Hythe Sound Mirror System* (1995).

Sean Sword's *Technical History of the Beginnings of Radar* (1986) is the best account of the early history of the technology. Robert Buderi's, *Radar: the Invention that Changed the World* (1996) provides a good narrative of the events at Bawdsey, although he has not used much primary evidence. His work is particularly good in linking events in the 1930s to the war and postwar periods. Still valuable is Henry Guerlac's *Radar in World War Two*, the secret official history of the Radiation Laboratory only published in 1987. David Fisher's rather eccentric account, *A Race on the Edge of Time: Radar – the Decisive Weapon of World War II* (1988), contains some useful material not found elsewhere. David Pritchard, *The Radar War: Germany's Pioneering Achievements, 1904-45* (1989), provided a important comparison with British developments.

Guy Hartcup's studies, *The War of Invention: Scientific Development, 1914-18*, (1988) and *The Challenge of War*, (1970) remain the best general studies of science and technology in the two world wars.

The three official biographies of the main protagonists in the Tizard-Lindemann dispute, the Earl of Birkenhead's, *The Professor and the Prime Minister* (1961), Ronald Clark's *Tizard* (1965), and Martin Gilbert's *The Prophet of Truth: Winston S. Churchill, 1922-39*, (1976), are essential reading. Clark's work is also invaluable in tracing the history of Tizard and his committee. C. P. Snow's *Science and Government* (1961) provides an emotive and inaccurate account of the famous feud. The biographical sketches of Watson-Watt, Bowen, Tizard, Lindemann, Blackett, Hill and Appleton in the *Biographical Memoirs of the Royal Society* were also very helpful.

The large number of studies on air force policy in this period all fail to place into proper perspective the importance of air defence. As such, few provide more than passing interest for this study. John Ferris' *Fighter Defence Before Fighter Command: The Rise of Strategic Air Defence in Great Britain, 1917-1934* (2000) provides a useful corrective to studies dominated by strategic bombing. J. P. Harris' *The 'Sandys Storm': the Politics of British Air Defence in 1938* (1989) was informative about Churchill's behavior in this period. N. H. Gibbs' official history, *Grand Strategy*, Volume I (1976), gives a good general overview of strategic policy making during the interwar era. Basil Collier's official history, *The Home Defence of Great Britain* (1957), is a good summary of air defence policy but contains little on the development of the radar chain.

The working of the air defence system is shown clearly in most studies of the Battle of Britain. The two best remain Derek Wood and Derek

Dempster's *The Narrow Margin* (1969), and Richard Hough and Denis Richards' *The Battle of Britain* (1989). John Ray's two studies of the battle, *The Battle of Britain: New Perspectives* (1994) and *The Night Blitz, 1940-41* (1996), contain an important re-examination of issues. The most important finding is that Dowding's dismissal ultimately rested on the failure of the night defences.

Sebastian Cox's *A Comparative Analysis of RAF and Luftwaffe Intelligence in the Battle of Britain, 1940* (1990), F. H. Hinsley's *British Intelligence in the Second World War*, Volume 1 (1979), and Ronald Lewin's *Ultra Goes to War* (1978) provide a good analysis of the role of intelligence in the battle.

Buderi and David Zimmerman's *Top Secret Exchange: The Tizard Mission and the Scientific War* (1996) furnish the best information on the long-term significance of the work at Bawdsey.

Notes

Notes to Chapter One

1. Christopher Cole and E. F. Cheesman, *The Air Defence of Great Britain* (London: Putnam, 1984), pp. 105-6, 151-3.
2. Cole and Cheesman, pp. 243-50. Twenty Gothas were dispatched , two turned back because of engine trouble, three others suffered engine trouble over England and dropped their bombs around Margate and Shoeburyness, and one further aircraft fell behind the main formation and dropped its bombs on the Royal Victoria Docks, London, sometime after the main attack.
3. Ibid, pp. 260-72.
4. As quoted in John Sweetman, 'The Smuts Report of 1917: Merely Political Window-Dressing?' *The Journal of Strategic Studies*, 4, (1981), p. 155.
5. Cole and Cheesman, pp. 219-21.
6. E. B. Ashmore, *Air Defence* (London: Longmans, Green and Co., 1929), p. 40.
7. Cole and Cheesman, pp. 276-86, 288-300.
8. Smuts as quoted in Ibid, p. 302.
9. John Ferris, 'Airbandit: C³ and Strategic Air Defenses during the First Battle of Britain, 1915-18,' *Strategy and Intelligence: British Policy During the First World War*, edited by Michael Dockrill and David French, p. 44.
10. Ashmore, p. 79.
11. Ibid, pp. 92-94.
12. Ibid. p. 95.
13. Cole and Cheesman, p. 447.
14. Hartcup, *The War of Invention: Scientific Developments, 1914-1918*, pp. 152-6.
15. Bernard Katz, 'A. V. Hill', *Biographical Memoirs of the Royal Society*, Volume 24 (1978), pp. 87-88.
16. Hartcup, pp. 161-2.
17. H. G. J. Holden to DMRS, 20 October 1917; Hugh Gamon, DMRS to Captain Holden DGMD 3 November 1917, PRO MUN 4/2487.

18. The Earl of Birkenhead, *The Professor and the Prime Minister* (Cambridge: Riverside Press, 1962), pp. 1-28; G. P. Thompson, Frederick Alexander Lindemann, Viscount Cherwell, *Biographical Memoirs of the Royal Society*, 1957, pp. 45-9.

19. Lindemann to Mr Richards, Naval Construction Department, 13 August 1914, Cherwell Papers, C/8.

20. Superintendent, Royal Aircraft Factory to Professor Callendar, 8 April 1915, Cherwell Papers, C 13/2; Admiralty to Lindemann, 16 July 1915, Cherwell Papers C 13/4; Lindemann, Royal Aircraft Factory Report No. 481, 'Note on the Possibility of Detecting Aircraft by Sound', 3 June 1915; Cherwell Papers, C 28/21.

21. T. Mather, J. T. Irwin, W. H. Cable, and H. C. Gibson, *Experiments on the Detection and Location of Hostile Aircraft*, n.d. but likely September 1915. Also see Superintendent RAF to ADMA, 29 June 1915; Mather to the Secretary, War Office, 14 September 1915 and 11 October 1915; Mather to the Secretary, Aircraft Department, War Office, 13 July 1916. (Note unless otherwise cited all documents are from the Public Records Office) AIR 1/121/15/40/105.

22. Captain W. B. Cadell, Deputy Assistant Director of Military Aeronautics, 6 October 1915, Ibid.

23. Major B. Hopkins, Deputy Assistant Director of Aircraft Equipment to Mather, 24 July 1916, Ibid.

24. The author wishes to thank Carol Ibrahim, of the University of London Library, and Anne Barrett, of the Imperial College archives, for this background information on Tucker. Tucker's DSc. was received on the basis of four articles: 'A High Potential Primary Battery', *The Proceedings of the Physical Society of London,* Vol. XXI; 'The Electrical Conductivity and Fluidity of Strong Solutions', ibid, Vol. XXV, Part II, February 1913; 'Heats of Dilution of Concentrated Solutions,' *Philosophical Transactions of the Royal Society of London,* Series A, Vol. 215; 'The Influence of Pressure on Convection Currents produced by a heated Cathode in a Vacuum Tube', no publication information available.

25. John Heilbron, 'Physics at McGill in Rutherford's Time', *Rutherford and Physics at the Turn of the Century*, (New York: Dawson and Science History Publications, 1979), p. 45.

26. W. S. Tucker, Statement of Qualifications, 11 February 1919, MUN 7/303.

27. W. S. Tucker, Officer Commanding Experimental Sound Ranging Section, 'Outline of Suggested Scheme for Location of Enemy Aircraft', Submitted to the Air Invention's Board, October 1917; Interview with Tucker 11 October 1917, MUN 7/303.

28. Tucker to The Controller, Munitions Inventions Department, 11 January 1918; RHS Bacon, Controller, MID, Acoustical Research Section, 22 February 1918, MUN 7/303.

29. See the Weekly Progress reports in AVIA 8/5.
30. Acoustical Research Section, Report for Week ending 28th September 1918, AVIA 8/5; Report for week ending 10th October 1918, MUN 7/308.

Notes to Chapter Two

1. Gibbs, N. H., *Grand Strategy*, Volume 1, London: HMSO, 1976, pp. 3-6; 35-49.
2. Ashmore, p. 114.
3. The three scientists and their peacetime positions were: Mr Jakeman, the National Physical Laboratory; Captain Ward, the Patent Office; and Lt. Milne, offered a fellowship at Trinity College, Cambridge. Major E. O. Henrici, DADFW to DDFW (c), 2 January 1919, MUN 7/309.
4. Ibid.
5. For Use of DSR at ADEE, 13 July 1938, AVIA7/3258; Also see AVIA 7/2792.
6. W. Tucker, Acoustical Research Section, 11 February 1919, MUN 7/303.
7. See the Reports of the Proceedings of the Royal Engineering Board, WO18.
8. Tucker to Controller, Munitions Inventions Department, 18 February 1918, MUN 7/308.
9. SEE Report No. 87, Preliminary Report on the Joss Gap Station of the Acoustical Section, 22 September 1920, AVIA 23/84.
10. 'Sound Mirrors for Anti-Aircraft Defence', 16 January 1924, AVIA 23/231.
11. SEE Report No. 73, Report of work carried out at Joss Gap, 20 March 1920, AVIA 2/70; SEE Report No. 87, op.cit.; SEE Report 67, Progress Report of the Acoustical Section, 27 April 1920, AVIA 23/64.
12. 'Sound Mirrors for Anti-Aircraft Defence', 16 January 1924, AVIA 23/231.
13. 'Sound Mirrors for Anti-Aircraft Defence', 16 January 1924, AVIA 23/231. Also see SEE Report No. 239 – January 1924, AVIA 7/3232.
14. Results of most of the mirror tests before June 1925 can be found in Notes on Acoustic Investigation in Hand at the Air Defence Experimental Establishment, Woolich, 6 July 1926, CAB 16/67.
15. SEE Report No. 285, 6 January 1925, AVIA 23/276; SEE Report No. 288, April 1925, AVIA23/279; SEE Report No. 298, May 1925; AVIA 23/289.
16. Conclusions of cabinet meeting 20 June 1923, PRO AIR 8/67 as quoted in Malcolm Smith, *British Air Strategy Between the Wars* (Oxford: Clarendon press, 1984), p. 32.
17. Lecture by the DCAS to the Naval Staff College on the 'Defence of England Against Air Attack', 10 March 1924, PRO AIR 9/69.
18. Basil Collier, *The Home Defence of Great Britain* (London: HMSO, 1957) pp. 8-18.
19. Ashmore, pp. 131-6.

20. Minutes of the First Meeting of the Anti-Aircraft Sub-Committee, 6 May 1925, CAB 16/67.

21. Lecture by the DCAS to the Naval Staff College on the 'Defence of England Against Air Attack', 10 March 1924; also see Expansion of the Royal Air Force for Home Defence, n.d. but 1923, PRO AIR 9/69.

22. Sir Geoffrey Salmond to AARC, 19 June 1925; Chatfield to AARC, 23 June 1925, CAB 16/67.

23. Colonel J. W. S. Sewell, 'The Present Position of Development of Methods of Locating and Plotting Position of Aeroplane, Sound Location', 23 June 1925, CAB 16/67.

24. Minutes of 2nd Meeting, 24 June 1925, CAB 16/67.

25. Minutes of 3rd Meeting of the ARC, 8 July 1925.

26. Minutes of 10th Meeting of the ARC, 28/4/26; Lindemann, 'Apparatus for Giving Warning at a Distance of the Approach of Aircraft', 19 April 1926, CAB 16/67.

27. F. A. Lindemann, 'The Efficiency of Sound-Gathering Devices for Detecting the Approach of Aircraft', 28 September 1925, CAB 16/67.

28. W. S. Tucker, Sound Detection of Aircraft, February 1926, CAB 16/67.

29. F. A. Lindemann, 'The Efficiency of Sound-gathering Devices for Detecting the Approach of Aircraft', 28 September 1925, CAB 16/67.

30. Minutes of the 4th Meeting of the ARC, 4/11/25; Anti-Aircraft Research Sub-Committee Interim Report, 23 December 1925, CAB 16/67.

31. The total approved funding for ADEE in fiscal 1925-26 was £34,680, of £19,330 was for operational expenses, the rest for special research projects. An additional £10,700 of requested funding for new building and the 200-foot mirror was rejected. MGOF, 'Research and Experiment in Connection with Anti-Aircraft Defence', 29 April 1926.

32. Geoffrey Salmond to Wing Commander Sir Norman Leslie, Secretary ARC, 31 December 1926; Tucker, 'Sound Detection of Aircraft', February 1926; Birch and Salmond, 'Acoustical Horizon of an Aircraft Flying at Various Heights', 27 April 1926, CAB 16/67.

33. The sub-committee members were F. E. Smith, DSR Admiralty; Colonel G. R. Pridham, President RE Board; and H. E. Wimperis, DSR Air Ministry. O. F. Brown, the Department of Scientific and Industrial Research, was the sub-committee's secretary.

34. Report of the Scientific Sub-Committee of the Anti-Aircraft Committee, 27 February 1928, CAB 4/17.

35. ARC, Report to the Committee of Imperial Defence, 24 March 1928, CAB4/17. Although the ARC remained in existence until Lord Haldane's death that summer this simple two page document endorsing the work of the scientific sub-committee is effectively the ARC's final report.

36. J. A. Webster, Air Ministry to ARC, 21 December 1926, CAB 16/67.

37. B. E. Holloway, to Air Ministry to Under Secretary of State, War Office, 19 April 1927, AIR 16/316 and AVIA 7/2764.

38. H. J. Creedy, War Office to The Secretary, Air Ministry, 16 July 1927; Secretary RE Board to Superintendent ADEE, 20 July 1927, AVIA 7/2764; B. E. Holloway, Air Ministry, to Under Secretary, War Office, AIR 16/316.

39. Air Commodore F. V. Holt, Chief of Air Staff Officer, ADGB to Colonel G. R. Pridham, RE Board, 6 November 1928, PRO AIR 16/316.

40. The Acoustics of Air Defence, n.d. but likely late 1934, AVIA 7/3186.

41. Ibid; ADEE Acoustical Report No. 77, October 1932, AVIA 17/47; The Acoustics of Air Defence, op.cit.; Notes on the 200-foot Strip Mirror, n.d. but likely 1933, AVIA 7/3185.

42. See weekly summaries for June 1930 in AVIA 7/2884; Report of the Acoustical by Service Personnel, June-July 1932, AIR 16/316.

43. Memorandum to Questionnaire put forward by ADGB on December 31st, 1932, AVIA 71/3184.

44. Group Captain R. E. Peirse for Director Operations and Intelligence to Air Officer Commanding ADGB, 28 April 1932, AIR 16/316.

45. ADEE Acoustical Report No. 77, October 1932, AVIA 7/47.

46. Salmond, ADGB to the Secretary Air Ministry, Acoustical Mirrors – Development and Policy, 22 October 1932, AIR 16/316.

47. To President RE Board, n.d. but report received at War Office on 20 December 1932, AVIA 71/3184.

48. Questionnaire by ADGB on ADEE Report No. 77, 31 December 1932, found as appendix A to Minutes of Meeting held on Tuesday, 7 February 1933 ... to discuss with H.Q. certain ADGB questions arising out of ADEE Secret Report No. 77, AVIA 71/3184.

49. Memorandum on the questionnaire put forward by ADGB, found as appendix B to Minutes of Meeting held on Tuesday, 7 February 1933.

50. Minutes of Meeting held on Tuesday, 7 February 1933, op. cit.

51. R. E. Peirse, Deputy Director of Operations and Intelligence to C-in-C ADGB, 7 June 1933, AIR 16/316.

52. Report of the Acoustical Range Test of the 200-foot mirror at Denge, June 1933, AVIA 17/24.

53. ADEE Acoustical Report No. 82, July 1933, AVIA 17/22.

54. Report of Air Exercises 17-20 July 1933, AIR 20/185.

55. Draft Minute of a Conference Held ... on 19 December, 1933, concerning the Extended Experiments to be carried out for the Developing the Acoustical Mirror Warning System, AIR 16/318; Memorandum on the Meeting Held at the Air Ministry on 19.12.33 in connection with the installation of Acoustic Mirrors in England, AVIA 12/133.

56. A copy of this article can be found in AVIA 12/133.

57. Report on Air Exercises 23-26 July 1934, AIR 20/186.

58. Air Council to War Office, 19 August 1935, AVIA 12/133. Emphasis provided.

59. Colonel F. J. C. Wyatt, President of the RE Board to MGO, AVIA 12/133.

60. Report on Combined RDF sound mirror station, 4 November 1935, AVIA 7/3258.

61. Air Ministry to Under Secretary of State, War Office, 26 August 1937, AVIA 12/133.

Notes to Chapter Three

1. Watson-Watt, Robert, 'Radar in War in Peace', *Nature*, September 15, 1945, p. 321.

2. Henry Guerlac, *Radar in World War Two*, Vol. I (New York: American Institute of Physics, 1987), pp. 32-4. Note: This is the first publication of the secret official history of the Radiation Laboratory originally completed in 1947.

3. Guerlac, Vol. I, pp. 38-44. Also see David Allison, *New Eye for the Navy: The Origins of Radar at the Naval Research Laboratory*.

4. Swords, pp. 52-68.

5. Clark, *Tizard*, p. 115.

6. Swords, pp. 101-5.

7. Ibid, p. 92.

8. Ibid, pp. 122-3, 135-9.

9. Ibid, p. 69. Emphasis added.

10. Ibid, pp. 69-74; Clark, *Tizard*, pp. 114-15.

11. Salmond to Tizard, 11 February 1924, as quoted in Clark, *Tizard*, p. 68.

12. Guy Hartcup, *The War of Invention: Scientific Development, 1914-18* (London: Brassey's, 1988), pp. 151-2; 'The Administration of Scientific Research at the Air Ministry and Ministry of Aircraft Production, 1918-1945', AVIA 46/158.

13. Robert Watson-Watt, *Three Steps to Victor*, (London: Oldhams Press, 1957), p. 80.

14. Bernal, p. 43.

15. A. P. Rowe, *One Man's Story of Radar* (Cambridge: Cambridge UP, 1948), p. 1.

16. 'The Administration of Scientific Research at the Air Ministry and Ministry of Aircraft Production, 1918-1945', AVIA 46/158.

17. G. H. Hardy as quoted in Gary Werskey, *The Visible College: A Collective Biography of British Scientists and Socialists of the 1930s*, (London: Free Association Books, 1988), p. 23.

18. David Wilson, *Rutherford: Simple Genius*, (London: Hodder and Stoughton, 1983), p. 538.

19. Hill as quoted in Werskey, p. 154.

20. *Nature*, 24 February 1934, 290.

21. William McGucken, *Scientist, Society and the State* (Columbus: Ohio State University Press, 1984), pp. 101-2.

22. Gibbs, pp. 48-9.

23. Ibid, pp. 80-1.

24. Wark, Wesley, *The Ultimate Enemy* (Oxford: OUP, 1986), pp. 28-33.

25. I. F. Clarke, *Voices Prophesying War* (Oxford: OUP, 1992), second edition, pp. 153-4.

26. Barry Powers, *Strategy Without Slide-Rule: British Air Strategy 1914-1939*, (London: Croom Helm, 1976), pp. 107-57; Clarke, pp. 145-62.

27. Martin Gilbert, *The Prophet of Truth: Winston S. Churchill, 1922-39* (Minerva: London, 1976), p. 151.

28. Churchill as quoted in Gilbert, pp. 51-2.

29. Gilbert, pp. 550-555.

30. Lindemann as quoted in Clark, pp. 107-8.

31. The Earl of Birkenhead, *The Professor and the Prime Minister* (Cambridge: Riverside Press, 1962), p. 180.

32. Gilbert, pp. 572-3, 576.

33. Ibid, p. 578.

34. U, Bialer, 'Humanization of Air Warfare in British Foreign Policy on the Eve of the Second World War', *Journal of Contemporary History*, pp. 13, 1 (1978).

35. David Omissi, *Air Power and Colonial Control* (Manchester: Manchester UP, 1990).

36. Smith, p. 67. Emphasis added.

37. John Ferris, 'Fighter Defence Before Fighter Command: The Rise of Strategic Air Defence in Great Britain, 1917-1934', unpublished manuscript provided by the author, pp. 22-3.

38. Handbook (Provisional) on Air Defence of Great Britain Scheme: Communications and Intelligence Systems, April 1930, AIR 5/768.

39. *Signals: Fighter Control and Interception*, Volume V, (A secret publication produced by the Air Historical Branch, 1950), p. 6, AIR 41/12.

40. Brooke-Popham, Report on Air Exercises – 17-20 July 1933, AIR 20/185; Air Defence of Great Britain Command Exercises, 1935, AIR 20/184.

41. Air Defence of Great Britain Command Exercises, 1935, AIR 20/184. For more on the effect of increasing aircraft speed on the defences see, The Sub Committee on the Re-orientation of the Air Defence System of Great Britain Interim Report, 31 January 1935, CAB 16/133.

42. Brooke-Popham, Report on Air Exercises – 17-20 July 1933, AIR 20/185.

43. Ibid.

44. Air Defence of Great Britain Command Exercises, 1935, AIR 20/184.

45. Sub-Committee on the Re-orientation of the Air Defence System of Great Britain, Interim Report, 31 January 1935, CAB 16/133.
46. Report by the A.O.C.-in-C. ADGB, 27 November 1928. AIR 9/69.
47. On Coastal vessels see CID Home Defence Committee Sub-Committee on the Re-Orientation of the Air Defence System of Great Britain 'Extension of Warning Period by Use of Surface Vessels' ADGB 17, 6 November 1934, CAB 13/18. For early warning recognisance see 'Air Defence of Great Britain Command Exercises, 1935', AIR 20/184.
48. CID Home defence Committee Sub-Committee on the Re-Orientation of the Air Defence System of Great Britain, Warning Period, ADGB paper 14 November 1934, CAB 13/18.

Notes to Chapter Four

1. The CID Home Defence Sub-Committee on the Re-orientation of the Air Defence System of Great Britain, Minutes of the 6th Meeting, 27 November 1934, PRO CAB 13/17. Also see the Earl of Birkenhead, *The Professor and the Prime Minister* (Boston: Houghton Mifflin Company, 1961), pp. 180-1; Brooke-Popham to Lindemann, 1 November 1934, Cherwell Papers, F 4/12; Lindemann to F. J. Hodson, Committee for Imperial Defence, 3 November 1934, Cherwell Papers, F5/1/3.
2. Londonderry to Lindemann, 28 November 1934, Cherwell papers, F5/2/1.
3. A. P. Rowe, *One Story of Radar*, ix, pp. 2-5. This account is, in part, confirmed in a letter Rowe sent to Tizard on 22 February 1936, HTT 79.
4. S. Russel Clarke, 'Suggestion for Producing a Heat Ray', n.d. but received at MID on 7 December 1917, PRO MUN 7/305.
5. James Swinburne to Col. Goold Adams MID, 10 January 1918, PRO MUN 7/305.
6. Press clipping from unspecified newspaper, PRO AVIA 7/2818.
7. Col. C. H. Silvester, Report on New Electrical Ray, 27 February 1933, PRO AVIA 7/2818.
8. Rowe, *One Story*, p. 6.
9. New York Times Oral History Program, Columbia University Oral History Collection. Part IV (1-219), 'Sir Robert Watson-Watt' (Sanford, North Carolina: Microfilming Corporation of America, 1979), p. 128. Note interview done July 1961.
10. Wimperis Diary Extract, 15 October 1934, HTT 700.
11. Henry Wimperis, 'Radiant energy methods of AA Defence. Note of a discussion with Professor A. V. Hill, FRS., 12 November 1934, HTT 700.
12. Hill to Secretary, Air Ministry, 12 December 1934, HTT 58. Also see Secretary of Air Ministry to Hill and Tizard, 12 December 1934, HTT 58; Dowding to Tizard, 17 May 1935, HTT 111.

13. Evan Davis, Patrick, 'Lord Blackett in the Royal Navy', unpublished paper presented at 'Patrick Blackett, Lord Professor and Lieutenant, Royal Navy: A Centenary Conference', Magdalene College, Cambridge, 24 September 1998.

14. Andrew Brown, 'Blackett at Cambridge, 1919-1933', unpublished paper presented at 'Patrick Blackett, Lord Professor and Lieutenant, Royal Navy: A Centenary Conference', Magdalene College, Cambridge, 24 September 1998.

15. Tizard as quoted in W. S. Faren, 'Henry Thomas Tizard', *Biographical Memoirs of Fellows of the Royal Society*, Volume 7 (1961), p. 324.

16. C. P. Snow, *Science and Government* (Cambridge: Harvard University Press, 1961), p. 6.

17. Tizard as quoted in W. S. Faren, 'Henry Thomas Tizard', pp. 320-1.

18. Cole and Cheesman, p. 167.

19. W. S. Faren, 'Henry Thomas Tizard', p. 330.

20. For more details on Tizard's early career see Clark, *Tizard*, 1-104. On Tizard and Wimperis' friendship see Henry Wimperis' Diary, National Archives of Canada.

21. A Note on the Problem of Air Defence, n.d. but late December 1934 or early January 1935. No author is provided, but it is almost certainly either written by or approved by Wimperis. AIR 20/145. Emphasis added.

22. J. A. Ratcliffe, 'Robert Alexander Watson-Watt', *Biographical Memoirs of Fellows of the Royal Society*, 21 (1975), pp. 549-68; Watson-Watt Oral History Interview.

23. Letter from R. Hanbury Brown as cited in Ratcliffe, p. 564.

24. The date of Wimperis' meeting with Watson-Watt is a matter of some conjecture. Wimperis' diary indicates that he met with him on 18 January to discuss his testimony before the Tizard Committee. However, he had also met with Watson-Watt on 7 November 1934, the contents of this meeting are unknown, but given its proximity to his meeting with Hill it seems likely to have included a discussion on death rays.

25. Watson-Watt Oral History Interview, pp. 127-8.

26. ibid, 128; Paper by A. F. Wilkins, no date but found with *Notes on the History of Bawdsey*, 12 November 1938, PRO AIR 20/195.

27. Robert Watson-Watt, *Three Steps to Victory* (London: Odhams Press, 1957) p. 95.

28. Robert Watson-Watt, Memorandum on the Damaging Effect of Radio Beams, January 1935, AVIA 10/349.

29. Wimperis Diary 23 January 1935.

30. Paper by A. F. Wilkins.

31. Watson-Watt Oral History Interview, p. 130; Wimperis Diary, 14 February 1935; Wimperis to Dowding, 15 February 1935, AIR 2/4483.

32. Basil Collier, *Leader of the Few: The Authorized Biography of Air Chief Marshal the Lord Dowding of Bentley Priory* (London: Jarrolds, 1957).

33. Dowding to Wimperis, 17 February 1935, AIR 2/4483. Emphasis provided.

34. Watson-Watt, *Three Steps*, p. 109. For a slightly different account see Watson-Watt Oral History Interview, p. 136.

35. Paper by A. F. Wilkins; Clark, *Tizard*, p.118; Watson-Watt, *Three Steps*, pp. 110-11; Rowe, Reflection of Radio Waves from Aircraft: Note on the results of a demonstration on 26 February 1935, AIR 2/4483. In Wilkins' account the name of the driver is given as Dyes.

36. Rowe, Reflection of Radio Waves from Aircraft: Note on the results of a demonstration on 26 February 1935, AIR 2/4483.

37. Wimperis Diary, 27-28 February 1935.

38. Detection and Location of Aircraft By Radio-Detection: Revised Note Prepared by Mr Watson-Watt, 28 February 1935, AIR 20/145.

39. Wimperis Diary, 1 March 1935.

Notes to Chapter Five

1. Sub-Committee on the Re-orientation of the Air Defence System of Great Britain, Interim Report, 31 January 1935, CAB 16/133.

2. Clark, *Tizard*, pp. 117.

3. Rowe to Tizard, 14 February 1935, HTT 79.

4. Rowe to Air Commodore Cunningham, 18 February 1935, AIR 16/182.

5. Rowe for the CSSAD, Notes on the committee's visit to ADGB, 21 February 1935, AIR 20/145.

6. Committee for the Scientific Survey of Air Defence. Interim Statement, 4 April 1935, AIR 20/145.

7. Committee for the Scientific Survey of Air Defence: First Interim Report, 16 May 1935, CAB 16/133.

8. Rowe to Tizard, 10 April 1935, HTT 79.

9. Lindemann to Londonderry, 3 December 1934, Cherwell Papers F 5/2/2.

0. Untitled summary of Lindemann's involvement in with air defence research, Cherwell F10/2.

1. Lindemann to Churchill, 7 January 1935; Letter to the Prime Minister 7 January 1935; Cherwell Papers 2/243.

2. MacDonald to Austen Chamberlain, 10 January 1935, Cherwell Papers F6/11.

3. Ramsay MacDonald to Austen Chamberlain, 15 January 1935, Cherwell Papers F 6/12.

4. Macdonald to Austen Chamberlain, 18 January 1935, Cherwell Papers F 6/15.

15. Lindemann to Austen Chamberlain, 18 January 1935, Cherwell Papers F 6/12; Churchill to Lindemann, 21 January 1935, Cherwell Papers F 6/17.
16. Lindemann to Churchill, 22 January 1935, Cherwell Papers 2/243.
17. Lindemann to Austen Chamberlain, 13 February 1935, Cherwell F 6/1/22.
18. Untitled summary of Lindemann's involvement in with air defence research, Cherwell F10/2.
19. Draft for Talk to 1922 Committee, n.d. but talk given on 18 February 1935, Cherwell Papers E30.
20. Marton Gilbert, *Prophet of Truth*, pp. 624-5.
21. House of Commons Debates, 27 February 1935, pp. 1117-8.
22. *The Manchester Guardian Weekly*, 8 March 1935, p. 183.
23. House of Commons Debate, 19 March 1935, p. 1003.
24. *The Manchester Guardian Weekly*, 22 March 1935, p. 221.
25. Lindemann to Hankey, 20 March 1935; Hankey to Lindemann, 25 March 1935, Cherwell Papers, F 6/3.
26. Tizard as quoted in Clark, *Tizard*, p. 17.
27. See various correspondence in the Cherwell Papers, D 243.
28. *The Times*, 6 April 1961.
29. Dowding to Tizard, 21 March 1935, HTT 111.
30. Committee of Imperial Defence Sub-Committee on Air Defence Research, 10 April 1935, CAB 16/133; Hankey to Tizard, 1 April 1935; 4 April 1935; Tizard to Hankey 4 April 1935, HTT99; Rowe to Tizard, 10 April 1934.
31. Air Defence Research Committee, Tentative Outline of Work for the Committee and Subjects on Which Information is Required, 11 April 1935; Tizard to the Secretary ADRC, 16 May 1935; Memorandum by Air Staff on Problems of Air Defence for the ADRC, 17 May 1935; Note Prepared by the Admiralty on the Problem and Location of Aircraft for the ADRC, May or June 1935, CAB 16/133; Minutes of the ADRC, CAB 16/132.
32. Lindemann, Letter to the Press, 25 May 1935, Cherwell Papers, E31/1, *The Daily Telegraph*, 25 May 1935.
33. House of Commons Debates, 7 June 1935, pp. 2247-2266.
34. *The Manchester Guardian Weekly*, 14 June 1935, pp. 462-3.
35. *The Daily Telegraph*, 8 June 1935.
36. Chiefs of Staff Sub-Committee, Re-orientation of the Air Defence of Great Britain, 14 May 1935; Note by Hankey for the Ministerial Committee on Defence Requirements, May 1935; Sub-Committee on Defence Policy Requirements, Re-orientation of the Air Defence of Great Britain, 18 May 1935; Sub-Committee on Air Defence Research, Interim Report, 25 June 1935, CAB 24/256.
37. Sub-Committee on Air Defence Research, Interim Report, 25 June 1935, CAB 24/256.
38. Minutes of Cabinet Meeting, 41 (35), July 1935, CAB 23/82.

39. Memorandum by Prof. F. A. Lindemann, n.d. but late June or early July 1935, HTT 67.
40. Rowe, Comments on Professor Lindemann's paper, 9 July 1935; Tizard to Lindemann; Tizard to Dowding 16 July 1935, HTT 67.
41. Lindemann to Tizard, 1 August 1935; Tizard to Lindemann, 4 August 1935, HTT 67.
42. Churchill to PM, 6 July 1935; Baldwin to Churchill, 8 July 1935, Cherwell Papers 2/244; Hankey to Churchill, 9 July 1935, Cherwell Papers 25/4.
43. Conclusions of the 4th Meeting of the ADRC, 25 July 1935, CAB 16/132.
44. Churchill to Cunliffe-Lister, 8 August 1935; Cunliffe-Lister to Churchill, 18 August 1935, Cherwell papers, 25/4; Gilbert, *Prophet*, pp. 657-9.

Notes to Chapter Six

1. National Trust Internet Site description of Orfordness.
2. Rowe, pp. 12-13; E. G. Bowen, *Radar Days* (Bristol: Adam Hilger, 1987), p. 16.
3. Watson-Watt, *Three Steps*, pp. 74-83, p. 168.
4. CSSAD, Minutes of 6th Meeting, 10 April 1935, AIR 20/181.
5. Bowen, pp. 7, 11, 14.
6. Bowen pp. 8-9; R. Hanbury Brown, H. C. Minnett, and F. W. G. White, 'Edward George Bowen', *Biographical Memoirs of Fellows of the Royal Society*, 1992, pp. 43-4; Visit of E. G. Bowen to F. E . Lutkin, 1 February 1951, AIR 53/331.
7. Bowen, pp. 9-10.
8. Rowe to the Royal Commission on Awards to Inventors, 24 May 1951, AVIA 53/350.
9. Bowen, 10; CSSAD, Minutes of 6th Meeting, 10 April 1935, AIR 20/181.
10. Bowen as quoted in Henry Guerlac, *Radar in World War II*, Vol 1 (New York: American Institute of Physics, 1987), p. 135.
11. Bowen, 11-19; Watson-Watt, *Three Steps*, pp. 126-30; Paper by A. F. Wilkins, no date but found with Notes on the History of Bawdsey, 12 November 1938, PRO AIR 20/195; Informal Report on Radio Work at Orfordness June 1935, 5 July 1935, AIR 20/474.
12. Watson-Watt, *Three Steps*, pp. 127-9; CSSAD, Notes on Visit of the Committee to Orfordness on 15 and 16 June 1935, AIR 20/145; Informal Report on Radio Work at Orfordness, June 1935, 5 July 1935, AIR 20/474; Wimperis Diary, 15-17 June 1935.
13. Notes on a Visit by the Secretary of the Committee to Orfordness on 16 July, 1935, AIR 20/145.
14. Watson-Watt Oral History Interview, 193-4, Watson-Watt, *Three Steps*, pp. 131-2.

15. Claim Before the Royal Commission on Awards to Inventors by A. F. Wilkins in Respect of Development in Radar Techniques, 28 March 1951, AVIA 53/347.

16. Paper by A. F. Wilkins, no date but found with Notes on the History of Bawdsey, 12 November 1938, PRO AIR 20/195; Watson-Watt, *Three Steps*, pp. 132-8; David Fisher, *A Race on the Edge of Time*, (New York: McGraw-Hill Books, 1988), pp. 121-2; Colin Latham and Anne Stobbs, *Radar a Wartime Miracle* (Stroud: Sutton Publishing, 1996), pp. 11-13.

17. Notes on a Visit by the Secretary of the Committee to Orfordness on 16 July, 1935, AIR 20/145.

18. Committee of Imperial Defence, Sub-Committee on Air Defence Research, Conclusions of the 4th Meeting, 25 July 1935, CAB 16/32.

19. WB1A to ADW, 1 October 1935; Bawdsey Manor, Suffolk, Description of Property, 4 October 1935, AIR 2/4485; Gordon Kinsey, *Bawdsey – Birth of the Beam* (Lavenham, Suffolk: Terence Dalton Limited, 1983), pp. 1, 9; Watson-Watt Oral History Interview, pp. 195-7.

20. Wimperis to Tizard, 22 August 1935, HTT 140.

21. Watson-Watt's Memorandum to the CSSAD on the State of RDF Research, 9 September 1935, in *Signals: Radar in Raid Reporting*, IV (Secret Publication of the Air Historical Branch, 1950), pp. 536-7, AIR 41/12.

22. Conclusions of the 5th Meeting ADRC, 16 September 1935, Cherwell Papers, 25/6; DSR to DWB, 19 September 1935, AIR 2/4485; Minutes of the 11th Meeting of the CSSAD, 25 September 1935.

23. R. V. Nind Hopkins to the Secretary of the Air Ministry, 19 December 1935, AIR 2/4485.

24. See for example Robert Buderi, *The Invention that Changed the World* (New York: Simon and Schuster, 1996), p. 58.

25. Watson-Watt Oral History Interview, pp. 202-3.

26. Wimperis to AMRD and Secretary Air Ministry, 19 September 1935, AIR 2/2724.

27. Wimperis to the Secretary Air Ministry, 30 November 1935, AIR 2/2724; Watson-Watt to Wimperis, 6 December 1935, AIR 2/2663.

28. Note by Wimperis, 2 January 1936, AIR 2/2724.

29. Wimperis to the Director, National Physical Laboratory, 29 October 1935, AIR 2/2663.

Notes to Chapter Seven

1. House of Commons Debates, 21 March 1936, pp. 259-400.

2. N. H. Gibbs, *Grand Strategy*, Vol 1, pp. 187-254; Malcolm Smith, *British Air Strategy Between the Wars*, p. 164.

3. Martin Gilbert, *Prophet of Truth*, pp. 647-714.

4. As quoted in Gilbert, *Prophet of Truth*, p. 716.

5. Cabinet 10 (36) K Cabinet Minutes, CAB 32/83.

6. Smith, pp. 159-65.

7. Rowe, Appreciation of the Present Position, 6 January 1936, AIR 20/145.

8. Rowe to Tizard, 9 January 1936, HTT 79.

9. Minutes of the 15th and 16th Meetings of the CSSAD, 3 and 25 February 1936, AIR 20/466.

10. Untitled, but Second Interim Report of the CSSAD, n.d. but March 1936, AIR 20/145.

11. Lindemann, Note on Air Defences, 7 February 1936, Cherwell Papers F13/1.

12. Churchill to Hankey, 26 February 1936; Lindemann to Churchill, 27 February 1936, Cherwell Papers, 25/9-10. There is no record of the Hankey letter in the official records, nor is it mentioned in Churchill's, Lindemann's or Hankey's authorized biographies.

13. Lindemann to Churchill, 5 March 1936; Lindemann's notes on Air Defence Problem, Cherwell Papers, 25/9.

14. Rowe to Tizard, 22 February 1936, HTT 79.

15. Wimperis Diary, 14 and 21 February 1936.

16. Memorandum by Watson-Watt, 27 April 1936, HTT 89; Tizard to Sir Frank Smith, 22 May 1936, HTT 90; Sir Frank Smith to Tizard, 22 May 1936, HTT 90; F. E. Smith to Sir Christopher Bullock, Air Ministry, 26 May 1936, HTT 90.

17. House of Commons Debates, 22 May 1936, 1398; Churchill to Inskip, 25 May 1936, CAB 64/5.

18. Tizard to Swinton, 12 June 1936, CAB 64/5.

19. Churchill to Hankey, 30 May 1936, Chartwell Papers 25/7; Churchill's Memorandum on Aerial Mines, 2 June 1936, CAB 64/5; Hankey to Churchill, 9 June 1936, Chartwell Papers 25/7.

20. Hankey to Swinton, 10 June 1936, CAB 21/426.

21. Tizard to Swinton, 12 June 1936, CAB 64/5; Note by Mr H. T. Tizard on Mr Winston Churchill's Memorandum, CAB 16/133.

22. Swinton to Inskip, 13 June 1936, CAB 64/5.

23. Watson-Watt to Sir Frank Smith, 13 June 1936, in Watson-Watt, *Three Steps*, pp. 148-9.

24. Lindemann to Churchill, n.d. but likely 12 June 1936, Chartwell Papers, 25/9.

25. Wimperis Diary, 15 February 1936; Swinton to Churchill, 18 June 1936, Chartwell Papers, 25/7.

6. Harcourt Johnstone to Churchill 15 June 1936; Churchill to Swinton, 16 June 1936, CAB 64/5; Swinton to Churchill, 18 June 1936, Chartwell Papers 25/7.

7. Churchill to Swinton, 22 June 1936, CAB 64/5.

28. Gilbert, *Prophet of Truth*, pp. 752-8.

29. Watson-Watt, *Three Steps*, pp. 149-50.

30. J. A. Webster, Air Ministry, to R. A. Watson-Watts [sic], 29 July 1936, T 166/80 (Part 1); Watson-Watt to the Secretary, Air Ministry, 31 July 1936, T 166/80 (Part 1).

31. Tizard to Lindemann, 17 June 1936, CAB 64/5.

32. Wimperis to Tizard, 19 June 1936, HTT 111.

33. The first draft of this letter was prepared on 23 June 1936, see Cherwell Papers, F8/6/4. The letter sent to Tizard on 25 June is in CAB 64/5. Lindemann to Churchill, 23 June 1936, Chartwell Papers, 25/9.

34. Tizard to Lindemann, 5 July 1936; Tizard to Swinton, 5 July 1936, CAB 64/5.

35. Lindemann to Rowe, 10 July 1936, HTT 67.

36. Minutes of the 20th Meeting of the CSSAD, 15 July 1936, AIR 20/181.

37. Note by F. A. Lindemann on the Progress Report of the Committee for the Scientific Survey of Air Defence, 20 July 1936, CAB 16/133.

38. Tizard to Swinton; Hill to Swinton; Blackett to Swinton, 15 July 1936, CAB 64/5.

39. Swinton to Hill, 16 July 1936, HTT 58; Swinton to Blackett, 16 July 1936; Hill to Swinton, 17 July 1936, CAB 64/5.

40. Hill to Tizard, 23 July 1936, HTT 67.

41. Conclusion of 10th Meeting of the ADRC, 24 July 1936, CAB 16/132; Wimperis Diary, 24 July 1936.

42. Cabinet 56(36) 8, 2 September 1936, CAB 23/85; Swinton to Hill, HTT 58.

43. Lindemann to Churchill, 23 October 1936, Chartwell 25/9.

44. Churchill to Baldwin, 22 July 1936, Chartwell Papers 2/266 b; Ismay to Churchill and transcript of Meeting with Prime Minster, 1 December 1936, Chartwell Papers 2/270.

45. Churchill to Swinton, 19 November 1936, Chartwell Papers 25/7.

46. Gilbert, *Prophet of Truth*, pp. 794-807.

47. Lindemann to Swinton, 23 September 1936, HTT 111; Wimperis to Lindemann, 5 October 1936, Cherwell Papers D123/8; Churchill to Hankey, 27 October 1936, CAB 21/426; Wimperis to Lindemann, 25 November 1936, Cherwell Papers D123/14.

Notes to Chapter Eight

1. AVM Joubert to Tizard, 31 March 1936, HTT 64.

2. Interception exercises, 18 November 1936, AIR 16/179.

3. *'Signals: Volume 5 – Fighter Control and Interception'*, Air Historical Branch Confidential History, 1950, pp. 8-9.

4. Minutes of the 13th Meeting of the CSSAD, 13 November 1935, AIR 20/181.

5. CSSAD – Note handed to the secretary by Joubert, 29 November 1935, AIR 20/2357.
6. Joubert to Tizard, 30 April 1936, HTT 131; Minutes of a meeting held at the Air Ministry on the co-ordination of DF and RDF and allied problems, 30 March 1936, AIR 2/1770.
7. Signals Report: No. 11 (Fighter) Group Final Sector Training, 1936, 22 August 1936, AIR 2/2728.
8. Notes of conclusions reached at an informal meeting held at the Air Ministry on 13/7/36 to discuss certain experiments suggested by Tizard AIR 16/45.
9. Air Ministry to Tizard, 27 July 1936, HTT 131; Ronald W. Clark, *The Rise of the Boffins* (London: Phoenix House, 1962), pp. 49-50.
10. Officer Commanding, Biggin Hill to CinC Fighter Command, 11 August 1936, AIR 16/45.
11. Minutes of conference held at Headquarters of Fighter Command on Friday 7 August 1936 to discuss questions concerning special interception experiments at Biggin Hill during August and September 1936, 10 August 1936, AIR 16/45.
12. Clark, *Tizard*, p. 150.
13. Interception Exercises, 18 November 1936, AIR 16/179.
14. The account of the Biggin Hill experiments was derived from: Tizard, Interception Experiments, 2 September 1936; 2nd to 6th Progress Reports of the Special Interception Experiments, 17 September 1936 to 15 January 1937. All quotes are from Interception Exercises, 18 November 1936, AIR 16/179. Also see: Tizard, CSSAD State of Investigations, October 1936, AIR 20/2357; Tizard to Swinton, 5 October 1936, CAB 64/5; Clark, *Tizard*, pp. 150-3; Sir Arthur McDonald, 'Biggin Hill', in Colin Latham and Anne Stubbs, *Radar: A Wartime Miracle* (Stroud: Sutton, 1996), pp. 69-71.
15. Grenfell, Signal Requirements for Operational Application of Interception Experiments, 18 January 1937, AIR 2/2728.
16. B. G. Dickins, Notes on Interception Experiments, February 1937, AIR 20/80.
17. Hankey to Tizard, 6 October 1936, CAB 21/426; Hankey to Swinton, 7 October 1936, CAB 21/621.
18. R. Hanbury Brown, *Boffin*, (Adam Hilgar: Bristol, 1991), pp. 14-15; Report on the Preliminary Special Communication Exercise – 17 September to 1 October 1936, 1 October 1936, AIR 2/2485.
19. Ibid; Bowen, pp. 24-5.
20. Tizard to Watson-Watt, 20 September, 1936, HTT 140.
21. DSD to CAS, 10 October 1936, AIR 2/2585.
22. Note by Professor E. V. Appleton, 2 October 1936, AIR 20/145.
23. Appleton to Tizard, 22 December 1936, HTT 30; Rowe to Tizard, 31 January 1937, HTT 79.
24. R. Hanbury Brown, pp. 1-4.

25. Donald Preist, in Colin Latham and Anne Stobbs, *Pioneers of Radar* (Stroud: Sutton, 1999), pp. 38-9.

26. Rowe, RDF Chain, 15 December 1935, AIR 2/2723; G. R. Brigstocke to The Secretary, Treasury, 31 March 1936, AIR 2/3103; Record of Discussion Held with Watson-Watt, 15 July 1936, AIR 2/2624.

27. Letter from Watson-Watt, 23 September 1936, AIR 2/12663; Abraham, Air Ministry to F. P. Robinson, Treasury, 27 September 1936, T161/973.

28. Bawdsey Research Station: Scientific and Technical Staff – January 1937, AVIA 53/356; A. P. Rowe, Notes on the History of Bawdsey Research Station 12 November 1938, AIR 20/195.

29. Bowen, p. 22; Brown, p. 17; Henry Guelrac, *Radar In World War II* (New York: American Institute of Physics, 1987), Volume I, pp. 140-2.

30. Edmund Dixon, Production Aspects of Investigation No. 100, 3 January 1937, AVIA 7/32; Watson-Watt to DSR, Air Ministry 15 February 1937 , AVIA 7/32; Edmund Dixon, Air Defence Communications, 18 February 1937, AVIA 10/47; D. R. Pye DSR to Watson-Watt, 13 March 1937, AIR 2/1980; Watson-Watt to Secretary Air Ministry, 30 March 1937, AIR 2/1980.

31. Watson-Watt, Superintendent to DSR AM, 15 February 1937, AVIA 7/32; DCAS to CAS, 5 March 1937, AIR 2/2653; CAS to DCAS, 9 March 1937, AIR 2/2653; D. R. Pye, DZR to Watson-Watt, 16 March 1937, AVIA 7/32.

32. W. S. Douglas, DSD to AMRD, 8 February 1937, AIR 2/2612; Dowding to the Secretary, Air Ministry, 13 March 1937, AIR 2/2612; R. H. Verney, Director of Technical Development to Chief Superintendent, RAE, Special Communication Exercises 1937, AVIA 7/394.

33. Watson-Watt, *Three Steps*, pp. 174-5.

34. Edmund Dixon, Technical Development RDF1, 15 January 1937, AVIA 10/47.

35. Headquarters, Fighter Command to Watson-Watt, 22 April 1937, AVIA 7/437; SA/DSR, Summary of Results of Special RDF Exercises, 19 to 30 April 1937, 18 June 1937, AVIA 7/394.

36. Watson-Watt, Note on April RDF Exercise, May 1937, AVIA 7/394; Rowe, Note on Bawdsey Exercises, April 1937, 8 June 1937, AIR 20/80; 86th Progress Meeting, Trials of RDF Equipment at Bawdsey, April 1937, 21 June 1937, AIR 6/49.

37. Joubert to Secretary, Air Ministry, 21 May 1937, AIR 2/2612; DCAS to CAS, 11 June 1937, AIR 2/4484; Watson-Watt to the Director Scientific Research, Air Ministry, 9 May 1937, AIR 2/1969; 86th Progress Meeting, Trials of RDF Equipment at Bawdsey, April 1937, 21 June 1937, AIR 6/49; Royal Air Force Expansion Measures, Secretary of State's Progress Meetings, Minutes of 86th Meeting, 22 June 1937, AIR 6/30; Watson-Watt, *Three Steps*, pp.180-1.

38. AMRD to DCAS, 5 February 1937, AIR 2/2618; Watson-Watt, Production of RDF Sets, 18 January 1937, AVIA 7/32.

39. Guerlac, 144-5; Watson-Watt, *Three Steps*, pp. 174-5; Rowe for Superintendent, to Secretary, Air Ministry, 24 August 1937, AIR 2/2653.

40. Edmund Dixon, Notes on the Function of Bawdsey Research Station in Communications Research, 9 July 1937, AVIA 7/230.

41. R. G. Hart, Training and Operational Development at Bawdsey Research Station, 27 August 1938, AVIA 7/410; Watson-Watt, *Three Steps*, pp. 173-4.

42. Co-operation during RAF Exercise on 9 and 11 August 1937, 4 September 1937, AVIA 7/395.

43. Bowen, *Radar Days*, pp. 30-46; E. T. Paris for Superintendent, Progress of RDF 2 Experiments, 9 September 1937, AIR 2/2624.

44. Paris, RDF and Coast Defences, 23 November 1937, AIR 2/2681.

45. Bawdsey Research Station: Scientific and Technical Staff, January 1938, AIR 53/356; Watson-Watt to the DSR, Air Ministry, Bawdsey Research Station – Staff Proposals, 2 November 1937, AIR 2/2724; Watson-Watt to Secretary, Air Ministry, War Office Staff at Bawdsey, 6 November 1937, AIR 2/4486.

Notes to Chapter Nine

1. A report written for the Prime Minister by the Air Staff giving him the situation if Britain were to be attacked on 1 January 1938, 22 October 1937, CAB 64/9.

2. The Air Staff as quoted in N. H. Gibbs, *Grand Strategy*, Volume I (London: HMSO, 1976), p. 567.

3. Gibbs, 565-89; Malcolm Smith, *British Air Strategy Between the Wars* (Oxford: Clarendon Press, 1984), pp. 198-226.

4. For Swinton's resignation see J. A. Cross, *Lord Swinton* (Oxford: Clarendon Press, 1982), pp. 191-219.

5. Gibbs, pp. 596-7.

6. Cabinet 53 (38), 7 November 1938, CAB 23/96.

7. H. C. Robbins, Secretary of the Admiralty, War Office, Air Ministry and Press Committee, 1 February 1938, AIR 2/2664.

8. Watson-Watt's claim is only found in the American version of his memoirs, *The Pulse of Radar* (New York: The Dial Press, 1959), pp. 177-8.

9. Viscount Templewood (Sir Samuel Hoare), *Nine Troubled Years* (London: Collins, 1954), p. 289.

0. Lindemann to Churchill, 16 May 1938, Chartwell Papers 25/16.

1. Churchill to Wood, 9 June 1938, CAB 21/630.

2. Wg/Cdr William Elliot to Hankey, 15 June 1938; Hankey to Minister, 16 June 1938; Hankey to Minister, Mr Churchill's Letter on the ADR Committee, 20 June 1938; Hankey to Elliot, 20 June 1938, CAB 21/634.

13. Hankey to Tizard, 16 June 1938; Tizard to Hankey, 16 June 1938; Mr Churchill's letter on the ADR Committee, 20 June 1938, CAB 21/634.

14. Memorandum by Sir Henry Tizard on Mr Winston Churchill's letter of June 9 to the Secretary of State for Air, 22 June 1938, CAB 64/5.

15. Kingsley Wood to Churchill, 24 June 1938; F. H. Sanford to H. G. Vincent, 24 June 1938, CAB 64/5.

16. Hankey to Prime Minister, 24 June 1938; Prime Minister's Office to Hankey, 27 June 1938, CAB 21/634.

17. J. P. Harris, 'The "Sandys Storm": the Politics of British Air Defence in 1938,' *Historical Research,* 62 Nu. 149 (October 1989), pp. 318-36.

18. Churchill to Wood, 26 July 1938, as quoted in Martin Gilbert, *Prophet of Truth* (London: Minerva, 1976), pp. 950-1.

19. Harris, p. 332.

20. Churchill to Wood, 30 October 1938, CAB 21/820.

21. Inskip to Tizard, 15 November 1938, CAB 21/820.

22. Tizard to Hill, 24 November 1938, Churchill College AVHL 2/5.

23. Tizard to Inskip, 24 November 1938; Inkip to Tizard, 24 November 1938, HTT 67.

24. Tizard to Wood, 29 January 1939, HTT 93.

25. Freeman to Tizard, 2 February 1939, HTT 93.

26. Hankey, Memorandum on Mr Churchill's letter to Sir Kingsely Wood of 9 June, 1938, 20 June 1938, CAB 21/634.

27. CSSAD, 27th and 28th Meetings, 21 and 24 December 1936.

28. Liddel-Hart to Tizard, 4 February 1937, with marginal note by Tizard dated 11 February; 'Suggestions for an Anti-Aircraft Projectile', n.d., HTT 66.

29. CSSAD 42nd Meeting, 13 July 1938.

30. CSSAD 45th Meeting, 3 January 1939; 48th Meeting, 17 May 1938.

31. ADRC, Report on the Progress of the Principal Researches Recommended by the CSSAD, June 1937, 22 July 1937, Chartwell Papers, 25/13.

32. R. V. Jones, *Most Secret War* (London: Hamish Hamilton, 1978), pp. 34-44; CSSAD 30th Meeting, 19 May 1937.

33. Balckett to Tizard, 16 December 1936, HTT 32.

34. Tizard to Dowding, 5 May 1938; Dowding to Tizard, 6 May 1938, HTT 131.

35. Ben Lockspeiser, RAE to Blackett, 14 April 1938, Blackett Papers D13.

36. CSSAD 46th Meeting, 9 February 1939.

37. RAE to Blackett, 4 February 1939, Blackett Papers D13; Time Fuses for Bombing the Bombers, n.d. but spring 1939, Blackett Papers D14; A note on the use of a stabilized sight for use with the PE Bomb, n.d. but spring 1939; RAE Instruction Leaflet No. H277, Blackett Papers D19; Dowding to Tizard, 15 September 1939, HTT 226.

38. CSSAD 44th Meeting, 23 November 1938; Guy Hartcup, *The Challenge of War,* 170-81; David Zimmerman, *Top Secret Exchange: The Tizard Missio*

and the Scientific War (Montreal and Stroud: McGill-Queens and Sutton, 1996).

39. Notes on Mr Churchill's Memorandum on Air Defence, 6 September 1938, AIR 19/25.

40. CSSAD 22nd Meeting, 16 October 1936.

41. SA to DSR, Investigations made in connection with Aerial Mines, in consultation with the Committee for the Scientific Survey of Air Defence, 22 June 1938, CAB 21/635.

42. SA to DSR, CSSAD, Long Aerial Mine Barrage, 15 November 1938, AIR 20/8.

43. D. R. Pye to Dowding, 24 April 1939, AIR 16/113.

44. CSSAD 49th Meeting, 21 June 1939.

45. Agenda for a Meeting for the purpose of discussing tactical trials of the long and short aerial mine, 22 February 1940; Long Aerial Mine – Experiments, 13 June 1940, AIR 16/113; Tizard to ACAS (T), Long Aerial Mines, 16 June 1940, HTT 15.

46. Hartcup, pp. 256-7; Long Aerial Mines Operations, 23 September 1940, AIR 16/113; Ground Control of Aerial Minelaying, 22 July 1941, AVIA 15/1546.

47. Hartcup, p. 235; UP Chronology, Appendix II of Memorandum on Mr Churchill's letter to Sir Kingsley Wood of 9 June, 1938, 20 June 1938, CAB 21/634.

48. Hartcup, pp. 236-7; Tizard, Note on War Office Memorandum on the Introduction of UP Weapon, 16 June 1938, CAB 21/631; R. H. Fowler, Rockets, 21 October 1938; Notes on Conversation with Wg/Cdr Elliott, 19 May 1939, HTT 10; Tizard, Defence Against Low Flying Aircraft, 29 March 1940, HTT 15; Tizard to Dowding, 1 March 1940, HTT 226; Ministry of Supply Executive Committee, UP Production, n.d. but likely early 1941, Beaverbrooke Papers, D160.

49. CSSAD 22nd Meeting, 16 October 1936; A. V. Hill, Aircraft Silhouettes: Area Illumination, 18 November 1936, CAB 21/1101; Tizard, State of Investigations of the CSSAD, October 1936, AIR 20/2357.

50. Secretary of State for Air, Memorandum for the Defence Policy Requirements Sub-Committee, April 1937, CAB 21/1101.

51. Secretary of State for Air to the Defence Policy Requirements Sub-Committee, Experiments in Silhouette Detection of Aircraft, CAB 64/3; CSSAD 29th Meeting, 10 March 1937.

52. CSSAD, 31st Meeting, 5 July 1937.

53. CSSAD, 50th Meeting, 16 August 1939; Air Ministry, Silhouette Detection of Aircraft, November 1939; H. L. Ismay, Silhouette Detection of Aircraft, 5 November 1939, CAB 21/1101.

54. On Balloon barrage research see the CSSAD's minutes of meetings; CSSAD, Position of Principal Researches recommended by the Committee, 30 June

1937, 30 June 1937, AIR 20/2357; Position of Principal Researches recommended by the CSSAD, 21 October 1937, CAB 16/134; research and development work on air defence problems carried out in consultation with the CSSAD, 22 June 1938, CAB 21/634; position of principal researches of the CSSAD, December 1938, CAB 16/134.

55. Churchill to Wood, 27 March 1939; Professor Lindemann's Memorandum, AIR 19/26.

56. Lindemann to Churchill, 26 April 1939; Wood to Pye, 29 March 1939, AIR 19/26.

57. Pye to AMDP, 17 April 1939, AIR 19/26.

58. Lindemann to Churchill, 18 May 1939, AIR 19/26.

59. Churchill to Wood, 27 June 1939, AIR 19/26.

Notes to Chapter Ten

1. DSR to AMRD, 18 November 1937, as quoted in 'The Administration of Scientific Research in the Air Ministry and Ministry of Aircraft Production, 1918-45,' AVIA 46/158.

2. PEO to AMRD, 5 May 1938; Pye to PEO, 19 May 1938, James Rae to Under Secretary State, Air Ministry, 2 August 1938, Air 2/2663.

3. Watson-Watt to Tizard, 4 April 1938, HTT 89.

4. Joubert, The Case of Wartime re-distribution of responsibilities, 13 December 1939, AIR 20/2269.

5. Bowen, *Radar Days*, pp. 49-50.

6. Colin Latham and Anne Stobbs, *Pioneers of Radar* (Stroud: Sutton, 1999), p. 74.

7. Ibid, p. 33.

8. Tizard to Freeman, 31 March 1938, HTT 140.

9. Interception Experiments at Biggin Hill, 1 February 1938; Rowe to DSR, RDF and Interception Experiments, 4 March 1938, AIR 2/2201; Interception Programme, 25 March 1938, HTT 10/1; CSSAD, Note on Interception Experiments with RDF, 20 April 1938, AVIA 7/229.

10. W. S. Douglas to Dowding, 28 March 1938; Signals to ACAS, 30 March 1938; Chandler for Air Commodore, Director of Signals to Dowding, 14 April 1938, AIR 2/2728.

11. *Signals Volume V: Fighter Interception and Control* (Air Historical Branch, classified monograph, 1952), pp. 16-19, AIR 10/5485. Also see Dowding to Tizard, 22 April 1937, HTT 39; Chief Superintendent RAE to CinC Fighter Command, 7 December 1937, AIR 16/180; Dowding to Secretary Air Ministry, 14 March 1938, AIR 2/2728.

12. Dowding to Secretary Air Ministry, 14 March 1938, AIR 2/2728.

13. Pye to Rowe, 27 May 1938, AIR 2/2201.

14. Dixon, 'Suggested Tactical Analysis of Large Scale Air Defence Operations in Relation to RDF,' 12 April 1938, AVIA 7/437.

15. DCD Radio Development Programme Monthly Conference, 11 July 1938; Rowe to Watson-Watt, 23 August 1938, AVIA 7/167.

16. Interception Experiments with RDF, 4/5/38-20/5/39, 23 May 1938. AIR 2/2201.

17. Organization 2 to SASO, 10 August 1938, AIR 16/178; Filter Room Techniques, 12 August 1938, AVIA 7/396; Dowding, Home Defence Excercise 1938, Preliminary Report, 18 August 1938, AVIA 7/396; Rowe, Interim Report on Home Defence Excercise 5- 7 August, 1938, 26 August 1938, AVIA 10/348; E. D. Whitehead, Home Defence Excercises, 5- 7 August 1938, 4 November 1938, AVIA 7/396.

18. Basil Collier, *The Defence of the United Kingdom* (London: HMSO, 1957), p. 65.

19. Rowe to Watson-Watt, 29 September 1938, AVIA 7/427; Rowe to Under Secretary of State, Air Ministry, 10 October 1938, AIR 2/3398.

20. Dowding to Under Secretary of State, Air Ministry, 17 October 1938, AIR 16/94; Watson-Watt, *Three Steps to Victory*, p. 188; *Fighter Interception and Control*, pp. 31-2.

21. *Fighter Interception and Control*, p. 51.

22. Rowe to Under Secretary of State, Air Ministry, Priority of Items for Bawdsey Research Station, 23 February 1939, AIR 2/3404.

23. Rowe to Tizard, 27 February 1939, HTT 140.

24. Rowe to DCD, 13 March 1939, AIR 2/2663.

25. See the various documents in the A. V. Hill Papers 2/1. The Admiralty's opposition to the proposal is found in ADM 1/10860.

26. Guy Hartcup and T. E. Allibone, *Cockcroft and the Atom*, (Bristol: Adam Hilger, 1984), pp. 85-7; Watson-Watt, *Three Steps*, pp. 211-3; Clark, *Tizard*, pp. 171-4; Sir Bernard Lovell, 'Wilfrid Bennett Lewis,' *Biographical Memoirs of Fellows of the Royal Society*, 34 (1992), pp. 464-6.

27. Rowe to Tizard, 27 February 1939, HTT 140.

28. Director of Signals to Air Member Personnel, 4 October 1938, AIR 16/186; S/L Tester to Rowe, 23 November 1938, AIR 7/411; *Fighter Interception and Control*, pp. 55-7.

29. Watson-Watt, Women Observers for RDF, 15 July 1938; Rowe to Watson-Watt, 27 July 1938; Tester to Rowe, 23 November 1938; Tester, Women Observers for RDF, 23 November 1938; H. J. Brooker, Visit to AMES Dover, 27 February 1939; Bawdsey Research Station, Women RDF Observers, 7 March 1939; Park to Rowe, 25 March 1939, AIR 7/411; Watson-Watt, *Three Steps to Victory*, pp. 176-7.

30. *Fighter Interception and Control*, pp. 53-4.

31. Sector and Supplementary D/F stations, 25 March 1939, AIR 2/2728.

32. Dowding to the Secretary Air Ministry, 11 March 1937, AIR 2/2587.

33. Draft Minutes of a Meeting held at Headquarters, Fighter Command on 22/12/38 to discuss the necessary arrangements to be made for a trial of the Challenger and Reply System of Identification; Procedure for Challenge; Procedure for Challenge and Reply System of Identification, 23 January 1939; Minor H.D. Exercises – Challenging and Recognition of Friendly Bombers, 23 February 1939, AIR 2/2587.

34. Director of Signals to ACAS, 22 April 1934, AIR 2/2615; Recognition in RDF Technique, AVIA 7/534; Watson-Watt, *Three Steps*, p.474.

35. Dowding to Tizard, 6 March 1939, HTT, 39; Rowe, Progress of certain research items at Bawdsey, 6 March 1939, AIR 2/3404; Tizard to Dowding, 7 March 1939, AIR 16/24; Watson-Watt, *Three Steps*, pp. 228-9.

36. Minutes of a conference held at Fighter Command Headquarters, 3/4/39: Home Defence Air Exercise, AIR 9/136; Air Commodore K. R. Park to The Under-Secretary of State, Air Ministry, 9 May 1939. AIR 16/24; Minutes of a conference held at the Air Ministry to discuss interception problems, 28 June 1939, AIR 20/222.

37. Rowe, Preliminary Report on the Bomber Command and Home Defence Exercises, August 1939, AVIA 7/397.

38. Woodward Nutt for the Director of Communications Development to the Superintendent, Bawdsey, 1 September 1938; Rowe to Watson-Watt, 14 September 1938, Air 2/2201.

39. Watson-Watt to Rowe, 2 June 1939; Rowe to Under Secretary of State, Air Ministry, 26 June 1939, AVIA 7/754.

40. Watson-Watt the Under Secretary Air Ministry, 27 June 1939, AVIA 7/754.

41. Colonel C. E. Colbeck, Royal Engineer Board, to M.4, 31 August 1939, AVIA 7/689.

42. D. P. Stevenson, DD Ops (H) to CAS, 28 June 1939, AIR 20/222.

43. Rowe to Dowding, 28 July 1939, AIR 16/251.

44. Report on a visit to Headquarters Fighter Command on 9 May 1939, AVIA 7/179; G. A. Roberts, memorandum on proposed experiments to be carried out during No. 11 (Fighter) Group Exercise, 9 July 1939, AVIA 7/181; G. A. Roberts, memorandum on experiments carried out at Hucknall, Duxford, Wittering and Digby on 13 and 14 July, 1939, 20 July 1939, AVIA 7/402; Harold Lardner, Progress Report, 16 June-15 July, 1939, Group I, 18 July 1939, AVIA 7/347; Lardner to Rowe, 4 August 1939, AVIA 7/401.

45. Dowding to Ludlow-Hewitt, C-in-C Bomber Command, 25 May 1939, AIR 16/109; Note on a visit to Biggin Hill, 6 June 1939, HTT 11/30; Dowding to Tizard, 12 June 1939, AIR 16/251; Minutes of a conference held at Fighter Command Headquarters, 3/4/39: Home Defence Air Exercise, AIR 9/136 Notes on a conference presided over by AVM Sholto Douglas at the Ai

Ministry, 28 June 1939, HTT 11/35; G. A. Roberts, memorandum on necessity for 'Lost Property Office' 25 July 1939, AVIA 7/180.

46. K. R. Park, Senior Air Staff Officer, Headquarters Fighter Command, Home Defence Air Exercise 8/11 August 1939, Instruction No. 1, July 1939; Air Officer Commanding No. 11 Group, Home Defence Exercise, 1939, August 1939; Dowding, Major Home Defence Exercise, 25 August 1939, AVIA 7/397; Dowding to HQ Nos 11, 12, 13 Groups, August Home Defence Exercise, 26 August 1939, AIR 16/271.

47. Bawdsey Research Station, Preliminary Report on the Bomber Command and Home Defence Exercise, August 1939, 15 August 1939, AVIA 7/397.

Notes to Chapter Eleven

1. David Fisher, *A Race on the Edge of Time* (New York: McGraw-Hill, 1988), pp. 84-91; *Signals: Volume IV, Radar in Raid Reporting*, pp. 79-80; W. S. Douglas to ACM Ludlow Hewitt Bomber Command, 7 July 1939, AIR 2/2587.

2. J. R. M. Butler, *Grand Strategy*, Volume II (London: HMSO, 1957), pp. 34-6.

3. Richard Hough and Denis Richards, *The Battle of Britain: The Greatest Air Battle of World War II* (London: Norton, 1989), pp. 81-91.

4. Basil Collier, *The Defence of the United Kingdom* (London: HMSO, 1957), pp. 93-5.

5. Churchill to Wood, 11 November 1939; Wood to Churchill, 13 November 1939; Churchill to Wood, 14 November 1939; Wood to Churchill, 15 November 1939, CAB 21/765.

6. Tizard to Dowding, 7 September 1939; Dowding to Tizard, 15 September 1939, HTT 226.

7. Operations Record Book, HQ No. 60 Group, AIR 25/679.

8. Denis Richards, *The Royal Air Force, 1939-45: The Fight at Odds* (London: HMSO, 1953), p. 67.

9. Dowding to Under Secretary of State, Air Ministry, 20 10 1939, AVIA 7/239; *Signals: Volume IV, Radar in Raid Reporting*, pp. 83-4; pp. 86-7.

10. Lardner to Rowe, 12 March 1940, AVIA 7/807; A. B. Jones to Under Secretary of State, Air Ministry, 27 May 1940, AVIA 7/602.

11. The minutes of the meetings of the chain observers are found in AVIA 7/243.

12. Lardner, 'Some lessons learned as a result of the differences between operating conditions of the RDF Chain in War as compared with Peace', 11 September 1939, AVIA 7/437.

13. Telling of Aircraft Tracks From Fighter Command RDF Filter Room, Instruction No. 1, 10 September 1939, AIR 16/343.

14. Lardner to Undersecretary of State Air Ministry for DCD, 23 November 1939, AVIA 15/81.

15. G. A. Roberts, Operational System Research Memorandum No. 11: Raid 18 of 19 January 1940, 20 January 1940, AIR 15/81.

16. G. A. Roberts, Operational System Research Memorandum No. 38: Investigation of Raid X6 of 30 May, 17 June 1940, AIR 16/343.

17. E. C. Williams, Filter Room Organization and Technique, 11 January 1940, AVIA 7/183.

18. *Signals Volume V Fighter Interception and Control*, pp. 40-2.

19. Harold Lardner, *The Origins of Operational Research*, ORA Report TL 89/5 Annex B.

20. Decisions Reached at a Production Conference on CHL Receivers, 11 January 1940, AVIA 7/689.

21. Operations Record Book HQ No. 60 Group, AIR 25/679.

22. Ibid, p. 42.

23. *Signals: Volume IV, Radar in Raid Reporting*, p. 85.

24. Photographic Record of Radar Stations (Ground), August 1943, AIR 10/4152; A. C. Mayman, Engineer-in-Chief, GPO, Electrical Calculators, 8 November 1939, AVIA 7/1012.

25. Report on a meeting held at TRE, 11 November, 1940; RDF Convertor Equipment, November 1940, AVIA 7/1012; Colin Latham and Anne Stubbs, *Radar: A Wartime Miracle*, pp. 21-4.

26. Dowding to Under Secretary of State Air Ministry, 8 November 1939; Draft minutes of a conference on RDF chain CH and CHL Sites, 21 December 1939 AIR 2/2920; Squadron Leader J. W. Gillan Sigs 1.a to W.2(a), 11 January 1940, AIR 20/ 2267; Stanmore Research Section, Mean Operational Coverage of CHL Stations April 1940, n.d., AVIA 7/900; Memorandum on relative position of RDF and other radio equipment: Comparison end of May 1940 with mid-April 1941, Beaverbrook Papers D/405; *Signals: Volume IV, Radar in Raid Reporting*, pp. 79-85.

27. Rowe to Under Secretary of State Air Ministry, 6 April 1940, AVIA 7/727.

28. Joubert to CAS, 12 December 1939; Interim Proposal for Extension of RDF Coverage of Great Britain, 22 January 1940, AIR 20/2267; Final Minutes of Meeting on RDF – Home Chain Policy, AVIA 15/255.

29. Dowding to Tizard, 13 November 1939, HTT 226. The other two troubles were the failure to fit AI radar (see chapter 13) and the lack of a heavy calibre armour piercing guns for the fighters.

30. F. S. Barton for Chief Superintendent Royal Aircraft Establishment to Under Secretary of State Air Ministry, 7 March 1940; Operations Record Book Fighter Command, 17 April 1940, AIR 24/507; Lardner to Rowe, 29 April 1940, AVIA 15/137; Lardner to AMRE, 30 May 1940, AVIA 7/807; *Signals: Volume IV, Radar in Raid Reporting*, pp. 74-8.

31. *Signals: Volume IV, Radar in Raid Reporting*, pp. 79-82.

32. *Signals: Volume V Fighter Interception and Control*, pp. 42-6.

33. J. A. Tester to the Under Secretary of State for Air, 9 October 1939, AVIA 7/410.

34. Watson-Watt, *Three Steps*, pp. 245-6.

35. Controller RAF Station Bawdsey, 12 May 1940, AIR 6/186; Dowding to Under Secretary of State Air Ministry, Height Finding as an Aid to Interception, 27 March 1940, AVIA 7/438.

36. E. C. Williams, Height Measurement by RDF, 3 June 1940, AVIA 7/183; Lardner, Memorandum on some causes which contribute to failure to intercept enemy aircraft, 3 June 1940, AVIA 15/474.

37. Notes on Calibration of RDF Stations, October 1940, AIR 16/877; *Signals: Volume IV, Radar in Raid Reporting*, p. 88.

38. Minutes of a Meeting Held in the CAS' room, 12 January 1940, AIR 20/2267.

39. Joubert, Draft Letter to C-in-C Fighter Command, n.d. but likely 22 January 1940, AIR 20/2267.

40. Dowding to Under Secretary of State Air Ministry, 31 January 1940, AIR 8/577.

41. E. C. Williams, Filter Room Organization and Technique, 11 January 1940, AVIA 7/183; Lardner to Rowe, Proposed Expansion of the Home Chain, 15 March 1940, AVIA 7/255; F. L. Sawyer, Scientific Officer, to Rowe, Scientific Observer's Report from AMES Canedown, 29 May 1940, AVIA 7/211.

42. Joubert to Sinclair, 11 January 1940, AIR 20/2268.

Notes to Chapter Twelve

1. John Ray, *The Battle of Britain: New Perspectives* (London: Arms and Armour, 1994), p. 42.

2. Watson-Watt to Secretary of State, 21 December 1940, AIR 20/2268.

3. Park to Rowe, 24 July 1940, AVIA 7/212.

4. The Capacity of the RDF System, OR Report No. 100, November 1940, AVIA 7/1007.

5. Dowding, The Battle of Britain, *Supplement to The London Gazette*, 11 September 1946, 4546.

6. *Signals: Volume IV, Radar in Raid Reporting*, p. 123.

7. Sebastian Cox, 'A Comparative Analysis of RAF and Luftwaffe Intelligence in the Battle of Britain, 1940', *Intelligence and National Security* 5, 2 (1990), pp. 425-442. Also see F. H. Hinsley, *British Intelligence in the Second World War*, Vol. 1 (London: HMSO, 1979), pp. 159-90; Ronald Lewin, *Ultra Goes to War* (Hutchinson: London, 1978), pp. 82-90.

8. W/C Thomas Lang, Reflections and Observations of a Group Controller, 1940, 5 May 1941, AIR 25/197.

9. Derek Wood and Derek Dempster, *The Narrow Margin* (London: Arrow Books, 1969) pp. 166-7.

10. *Signals: Volume IV, Radar in Raid Reporting*, p. 125.

11. AMES Serviceability, n.d. but August-September 1940, AIR 20/2267; Memorandum on relative positions of RDF and other Radio Equipment – Comparison end of May 1940 with mid-April 1941, Beaverbrook Papers D/405.

12. Park to Rowe, 24 July 1940, AVIA 7/212.

13. Rowe to Lewis 27 July 1940; Lewis to Rowe, 28 July 1940, AVIA 7/212.

14. D. R. W. Thompson to Headquarters No. 26 Group, 29 July 1940, AVIA 7/410.

15. E. C. Williams to Sigs 1, 6 August 1940, AVIA 7/410.

16. Lardner to Rowe, 23 July 1940, AVIA 15/474.

17. The Capacity of the RDF System, OR Report No. 100, November 1940, AVIA 7/1007.

18. Keith Park, HG Fighter Group No. 11, German Air Attacks on England – 8/8-10/9/40, 12 September 1940, AIR 16/1067.

19. Draft of a Letter to the Prime Minister, 9 November 1940, AIR 19/476.

20. Dowding to Headquarters No. 11 Group and Commandant, Observer Corps, 15 September 1940, AVIA 7/178. Also see Lardner to Rowe, 14 September 1940, AVIA 7/178.

21. The Capacity of the RDF System, OR Report No. 100, November 1940, AVIA 7/1007.

22. Park, Report of Air Fighting in No. 11 Group Area, 11 September-31 October 1940, 7 November 1940, AIR 25/198.

23. *Signals Volume V: Fighter Control and Interception*, p. 51.

24. Park to Joubert, 29 October 1940, AIR 20/2267.

25. No. 11 Group Report on Battles, 0816/1030 hours, 10 October 1940; Group Conference, No. 11 Group, 18 October 1940, AIR 25/198; Group Captain Hart to Group Captain C. K. Chandler, Ministry of Aircraft Production, 2 December 1940, AVIA 15/89.

26. Park, Report of Air Fighting in No. 11 Group Area, 11 September-31 October 1940, 7 November 1940, AIR 25/198.

27. Zimmerman, *Top Secret Exchange*, p. 69.

28. Colin Latham and Anne Stubbs, *Radar: A Wartime Miracle*, pp. 1-3.

29. Ibid, 7; Cox, 'A Comparative Analysis of RAF and Luftwaffe Intelligence'; Wood and Dempster, pp. 43-5.

30. Richard Hough and Denis Richards, *The Battle of Britain* (New York: Norton, 1989), pp. 141-3.

31. *Signals: Volume IV, Radar in Raid Reporting*, pp. 118-9; Wood and Dempster, pp. 165-74; Hough and Richards, pp. 146-8.

32. Park, Group Controllers Instruction No. 4, 25 August 1940, AIR 25/197.

33. Park, German Air Attacks on England – 8/8-10/9/40, 12 September 1940, AIR 16/1067.

34. *Signals: Volume IV, Radar in Raid Reporting*, pp. 121-3.

35. Wood and Dempster, pp. 43-5.

36. As quoted in *The Rise and Fall of the German Air Force, 1933-1945* (Poole: Arms and Armour Press, 1983), p. 80. This book is a reprint of a postwar secret Air Ministry historical study.

37. Keith Park, HG Fighter Group No. 11, German Air Attacks on England – 8/8-10/9/40, 12 September 1940, AIR 16/1067.

38. Appendixes to Battle of Britain History, n.d., AIR 41/16.

39. Dowding to Under Secretary of State, 27 September, AIR 16/387; Extracts from Draft Conclusions of 6th Air Council Meeting on 2.10.40, AIR 8/577.

40. E. E. B. to Churchill, 11 October 1940; Churchill to Sinclair, 12 October 1940, AIR 19/476.

41. Draft by Joubert; Draft by Private Secretary to the Secretary of State for Air, 18 October 1940, AIR 19/476.

42. Churchill to Sinclair, 27 October 1940; Joubert to Assistant Private Secretary and Chief of Air Staff, 30 October 1940; Draft of letter to Prime Minister 9 November 1940; Assistant Private Secretary to Deputy Chief of the Air Staff, 9 November 1940; Private Secretary to Vice Chief of Air Staff to Private Secretary of the Secretary of State, 10 November 1940, AIR 19/476.

Notes to Chapter Thirteen

1. John Ray, *The Battle of Britain: New Perspectives*, (London: Arms and Armour, 1994), pp. 124-169.

2. Ibid, p. 141.

3. Phillips to Churchill, 16 October 1940, AIR 16/676.

4. Dowding to Churchill, 17 October 1940, AIR 16/676.

5. Kenley Night Fighter Operations , 1st October to 16th October, 1940, 19 October 1940, AVIA 7/946.

6. Operations Record Book, Fighter Command, 21-22 December 1940, AIR 24/507.

7. Bowen, *Radar Days*, pp. 60-4.

8. Report on the Progress at Bawdsey Research Station, 16 June, 1939, AIR 2/3404.

9. Bowen, *Radar Days*, pp. 70-2.

10. Dowding to Rowe, 7 July 1939, AIR 16/251.

11. Dowding to the Under-Secretary of State, Air Ministry, 10 July 1939; Deputy Director Operations (H) to Deputy Chief of the Air Staff, 13 July 1939, AIR 20/222; DCAS to AMDP and AMSO, 14 July 1939, AIR 2/3377.

12. Bowen, pp. 75.

13. David Zimmerman, *The Great Naval Battle of Ottawa*, (Toronto: University of Toronto Press, 1989), pp. 69,131.

14. Bowen, pp. 75-82.

15. Rowe, The Evacuation of Bawdsey Research Station, 13 December 1938, AVIA 7/600.

16. Bowen, p. 84.

17. Bowen, pp. 83-9.

18. Lovell to Blackett, 14 October 1939, Imperial War Museum, Henry Tizard Papers (HTT) 32.

19. Rowe to Tizard, 28 October 1939, HTT 32.

20. Tizard Diary, 30 October 1939, HTT 11. Also see, Tizard to Watson-Watt, 24 October 1939, HTT 12; Edmund Dixon to Wg/Cdr R. G. Hart, HQ Fighter Command, 7 November 1939, AIR 2/3377.

21. Rowe to the Under Secretary of State for the Attention of the DCD, 30 October 1939, HTT 79.

22. DGRD to AMDP, 8 November 1939, HTT 236.

23. Tizard to Dowding, 11 November 1939, HTT 226.

24. Dowding to Tizard, 13 November 1939, HTT 226.

25. Operation Record Book, Fighter Command 12 December 1939, AIR 24/507.

26. Dowding to the Under Secretary of State, Air Ministry, 18 December 1939, AIR 2/3377; Tizard, Notes on visit to St Athans on Thursday, December 14, 1939, HTT 232.

27. R. Hanbury Brown, Suggestions for Fighter Control by RDF, 24 November 1939, Churchill College, E. G. Bowen Papers (EGBN), 1/7.

28. R. Hanbury Brown, *Boffin* (Bristol: Adam Hilger, 1991), pp. 62-3.

29. Dewhurst, Report of Visit to No. 10 Department RAE, 23 September 1940; Signals, Air Ministry to Fighter Command, No. 60 Group and RAE, 13 October 1940; Report on GCI Position, 11/40, 20 November 1940, AIR 16/885; GCI Information Pamphlet, 29 December 1940, AVIA 7/1132; Minutes of a Conference on Night Interception held in the Air Council Room, 1 January 1941, AIR 8/577.

30. Joubert, The Case of Wartime re-distribution of responsibilities, 13 December 1939, AIR 20/2269.

31. Operations Record Book, No. 60 Group, 23 February 1940, AIR 25/679.

32. Tedder to Tizard, 24 January 1940, HTT 14; Minutes of the 19th Meeting of the Air Fighting Committee, 12 February 1940, AIR 16/1024.

33. 'Wilfrid Bennett Lewis', *Biographical Memoirs of Fellows of the Royal Society*, V. 34, 1988, pp. 467-76; Bowen, pp. 83-6, 123-31, 136-140; R. Hanbury Brown, *Boffin*, pp. 60-1.

34. Minutes of the Night Interception Committee, 1st to 8th Meeting, 14 March-4 July 1940, AIR 16/427.

35. F/L R. Hiscox to Dowding, 16 April 1940; Watson-Watt, Summary of Results with Improved AI Equipment; AIR 16/427.
36. FIU Report No. 7, Night Fighting Exercises Carried out after Lecture, 24 May 1940; Wg/Cdr R. G. Hart to Dowding, 24 May 1940, AIR 16/427.
37. Tizard to Dowding, 31 May 1940, HTT 226; Tizard, Notes on 'AI', 31 May 1940, Air 2/3377; Ismay to Churchill 23 May 1940 and 3 June 1940, CAB 21/1103.
38. David Zimmerman, *Top Secret Exchange: The Tizard Mission and the Scientific War* (McGill-Queens & Sutton, 1996).
39. Sqn Ldr R. Hiscox, FIU Report No. 17, Test of Experimental AI Mark IV Apparatus, 3 July 1940, AVIA 7/946; Wg/Cdr Chamberlain, FIU Report No. 23, Further Test of Experimental AI Mark IV Apparatus, AVIA 7/774.
40. Operational Research Section Fighter Command, Report No. 173, A Statistical Report on the interception of enemy bombers by night over Great Britain, 2-3 March to 12-13 April 1941, AIR 16/1043.
41. Tedder to Tizard, 24 January 1940, HTT 15.

Notes to Conclusion

1. C. P. Snow, *Appendix to Science and Government* (Cambridge, Mass.: Harvard University Press, 1962), p. 7.
2. The Oxford English Dictionary first lists the word boffin as being used in 1945. Undoubtedly it was common slang several years before that, although there is no evidence that boffin was used prior to the summer of 1941.
3. Tizard to R. A. Butler, 8 December 1953, HTT 696.
4. Cockburn and Ellyard, *Oliphant*, pp. 84-9; Bowen, *Radar Days*, pp. 152-3.
5. David Zimmerman, *Top Secret Exchange.*
6. Tizard to Appleton as quoted in Clark, *Tizard*, p. 322.
7. Rowe to the Royal Commission of Awards to Inventors, 7 May 1951, AVIA 53/350.
8. Watson-Watt, *Three Steps*, pp. 461-2.
9. Obituary of Dr A. P. Rowe, *The Times*, 31 May 1976.
10. Paul Crook, 'Science and War: Radical cientists and the Tizard-Cherwell Area Bombing Dispute in Britain, *War and Society*, 12 No. 2 (October 1994), pp. 69-101.
11. Clark, *Tizard*, pp. 272-419.
12. Birkenhead, *The Professor and the Prime Minister*, pp. 275-360.
13. C. P. Snow, *Science and Government* (Cambridge, Mass.: Harvard University Press, 1960), The Earl of Birkenhead, *The Professor and the Prime Minister*. In the Blackett papers at Royal Society there is a file containing a series of private letters between Blackett and Jones about Tizard and Cherwell. Jones wrote an article praising Churchill as a brilliant scientific leader in a collection of essays on Churchill printed in 1989.

Index